METHODS IN MOLECULAR BIOLOGY™

Series Editor
John M. Walker
School of Life Sciences
University of Hertfordshire
Hatfield, Hertfordshire, AL10 9AB, UK

For further volumes:
http://www.springer.com/series/7651

Mammalian Oocyte Regulation

Methods and Protocols

Edited by

Hayden A. Homer

Department of Cell and Developmental Biology, Institute for Women's Health,
University College London, London, UK

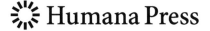 Humana Press

Editor
Hayden A. Homer
Department of Cell and Developmental Biology
Institute for Women's Health
University College London
London, UK

ISSN 1064-3745   ISSN 1940-6029 (electronic)
ISBN 978-1-62703-190-5   ISBN 978-1-62703-191-2 (eBook)
DOI 10.1007/978-1-62703-191-2
Springer New York Heidelberg Dordrecht London

Library of Congress Control Number: 2012950364

Printed on acid-free paper

Humana Press is a brand of Springer
Springer is part of Springer Science+Business Media (www.springer.com)

Preface

Mammalian oocytes occupy a critical nexus in reproduction as they not only contribute half the genetic makeup of the embryo but also provide virtually all of the cytoplasmic building blocks required for sustaining embryogenesis. The journey that transforms a primordial germ cell into a mature oocyte (or egg) capable of fertilization and embryonic development is of unrivalled complexity characterized by a discontinuous stop–start tempo that spans an astounding period lasting up to four to five decades in humans. Built into this protracted process of oogenesis are two consecutive meiotic nuclear divisions, meiosis I and meiosis II, which proceed without an intervening phase of DNA replication thereby ultimately halving the chromosomal content.

Oogenesis begins in the fetal ovary when the earliest oocyte precursors commit to meiosis and undergo reciprocal recombination. Recombination leads to the reshuffling of genes important for diversity within the species and also for the establishment of a link between homologous chromosome pairs critical for setting up the correct pattern of chromosome segregation during meiosis I. Oocytes then enter their first arrest state at the dictyate stage of prophase I with an intact nucleus (or germinal vesicle as it is known in the oocyte) and surrounded by a somatic cell investiture (or follicle) that acts as a private boudoir for each oocyte within the ovary. This suspended state is maintained until cohorts of prophase I-arrested oocytes are induced to undergo a protracted growth phase lasting ~2 weeks in mice and ~10–12 weeks in humans during which the oocyte builds up its repository of mRNAs and proteins. As a consequence, oocyte volume increases over 100-fold to produce some of the largest cells known—the volume of a fully grown mouse oocyte is ~270 pL, whereas that of a PtK1 somatic cell is ~6 pL—all the while remaining in intimate contact with the surrounding follicle, which is also undergoing development in parallel. Fully grown oocytes located within dominant antral follicles respond to hormonal cues and resume meiosis I marked by breakdown of the GV. Over the course of several hours (6–10 h in mice and 24–36 h in humans), oocytes complete meiosis I and progress immediately to meiosis II only to become arrested for a second time at metaphase II. The end-product of this remarkable developmental process is the "mature" egg, now replete with macromolecules and organelles, and poised to support embryonic development if fertilization should occur to break the metaphase II-arrest state.

Given the importance of the oocyte for embryonic development, it is not surprising that the oocyte holds centre stage in fertility clinics and greatly influences the outcome of assisted reproductive treatments (ARTs). In the current economic and social climate in which child-bearing occurs at increasingly older ages, women in their late 30s and early 40s comprise an ever-expanding proportion of fertility patients. For such patients, the single biggest impediment to successful fertility treatment and pregnancy outcome is poor oocyte quality, a biological phenomenon that at present remains elusive at the molecular level and for which there is no effective therapy apart from the use of "good-quality" eggs provided by donors. Poor oocyte quality is also at the heart of other adverse reproductive sequelae such as chromosomally abnormal pregnancies (e.g., Down's syndrome) and miscarriage both of which most often stem from chromosome segregation errors arising during oocyte maturation. Understanding the biology of the mammalian oocyte is therefore at the heart of mammalian reproductive performance and a priority agenda for reproductive biologists. Notably, due to its remarkable property for nuclear reprogramming, the mammalian oocyte is also a central focus of stem cell research.

The paucity of human oocytes for research has meant that surrogate mammalian models, most notably the mouse, have been indispensable for elucidating oocyte mechanisms. The aim of *"Mammalian oocyte regulation: Methods and Protocols"* is to provide a highly diverse compendium of detailed methodologies—primarily focusing on the murine model but also including chapters on human oocytes—for enabling researchers to interrogate every aspect of mammalian oocyte development including recombination, meiotic maturation, oocyte substrate uptake, chromosomal segregation, and fertilization.

The chapters cover generic techniques such as those required for obtaining oocytes from female mice (oocytes have to be obtained from hormonally primed live hosts and cannot be propagated in culture in the lab), for stripping oocytes of their surrounding cumulus cells, for oocyte handling and microinjection, and for the in vitro manufacture of cRNAs (important for experimental protein overexpression in the transcriptionally quiescent oocyte). Superimposed on these generic technical descriptions is a diverse range of protocols having more specific focus. These include protocols for studying recombination using allele-specific PCR; for analyzing high-resolution details within the voluminous oocyte using immunofluorescence and electron microscopy; for studying substrate uptake and protein turnover using radioactive reagents, kinase assays, and timelapse fluorescence imaging; for studying chromosomal structure, chromosomal compliment, and chromosome-centred spatial regulation using chromosome spreads, immunofluorescence, FISH, microarray-CGH, FRET technology, and oocyte bisection; for studying germ cell development in relation to the unique ovarian niche using immunohistochemistry of ovarian sections; for studying gene function using long double-stranded RNA for RNA interference and morpholino antisense technology; and for exploring nuclear reprogramming and generating ES cells using nuclear transfer techniques.

This book has strong appeal for the novice oocyte researcher as leading oocyte labs share with readers their unique experience acquired over many years regarding techniques that are core to work in the field; this book therefore represents one of the largest individual collections of "insider secrets" for circumventing unique obstacles within this challenging experimental system. With the wide range of protocols described herein, this book will also be of great interest to experts who wish to add another dimension to their established portfolio. By describing techniques of wide interest such as reporter gene technology, gene silencing, chromosomal analyses, and nuclear transfer, this book will also be a great companion to the wider research community.

Above all, I would like to express my heartfelt gratitude to all the authors who made this book possible by sharing their wealth of expertise in such a selfless, thoughtful, and meticulous manner.

London, UK   *Hayden A. Homer*

Contents

Contributors

RICHARD A. ANDERSON • *MRC Centre for Reproductive Health, The Queen's Medical Research Institute, The University of Edinburgh, Edinburgh, UK*

LYNNE ANGUISH • *Baker Institute for Animal Health, Cornell University, Ithaca, NY, USA*

JAY M. BALTZ • *Departments of Obstetrics & Gynecology and Cellular & Molecular Medicine, Ottawa Hospital Research Institute, University of Ottawa, Ottawa, ON, Canada*

SUZANNE CAWOOD • *Centre for Reproductive and Genetic Health, Eastman Dental Hospital, London, UK*

JEAN-PHILIPPE CHAMBON • *CNRS UMR7622 Biologie du Développement and UPMC Paris 6, Paris, France*

TERESA CHIANG • *Institute of Biomedical Sciences, Academia Sinica, Taipei, Taiwan*

ANDREW J. CHILDS • *MRC Centre for Reproductive Health, The Queen's Medical Research Institute, The University of Edinburgh, Edinburgh, UK*

PAULA E. COHEN • *Department of Biomedical Sciences, Cornell University, Ithaca, NY, USA*

SCOTT COONROD • *Baker Institute for Animal Health, Cornell University, Ithaca, NY, USA*

HANNAH E. CORBETT • *Departments of Obstetrics & Gynecology and Cellular & Molecular Medicine, Ottawa Hospital Research Institute, University of Ottawa, Ottawa, ON, Canada*

ESTHER DE BOER • *Institut de Génétique et Microbiologie, Université Paris-Sud, Orsay, France*

JULIEN DUMONT • *Institut Jacques Monod, CNRS, UMR 7592, Univ Paris Diderot, Sorbonne Paris Cité, Paris, France*

LIMING GUI • *Mammalian Oocyte and Embryo Research Laboratory, Division of Biosciences, Department of Cell and Developmental Biology, University College London, London, UK*

KHALED HACHED • *CNRS UMR7622 Biologie du Développement and UPMC Paris 6, Bernard, Paris, France*

JING HE • *MRC Centre for Reproductive Health, The Queen's Medical Research Institute, The University of Edinburgh, Edinburgh, UK*

JANET E. HOLT • *School of Biomedical Sciences and Pharmacy, University of Newcastle, Callaghan, NSW, Australia*

HAYDEN A. HOMER • *Department of Cell and Developmental Biology, Institute for Women's Health, University College London, London, UK*

KEI-ICHIRO ISHIGURO • *Laboratory of Chromosome Dynamics, Institute of Molecular and Cellular Biosciences, University of Tokyo, Tokyo, Japan*

SOURAYA JAROUDI • *Reprogenetics UK, Institute of Reproductive Sciences, Oxford, UK*

MARIA JASIN • *Developmental Biology Program, Memorial Sloan-Kettering Cancer Center, New York, NY, USA*

KEITH T. JONES • *School of Biomedical Sciences and Pharmacy, University of Newcastle, Callaghan, NSW, Australia*

SCOTT KEENEY • *Molecular Biology Program, Memorial Sloan-Kettering Cancer Center, New York, NY, USA; Howard Hughes Medical Institute, New York, NY, USA*

JI-HYE KIM • *Laboratory of Chromosome Dynamics, Institute of Molecular and Cellular Biosciences, University of Tokyo, Tokyo, Japan; Graduate School of Agricultural and Life Science, University of Tokyo, Tokyo, Japan*

JACEK Z. KUBIAK • *CNRS, UMR 6290, Institute of Genetics and Development of Rennes (IGDR), Cell Cycle Group, Rennes, France; University Rennes 1, UEB, UMS 3480, Faculty of Medicine, Rennes, France*

NOBUAKI KUDO • *Institute of Reproductive and Developmental Biology, Imperial College London, London, UK*

MICHAEL A. LAMPSON • *Department of Biology, University of Pennsylvania, Philadelphia, PA, USA*

SIMON I.R. LANE • *School of Biomedical Sciences and Pharmacy, University of Newcastle, Callaghan, NSW, Australia*

MARK LEVASSEUR • *Institute for Cell and Molecular Biosciences, The Medical School, Newcastle University, Newcastle, UK*

SHU-KUEI LI • *Institute of Biomedical Sciences, Academia Sinica, Taipei, Taiwan*

YI-NAN LIN • *Institute of Biomedical Sciences, Academia Sinica, Taipei, Taiwan*

PETROS MARANGOS • *Department of Biological Applications and Technology, University of Ioannina, Ioannina, Greece; Division of Biosciences, Department of Cell and Developmental Biology, University College London, London, UK*

EIJI MIZUTANI • *Center for Developmental Biology, RIKEN Kobe institute, Kobe, Japan; Bioresource Center, RIKEN Tsukuba Institute, Tsukuba, Japan; Faculty of Life and Environmental Sciences, University of Yamanashi, Kofu, Japan*

ATSUO OGURA • *Bioresource Center, RIKEN Tsukuba Institute, Tsukuba, Japan*

ZBIGNIEW POLANSKI • *Department of Genetics and Evolution, Institute of Zoology, Jagiellonian University, Cracow, Poland*

SAMANTHA RICHARD • *Departments of Obstetrics & Gynecology and Cellular & Molecular Medicine, Ottawa Hospital Research Institute, University of Ottawa, Ottawa, ON, Canada*

SOLON RIRIS • *Reproductive Medicine Unit, EGA Wing, University College London Hospitals NHS Trust, London, UK; Institute for Women's Health, University College London, London, UK*

RICHARD M. SCHULTZ • *Department of Biology, University of Pennsylvania, Philadelphia, PA, USA*

PAUL SERHAL • *Centre for Reproductive and Genetic Health, Eastman Dental Hospital, London, UK*

PAULA STEIN • *Department of Biology, University of Pennsylvania, Philadelphia, PA, USA*

XIANFEI SUN • *Department of Biomedical Sciences, Cornell University, Ithaca, NY, USA*

PETR SVOBODA • *Institute of Molecular Genetics, Academy of Sciences of the Czech Republic, Prague, Czech Republic*

KARL SWANN • *School of Medicine, Cardiff University, Heath Park, Cardiff, UK*

TANG K. TANG • *Institute of Biomedical Sciences, Academia Sinica, Taipei, Taiwan*

MARIE-HÉLÈNE VERLHAC • *Collège de France, Center for Interdisciplinary Research in Biology (CIRB), UMR-CNRS7241/INSERM-U1050, Paris, France; Memolife Laboratory of Excellence and Paris Science Lettre, Paris, France*

TERUHIKO WAKAYAMA • *Center for Developmental Biology, RIKEN Kobe institute, Kobe, Japan; Faculty of Life and Environmental Sciences, University of Yamanashi, Kofu, Japan*

KATJA WASSMANN • *CNRS UMR7622 Biologie du Développement and UPMC Paris 6, Paris, France*

YOSHINORI WATANABE • *Laboratory of Chromosome Dynamics, Institute of Molecular and Cellular Biosciences, University of Tokyo, Tokyo, Japan; Graduate School of Agricultural and Life Science, University of Tokyo, Tokyo, Japan*

DAGAN WELLS • *Nuffield Department of Obstetrics and Gynaecology, Women's Centre John Radcliffe Hospital, University of Oxford, Oxford, UK*

KUO-TAI YANG • *Institute of Biomedical Sciences, Academia Sinica, Taipei, Taiwan*

JIEQIAN ZHOU • *MRC Centre for Reproductive Health, The Queen's Medical Research Institute, The University of Edinburgh, Edinburgh, UK*

Chapter 1

Studying Recombination in Mouse Oocytes

Xianfei Sun and Paula E. Cohen

Abstract

Meiosis is the specialized cell division in sexually reproducing organisms in which haploid gametes are produced. Meiotic prophase I is the defining stage of meiosis, when pairing and synapsis occur between homologous chromosomes, concurrent with reciprocal recombination (or crossing over) events that arise between them. Any disruption of these events during prophase I can lead to improper segregation of homologous chromosomes which can cause severe birth defects in the resulting progeny, and this occurs with alarming frequency in human oocytes. Thus, while the pathways that regulate these events in prophase I are highly conserved in both males and females, the stringency with which these events are monitored and/or controlled appears to be dramatically lower in females. These observations underscore the need to examine and compare meiotic mechanisms across the sexes. However, the study of female meiosis is impeded by the early start of meiosis during fetal development and the very limited amount of ovarian tissue available for meiotic analyses. Here we describe three different techniques which are useful for meiotic prophase I analysis in mouse/human oocytes, ranging from early prophase I events through until the resolution of crossing over at the first and second meiotic divisions.

Key words: Meiosis, Meiotic prophase I, Chromosome spreading, Chiasmata preparation, Spindle preparation, Oocyte

1. Introduction

Meiosis is a special type of cell division in which diploid precursor cells give rise to haploid daughter cells through one round of DNA replication followed by two rounds of cell division. The meiotic division is an integral part of the reproduction in sexually reproducing organisms and haploid gametes are generated from this process.

Meiosis is divided into two stages, based on the two rounds of cellular division: meiosis I and meiosis II. Each stage can be further divided into prophase, metaphase, anaphase, and telophase. Among all of these substages, prophase I is the most remarkable since this stage encompasses the defining and unique features of meiosis.

Hayden A. Homer (ed.), *Mammalian Oocyte Regulation: Methods and Protocols*, Methods in Molecular Biology, vol. 957,
DOI 10.1007/978-1-62703-191-2_1, © Springer Science+Business Media, LLC 2013

These events include pairing of homologous chromosomes through the proteinaceous scaffold structure known as synaptonemal complex (SC), recombination between chromosome pairs and formation of chiasmata between homologs at sites of crossing over. The importance of these events is underscored by the observation that nondisjunction errors occurring in first meiotic division lead to about 50% of all spontaneous miscarriages in human (1, 2).

According to the structure and status of synaptonemal complex (SC), prophase I is further divided into five substages: leptonema (adjective: leptotene), zygonema (adjective: zygotene), pachynema (adjective: pachytene), diplonema (adjective: diplotene), and diakinesis (see Fig. 1). The replicated chromosomes condense and search for their homologous partners in leptonema. The lateral element of SC formed along the axes of homolog chromosomes at this stage. The lateral element is also called axial element prior to synapsis (3). During zygonema, the central element of the SC element forms and "zip" chromosome pairs together. By pachynema, the chromosome pairs are fully synapsed. The SC starts to degrade after this, a process that results in the separation of chromosome pairs in diplonema. By the time the cell has progressed to diakinesis, the chromosomes remain attached only at sites of chiasmata where crossing over, or reciprocal recombination, has occurred.

Although meiosis is an essential process common to all sexually reproducing organisms, it shows tremendous variation between eukaryotic species in terms of its timing, regulation, and success rate. In mammals, this variability is also observed between sexes, and such sexual dimorphism leads to significant differences in the aneuploidy rates observed in male and female gametes. In humans, oocyte aneuploidy rates are as high as 25%, compared to the rates of around 2% reported for human sperm (1, 4), and the majority of these errors (90%), at least in females, are the result of errors in maternal meiosis I (5, 6). Trisomy, as one type of aneuploidy, involves an extra chromosome copy within the nuclei of affected offspring. While the parental origin of the additional chromosome varies in different human trisomic conditions, maternal errors account for more than 60% of these abnormalities. Once again, nondisjunction events in maternal meiosis I are the major cause of these trisomies (1). Trisomy 21, also known as Down syndrome, is one of the most common aneuploid abnormalities in human newborns, and maternal errors account for 88% of all Down syndrome cases, >70% of them occurring during meiosis I (1, 7). These errors correlate well with the heterogeneity of prophase I events observed in human fetal oocytes (8).

The sexual dimorphism observed in mammalian meiosis is also illustrated in transgenic mice bearing mutations in genes essential for meiotic progression. *Sycp3* encodes a protein essential for SC formation and homolog pairing (9). In females homozygous for an

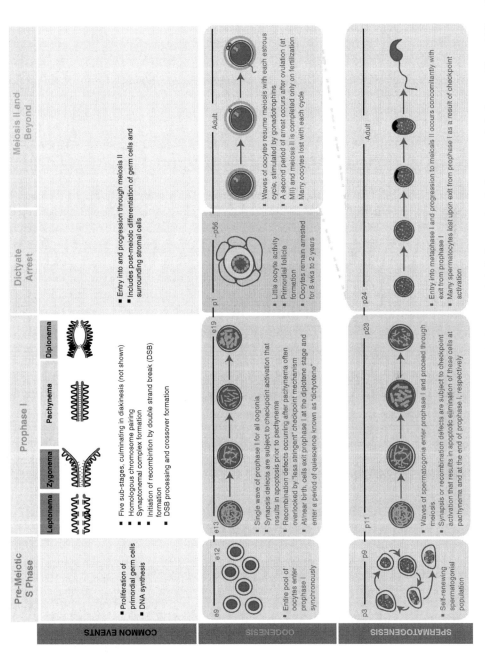

Fig. 1. Gametogenesis and meiosis in mice. The *top panel* shows the sequential events during pre-meiotic S phase and meiosis common in both sexes. The *middle and bottom panels* illustrate the gametogenesis in female and male mouse separately, demonstrating obvious sexual dimorphism not only in their different meiotic products but also in timing and checkpoint control during meiosis.

Sycp3 deletion, synapsis and chiasmata formation are reduced, leading to a high aneuploidy rate in oocytes and a subfertility phenotype (10, 11). By contrast, *Sycp3–/–* males are totally sterile. The spermatocytes in *Sycp3* knockout males show seriously disrupted synapsis and undergo apoptosis at a cytologically defined zygotene stage (9, 12).

A major difference in male and female meiosis, and one of the principal roadblocks in understanding female meiotic errors, comes from the fact that female meiosis starts during fetal development, whereas males initiate meiosis during early postnatal life (in the mouse). Mouse oocytes initiate prophase I entry soon after the oogonia populate the fetal ovary at embryonic day 10.5–12.5 (13). The oocytes progress into leptonema from embryonic day 13.5 (14). At around the time of birth, the oocytes start to enter diakinesis and arrest at a prolonged quiescent state known as dictyate (13). By the fifth day after birth, all oocytes are arrested in dictyate, remaining dormant until puberty when only a few will be selected to resume meiosis at each estrous cycle. Even for these selected oocytes, they will not complete meiosis but get arrested again in metaphase II unless they can be fertilized by sperm after ovulation. By contrast, male germ cells enter meiosis after birth. At puberty, a group of male germ cells are recruited to enter meiosis and produce mature sperm. Meiotic entry then proceeds in waves throughout the lifetime of the animal (see Fig. 1).

The temporal differences mentioned above illuminate the inherent difficulty in studying female meiosis. In mouse, while a small piece of testicular tissue from an adult male could provide enough meiotic and post-meiotic germ cells for most experimental purposes, and would certainly generate hundreds of "chromosome spread" preparations for prophase I analysis, the meiotic material obtained from an individual female is very limited. Thus the efficient processing of the scant tissue samples is extremely important for female study.

The following protocols are designed to provide three different methods that can be used in the female meiotic studies with restricted amounts of experimental material. Subheading 3.1 will introduce a method for preparing chromosome spreads from late stage mouse fetal ovaries and labeling key meiotic markers of prophase I with fluorescent antibodies, often termed indirect immunofluorescence. In Subheading 3.2, a technique will be described for staining diakinesis stage mouse oocytes and detecting chiasmata in vitro. Often, the abnormality in meiotic recombination during prophase I may not show up until later, which could display as a disrupted spindle structure in metaphase I and/or a lower rate for the first polar body extrusion. Subheading 3.3 will demonstrate how to make spindle preparations. Collectively, and in conjunction with fertility assessment studies, these techniques allow for a comprehensive analysis of prophase I events from beginning to end in mouse/human oocytes.

2. Materials

2.1. Ovarian Chromosome Spreads

1. Mice: Mouse embryos are used. To set up timed mating, female mice (2 months to 6 months old) are bred to males (2 months to 1 year old). Start checking vaginal plug the next day. Noon on the day of plug is 0.5 day post coitum (dpc) (see Table 1).

2. Dissecting equipment and dissecting tools: dissecting scope, microdissecting scissors (11-1165 and 11-1025, Biomedical Research Instruments, Inc., Silver Spring, MD 20914), fine forceps (watchmaker No. 5), surgery scalpel (B-P blade handle No. 5 and compatible scalpel blade No. 11), disposable 1 ml syringe and needle (30G1/2, 305106, Becton Dickinson & Co., Franklin lakes, NJ 07417).

3. Culture plates: Sterile 100 mm and 60 mm culture.

4. 15 ml Conical centrifuge tubes.

5. Two-well Concavity slides (71878-08), Electron Microscopy Sciences (EMS), 1560 Industry Road (Hatfield, PA 19440).

6. Six-well microscope slides (63423-08, EMS) (see Note 1).

7. Microscope coverslips (24 mm × 60 mm).

8. Humid chamber: Put wet towel papers on both sides of a square plastic bioassay plate (431272, Corning Incorporated), One Riverfront Plaza (Corning, NY 14831), leaving the middle section dry for slides.

9. Coplin staining jars (70315, EMS).

Table 1
Percentage of oocytes in different prophase I stages from the 14th day of fetal life to the first day after birth[a] (adapted from Evan et al., 1982 (14))

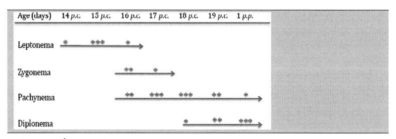

p.c. post coitum, *p.p.* post partum
[a]The number of "+" indicates the relative percentage of each population in a particular developmental stage

10. 1× PBS: 8 g NaCl, 0.2 g KCl, 0.92 g Na_2HPO_4, 0.2 g KH_2PO_4. Adjust volume to 1 L with ddH_2O. Adjust pH to 7.4 with NaOH.

11. 10× Sodium borate buffer (pH 8.2, 1 M): Dissolve 61.83 g boric acid in 500 ml ddH_2O. Adjust pH to 8.2 with solid NaOH pellets. Bring final volume to 1 L with ddH_2O. Use filter to sterilize.

12. Hypotonic extraction buffer (HEB): 100 µl of 500 mM EDTA (final concentration 5 mM), 300 µl of 1 M Tris–HCl (pH 8.5) (final concentration 30 mM), 10 µl of 500 mM dithiothreitol (DTT) (final concentration 0.5 mM), 50 µl of 100 mM phenylmethylsulfonylfluoride (PMSF) (final concentration 0.5 mM), 0.051 g trisodium citrate (final concentration 20 mM), and 0.172 g sucrose (final concentration 50 mM).

    Combine all ingredients and bring the final volume to 10 ml (see Note 3). The pH should be between 8.2 and 8.4. If not, adjust pH with sodium borate buffer. Stocks of DTT and PMSF should be stored at –20°C in small aliquots.

13. Hypotonic sucrose solution (100 mM): Dissolve 0.3423 g sucrose in 10 ml ddH_2O. Adjust pH to 8.2 (see Note 2).

14. 1% Paraformaldehyde (PFA): Add 0.5 g of PFA (19200, EMS) to 50 ml of pre-warmed ddH_2O (at approximately 65°C). Add one drop of 1 N NaOH and shake to dissolve. Adjust pH to 9.2 with sodium borate buffer. Add 75 µl Triton X-100 and mix well but gently (see Note 2).

15. Antibody dilution buffer (ADB): 10 ml Normal goat serum (NGS), 3 g bovine serum albumin (A9647, Sigma), 50 µl Triton X-100, and 90 ml 1× PBS. Combine the ingredients above and mix well. The solution can be stored at 4°C up to 1 week (see Note 3).

16. 0.4% Photoflo/PBS: 400 µl Photoflo in 100 ml of PBS.

17. 0.1% Triton/PBS: 100 µl Triton X-100 in 100 ml of PBS.

18. 10% ADB/PBS: 10 ml ADB in 90 ml of PBS.

19. 0.4% Photoflo/H_2O: 400 µl Photoflo in 100 ml of ddH_2O.

20. Primary antibodies: Many antibodies are compatible with prophase I chromosome spreads. These include:

    (a) Mouse or rabbit SYCP3 (available from many labs or commercial resources (8, 15–18)).

    (b) Goat SYCP3 (SC-20845, Santa Cruz).

    (c) Mouse Monoclonal MLH1 (550838, BD Biosciences).

    (d) Mouse or rabbit γH2AX (05-636 or 07-164, Upstate).

    (e) Rabbit RAD51 (PC 130, Calbiochem).

    (f) Rabbit SYCP1 (ab 15090, Abcam (17)).

21. Secondary antibodies:

    (a) Goat anti-mouse Alexa Fluor 488 (A11017, Invitrogen; 1:1,000 dilution).

    (b) Goat anti-rabbit Alexa Fluor 488 (A11070, Invitrogen; 1:1,000 dilution).

    (c) Goat anti-mouse Alexa Fluor 555 (A21425, Invitrogen; 1:1,000 dilution).

    (d) Goat anti-rabbit Alexa Fluor 555 (A21430, Invitrogen; 1:1,000 dilution).

22. Antifade with DAPI: Mix 5 ml of 1× PBS, 50 µl of 1 mg/ml DAPI stock (1 µg/ml final), 1.2 g 1,4-Diazabicyclo[2.2.2] octane (DABCO) (24 mg/ml final), and 45 ml glycerol (90% final). Dissolve well. Store at –20°C in small (~1 ml) aliquots in brown 1.5 ml tubes to protect from light. Alternatively, use commercial antifade mounting medium, such as Prolong Gold antifade reagent with DAPI (P36935, Invitrogen).

23. Parafilm (13-374-10, Fisher Scientific).

2.2. Chiasmata Prep

1. Sterile 60 mm culture dish.

2. Watch glasses.

3. Collection Medium: 45 ml Waymouth's medium with 1% penicillin–streptomycin solution, 5 ml fetal bovine serum. Add 2.5 mg/ml sodium pyruvate before use. Medium without sodium pyruvate can be stored at 4°C in the dark for up to 1 month.

4. KSOM medium (MR-020P-5F, Millipore).

5. Light mineral oil (ES-005-C, Millipore).

6. Hand-pulled micropipettes and suction mouth piece: use a fine flame and pull 100 µL capillary glass pipettes (53432-921, VWR). Snip the glass pipette in the middle. Micropipettes with small openings (about 0.3 mm in diameter) should be obtained in this way. Then attached the micropipettes to the mouth piece (A5177-5EA, Sigma).

7. Microscope slide preparation: Use china marker pen to mark the area on the underside of a microscope slide.

8. Hypotonic solution: Dissolve 1 g sodium citrate in 100 ml ddH$_2$O (1% final).

9. Carnoy's fixative: Mix three parts methanol and one part glacial acetic acid. Make fresh immediately prior to use.

10. Giemsa stain (GS500, Sigma-Aldrich).

11. Histological mounting medium (Permount SP15-100, Fisher Scientific).

12. Coplin staining jars (70315, EMS).

13. Microscope coverslip (24 mm × 60 mm).

2.3. Spindle Prep

1. Sterile 60 mm culture dish.

2. Watch glasses.

3. Lysine coated slide preparation: Make a stock solution of 2 mg/ml Poly-d-lysine hydrobromide (P1024, Sigma-Aldrich). Boil microscope slides in soapy water, rinse well, wash in 95% ethanol, and then dry either in the air or with a lint-free cloth. Immerse the bottom half of the slides in poly-d-lysine solution for 15 min at room temperature and rinse well in ddH$_2$O. Let the slides dry in the air. Then use rubber cement to make separate wells on the slides to hold medium drops.

4. Slide mailer (HS15986, Fisher Scientific).

5. Hand-pulled micropipettes and suction mouth pieces: Please refer to Subheading 2.2, item 6 for details.

6. Collection medium: Please refer to Subheading 2.2, item 3 for details.

7. KSOM medium (MR-020P-5F, Millipore).

8. Light mineral oil (ES-005-C, Millipore).

9. 5× Stabilization buffer (SB) stock: Add 7.55 g of PIPES, 0.25 g of MgCl$_2$, and 0.235 g of EGTA to 50 ml ddH$_2$O. Can be stored at 4°C for 1 month (see Note 4).

10. Spindle Fixative: Dissolve 0.2 g of PFA to 5 ml of ddH$_2$O (add one drop of 1 N NaOH and heat to 65°C to dissolve). Add 2 ml of 5× SB stock, 50 µl 20% Triton X-100 to the dissolved PFA. Bring the total volume to 10 ml with ddH$_2$O. Make fresh prior to use.

11. 0.1% NGS wash: 50 µl NGS with 50 ml 1× PBS. Can be stored at 4°C for 1 month.

12. 10% NGS wash: Add 50 µl of Triton X-100 and 5 ml of NGS to 45 ml of 1× PBS. Can be stored at 4°C for 1 month. Add 0.01 g sodium azide if long time storage is desired.

13. Ringer solution: Add 0.45 g of NaCl, 0.021 g KCl, and 0.0125 g CaCl$_2$ to 50 ml of ddH$_2$O. Sterilize with syringe filter.

14. Fibrinogen solution: Add 400 µl Ringer's solution to 0.005 g fibrinogen. Incubate at 37°C for 10 min to dissolve.

15. Thrombin: Reconstitute 250 U of lyophilized thrombin with 1 ml ddH$_2$O. Add 1.5 ml 1× PBS for a final volume of 2.5 ml. Store in 20 µl aliquots at −20°C.

16. Primary antibody: Monoclonal mouse anti-β-tubulin (T4026, Sigma-Aldrich).

17. Secondary antibody: Goat anti-mouse AlexFluor 488 (A11017, Invitrogen).

18. 2% Triton/PBS: 200 µl Triton-X-100 in 10 ml PBS.

19. 0.4% Photoflo/ddH$_2$O: 40 µl Photoflo in 10 ml ddH$_2$O.

20. Antifade with DAPI: Please refer to Subheading 2.1, item 22 for details.

21. Microscope coverslips.

3. Methods

3.1. Chromosome Spreads from Mouse Fetal Ovaries for Prophase I Analysis

Since meiosis begins during fetal development in females, ovaries obtained from fetuses at the appropriate developmental stage will be used. Timed matings need to be set up ahead of time. Noon on the day of a copulation plug is counted as embryonic day (E) 0.5 (see Note 5). The timing of meiotic progression may vary among different mouse strains. Generally, leptotene- and zygotene-stage cells can be found at E15.5-16.5. Pachytene- and diplotene-stage cells predominate at E18.5-19.5 (see Table 1).

3.1.1. Chromosome Spreading Procedure

1. Euthanize pregnant female and remove the gravid uterine horns. Place uteri containing embryos into a 100 mm culture dish containing 1× PBS on ice.

2. Dissect ovaries from female mouse embryos. Do one embryo at a time and maintain the rest in 1× PBS on ice. Remove the embryo from uterus, decidual tissue, and yolk sac. Sacrifice the embryo by removing the head and tape the torso down on a paper towel, ventral surface up. A testis should be easily distinguished from an ovary at this stage. Testes are located in the caudal region of the pelvis while ovaries are located near the kidneys (see Fig. 2). To locate the ovaries, first identify the bladder and, protruding out from there, identify each uterine horn traveling outwards and anteriorly from the bladder. Follow the

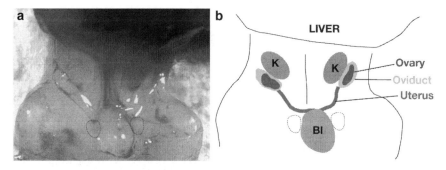

Fig. 2. Illustration of late stage mouse fetal ovaries and testes. (**a**) The abdominal part of a 16.5 dpc female embryo is shown with the abdominal cavity open and liver/intestines flipped to the side. The *dotted circles* indicate the location of the testes in a male embryo at this age. Panel (**b**) represents the cartoon version of (**a**) with the neighboring organs labeled. *K* kidney, *Bl* bladder.

uterine horns up towards the kidney, being careful not to tear each horn, until you locate the larger clear "jelly like" structure at the end of each horn. This is the ovary and oviduct. Pinch off entire structure with forceps or scissors. You can dissect the ovary away from the other tissues in a drop of PBS.

3. Place ovaries in HEB and incubate on ice for 30 min (see Note 6).

4. Add 40–60 µl hypotonic sucrose solution to the concave well of the concavity slide and transfer two to three pairs of fetal ovaries to the sucrose drop (or 20 µl hypotonic sucrose solution to one pair if embryos are being treated separately). Use one surgery scalpel blade (size #11) and one 30-guage needle to tease the ovaries apart and release cells. Remove the larger pieces of ovaries from the sucrose solution and pipette sucrose solution up and down to resuspend cells.

5. Place 35 µl PFA drops on each well of the six-well slides. Pipette 5 µl of the sucrose solution to each PFA drop until the sucrose solution is exhausted. You should have enough cell slurry to prepare 2 six-well slides from three pairs of fetal ovaries. Again, if each pair of ovaries is being treated separately, apply cell slurry from embryo 1 to well 1 of each of your slides. Then apply the cell slurry from embryo 2 to well 2 of each slide, and so on. In this way, each of the slides will have 6 wells, each containing cells from a distinct animal. The entire slide will be processed together, such that each well will be subjected to the same treatment, and hence each slide will contain a mixture of genotypes ALL receiving the same antibodies.

6. Incubate the six-well slides in a closed humid chamber overnight at 4°C.

7. Open the chamber and let the slides dry completely.

8. Wash slides in 0.4% Photoflo/H_2O for 5 min. Repeat the wash two more times and then air-dry. At this point, the slides can be either used for immunostaining or stored at −80°C (see Note 7).

3.1.2. Indirect Immunofluorescence of Chromosomes from Prophase I Oocytes

1. Blocking for primary antibody: Soak the slides in 0.4% Photoflo/PBS for 10 min, then in 0.1% Triton/PBS for 10 min, followed by another 10 min in 10% ADB/PBS. From this point onwards, never allow slides to dry out.

2. Dilute primary antibodies in ADB (see Note 8). Maintain antibodies on ice until needed.

3. Cut one piece of parafilm and place it in the central dry area of the humid chamber. Use strips of parafilm to lift the slides off the surface of the underlying film and create room for the antibody dilutions. Add 80 µl of antibody dilution on the parafilm.

Put the slides face down to the antibody drop on the parafilm and expel air bubbles out. Seal the humid chamber with saran wrap.

4. Incubate overnight at room temperature.

5. Remove the slides from the humid chamber and repeat the blocking steps: 10 min in 0.4% Photoflo/PBS, 10 min in 0.1% Triton/PBS, and 10 min in 10% ADB/PBS.

6. Dilute secondary antibodies in ADB.

7. With the same method described in step 3 of this section, add 80 μl of secondary antibody dilution on the parafilm in a humid chamber and place the slides face down to the antibody dilution drop (see Note 9).

8. Seal the humid chamber and wrap the box with aluminum foil as well.

9. Incubate in the dark for 2 h at 37°C.

10. Carefully lift the slides off the parafilm. It is often useful to flood the parafilm with 0.4% Photoflo/PBS first. Transfer slides to Coplin jars and wash twice in 0.4% Photoflo/PBS for 5 min and twice in 0.4% Photoflo/H_2O for 5 min.

11. Air-dry the slides thoroughly in the dark and add antifade with DAPI. Then mount with glass coverslip and blot excess liquid (see Note 10).

12. Dry on the bench for at least 4 h before visualizing. The slides can be stored at 4°C in the dark (before or after drying) or immediately visualized using an epifluorescence microscope (see Fig. 3, for example).

3.2. Chiasmata Preparation Using Diakinesis Stage Mouse Oocytes

The following protocol is modified from the method described previously by Tarkowski and others (15, 19–21). A nice scattering of oocyte nuclei and spreading of chromosomes can be obtained using this technique. It can be applied not only to chiasmata analysis on diakinesis-stage oocytes but also to studies for meiosis I disjunction study on metaphase II arrested oocytes.

1. Remove ovaries from unstimulated female mice at 24–28 days of age (see Note 11), remove fat and other tissue, and transfer them into the oocyte collection media.

2. Grasp ovary with blunt forceps in one hand and release the oocytes from the ovaries by puncturing the ovaries with one 30-gauge needle in the other hand.

3. Transfer the oocytes with germinal vesicle (nuclear membrane) intact into 20 μl KSOM drop under light mineral oil using mouth pipette and hand-pulled pipette (see Note 12).

4. Incubate the oocytes in a cell culture incubator (37°C, 5% CO_2) until the oocytes enter metaphase I (see Note 13).

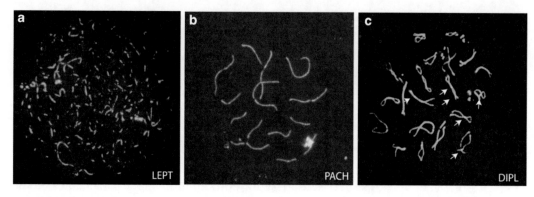

Fig. 3. Examples of female chromosome spreads made from fetal mouse ovaries. Shown are chromosome spreads immunostained for SYCP3 (*green*), centromeres (using CREST autoimmune serum, *green*), and MLH1 (*red*) at leptonema (LEPT), pachynema (PACH), and diplonema (DIPL). The structure of the synaptonemal complex (SC), which is stained with SYCP3 antiserum in this figure, changes during the progression of prophase I. (**a**) The components of SC start to accumulate on the chromosome core during leptonema, shown as fragmented SYCP3 signals. (**b**) In pachynema, the homologous chromosomes are fully synapsed illustrated by associated SYCP3 at full length. MLH1, a key marker of the homologous recombination site, is detected in this stage, shown as *red/orange* foci on the chromosome core. (**c**) Homologous chromosomes start to desynapse during diplonema. MLH1 foci (*yellow dots* indicated by *arrow*) are gradually lost from the crossing over site. SYCP3 is represented by the *green* signal in (**a–c**). The centromeres are detected by CREST autoimmune serum (*blue* in all three).

5. Move oocytes to hypotonic solution and incubate for 15 min at room temperature.

6. Use a fine hand-pulled micropipette and transfer a small drop (1–2 μl) of hypotonic solution onto the upper surface of the slides with markers underneath.

7. Transfer a few oocytes from hypotonic solution to the drop on the slides. Remove excess fluid from the drop until the oocytes appear to be attached to the slides (see Note 14).

8. Add one drop of Carnoy's fixative onto the top of the oocytes without dispersing them. Pause briefly to let the fixative disperse and quickly add another two drops of fixative (see Note 15, Fig. 4).

9. Allow the slides to air-dry. The spreading of chromosome will be achieved during this drying process (see Note 16).

10. Stain the slides with Giemsa solution for 3 min. Wash the slides with water for three times, 3 min each.

11. Air-dry completely. Load mounting medium and put on coverslip carefully to avoid air bubbles. When the slide is dried, the oocytes can be visualized under the microscope.

3.3. Oocyte Spindle Preparation

This method is designed to observe the metaphase spindle configuration in oocytes following meiotic resumption. The three-dimensional structure of spindles and the developing oocytes are

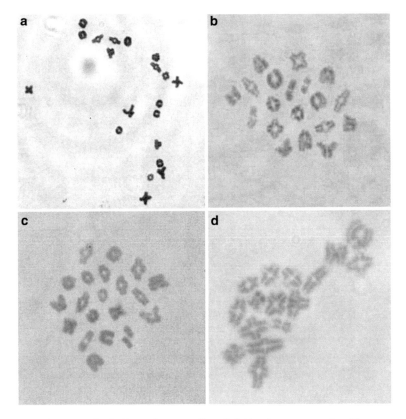

Fig. 4. Giemsa-stained chiasmata preparation of diakinesis mouse oocytes. Shown are giemsa-stained chiasmata preparations of wild type mouse oocytes under different evaporation conditions. The density of the chromosomes will vary widely responding to subtle changes of the local humidity level. The slower the fixative evaporates, the more condense the chromosomes appear. When the evaporation is sped up, the chromosomes tend to be more spread out. (**a–d**) Oocytes at diakinesis stage are fixed under slow to higher evaporation conditions.

well preserved using this technique, which makes the temporal analysis of chromosome behavior during this process possible (21) (see Fig. 5).

1. Follow steps 1–4 in Subheading 3.2 to collect oocytes from ovaries of 24–28-day-old mice and culture them in KSOM medium until the oocytes reach metaphase I or metaphase II (see Note 13).

2. Place four to six 2 μl drops of fibrinogen solution within each rubber cement well on the poly-lysine coated slides and cover with oil. Incubate the slides at 37°C for 10 min.

3. Transfer five to ten oocytes to each fibrinogen drop (see Note 17).

4. Add 2 μl thrombin to the fibrinogen drop using fine hand-pulled micropipette (see Note 18).

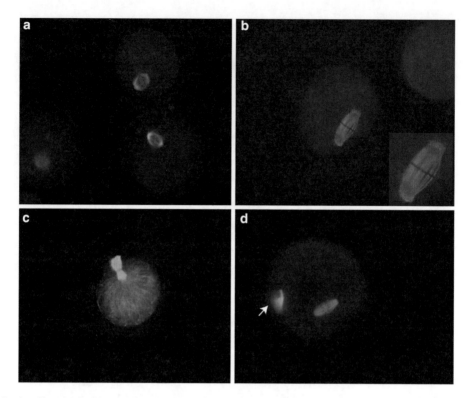

Fig. 5. Confocal images of wild type mouse oocytes from spindle preparation with immunofluorescence staining of β-tubulin. Shown are oocyte spindle preparations immunostained for β-tubulin (*green*) and DNA (*magenta*). The transition from metaphase I to metaphase II can be observed by culturing the mouse oocytes and then fixing them at different time points after the resumption of meiosis. (**a**) 7 h after germinal vesicle break down (GVBD), the oocytes are in early metaphase I with chromosomes align on the spindle equator. (**b**) An oocyte with a characteristic anaphase I configuration with chromosomes on the opposite poles of the spindle is shown. (**c**) The first polar body is being extruded out in telophase I. (**d**) The oocytes get arrested again in metaphase II with the first polar body (*arrow*). Spindle is visualized by β-tubulin staining (*green*); chromosomes are detected by DAPI (*magenta*).

5. Keep the slides at 37°C for 1 min and check the formation of the clot.

6. Carefully drain off oil and rinse the slide well in PBS containing 2% Triton X-100 for 3 min.

7. Put the slide in Spindle Fixative solution for 30 min in a 37°C incubator.

8. Wash in 0.1% NGS for 15 min at 37°C.

9. Transfer the slide to 10% NGS at 37°C for 1 h. Alternatively, the slide can be stored at 4°C in 10% NGS overnight.

10. Dilute mouse anti-β-tubulin antibody (1:800) in 10% NGS (see Note 19). Add the diluted antibody to the slides and incubate at 37°C for 1 h.

11. Wash the slides in 10% NGS for 10 min at 37°C. Repeat the washing two more times.

12. Apply diluted secondary antibody in 10% NGS (1:1,000) to the slides for 1 h at 37°C.

13. Wash the slide in 10% NGS for 10 min at 37°C. Repeat the washing two more times.

14. Briefly immerse the slide in ddH$_2$O containing 0.4% Photoflo.

15. Carefully remove the rubber cement well around the clots.

16. Mount antifade with DAPI to the slide and put on coverslip. Blot the excess mounting medium and seal with rubber cement.

17. Dry in the dark for at least 2 h at room temperature or overnight at 4°C before analyzing under fluorescence microscope.

4. Notes

1. These slides can be reused by washing them well with soapy water, followed by 70% ethanol and distilled water consecutively. Let the slides dry in the air and store them in a clean box.

2. The HEB, hypotonic sucrose solution, and 1% PFA should be made fresh within 2 h of use and maintain on ice.

3. If anti-goat primary antibody will be used later, replace NGS with serum from the same species in which the secondary antibodies are raised (e.g., donkey or chicken serum).

4. Notice: PIPES will not dissolve until pH reaches 6.1–7.5. Add solid NaOH a few pellets a time until dissolve.

5. Plug date is not always an accurate indicator of gestational age. The actual date may vary ±0.5 day.

6. If mice are of a specific genotype, keep each pair of ovaries separate from those of other embryos in the litter. If tissue is required for genotyping, obtain this from inside the peritoneal cavity of the embryo because the outside surface of each embryo will be contaminated with maternal tissue/fluids.

7. The slides can be stored in a −80°C freezer up to 1 week. Since some epitopes are not very stable, immediate staining of the slides is highly recommended.

8. Antibodies from different vendors or different batches may vary significantly. The dilution of antibodies may require adaptation from time to time.

9. It is ESSENTIAL not to let the slides dry out in this step. So 80 µl is the minimum amount you should use. If in doubt, use more (up to 120 µl or so).

10. It is very important to press the coverslip down firmly at several points and to "wiggle" the coverslip gently across the surface of the slide. This helps to flatten the coverslip on the slide, prevents uneven build-up of antifade, and removes excess antifade.

11. Although one should be able to collect enough oocytes from female mice at this age, stimulation by exogenous hormone can also be used to increase the yield of the eggs harvested. Intraperitoneal injection of two hormones will be needed in this procedure: pregnant mare serum gonadotropin (PMSG) and human chorionic gonadotropin (hCG). These two hormones are used to mimic the actions of FSH and LH, respectively. The injection timing is dependent on the light/dark cycle of the mouse room. For example, in a room following a 7 am to 7 pm light cycle, inject 5 IU PMSG to each female in the late afternoon on day 1. After 46–48 h, inject each animal with 5 IU hCG (day 3). Oocytes are supposed to be ovulated 10–13 h after hCG injection (day 4). Note that oocytes obtained by superovulation are often more difficult to handle and more susceptible to damage. To increase the efficiency of the hormonal regimen, always use mice as young as possible (3 weeks of age is optimal).

12. Move the oocytes through two to three 20 μl KSOM drops to dilute the somatic cells transferred with oocytes. Each 20 μl KSOM drop can hold up to 30 oocytes.

13. The cell cycle progression may vary among different mouse strains. For the most commonly used C57B6 strain, it usually takes about 2.5 h for the oocytes to loose their germinal vesicle, which is a sign of meiotic resumption. 8–10 h after break down of germinal vesicle, most oocytes will in metaphase I. For observation of oocytes at metaphase II, overnight culture (16–18 h) can be used.

14. It is critical to have the right amount of fluid with the oocytes. If too much fluid remains, the oocytes will not adhere onto the surface of the slide and easily get washed away later by fixative. Meanwhile, it is also necessary to have enough fluid with the oocytes to prevent drying before fixative is applied.

15. Three drops of freshly prepared fixative are recommended for most conditions. However, the total number of the drops added to the oocytes also depends on the humidity condition of the room where the experiment is performed (see Fig. 4). Thus, the optimal amount of the fixative used for a particular experiment needs to be tested individually.

16. The chromosomes can be observed as light reflecting spots under dissecting microscope during drying but become much less visible when the slide is totally dry. So it will be helpful for

later visualization to remember the location of the oocytes using markers underneath the slides before the fixative on the slide is completely dried.

17. If desired, wash the oocytes through two to three fibrinogen drops. This could help a lot in later staining step since extra somatic cells will increase background staining.

18. Inject thrombin into fibrinogen drop slowly and evenly throughout the drop from different directions. Don't leave the tip of glass pipette with thrombin in the same spot for too long. The tip may get stuck!

19. Depending on the batch and freshness of this antibody, the dilution factor may vary from 200 to 1,000 in our hands.

Acknowledgements

This work was supported by funding from the NICHD to P.E.C. (HD041012) and by a student fellowship from the Cornell Center for Vertebrate Genomics to X.S. We thank members of the Cohen lab for their helpful comments and suggestions during the preparation of this manuscript.

References

1. Hassold T, Hunt P (2001) To err (meiotically) is human: the genesis of human aneuploidy. Nat Rev Genet 2:280–291

2. Hunt PA, Hassold TJ (2008) Human female meiosis: what makes a good egg go bad? Trends Genet 24:86–93

3. Page SL, Hawley RS (2004) The genetics and molecular biology of the synaptonemal complex. Annu Rev Cell Dev Biol 20:525–558

4. Morelli MA, Cohen PE (2005) Not all germ cells are created equal: aspects of sexual dimorphism in mammalian meiosis. Reproduction 130:761–781

5. Hassold T, Hall H, Hunt P (2007) The origin of human aneuploidy: where we have been, where we are going. Hum Mol Genet 16 Spec No. 2: R203–R208

6. Cohen PE, Holloway JK (2010) Predicting gene networks in human oocyte meiosis. Biol Reprod 82:469–472

7. Hulten MA, Patel S, Jonasson J, Iwarsson E (2010) On the origin of the maternal age effect in trisomy 21 Down syndrome: the Oocyte Mosaicism Selection model. Reproduction 139:1–9

8. Lenzi ML, Smith J, Snowden T, Kim M, Fishel R, Poulos BK, Cohen PE (2005) Extreme heterogeneity in the molecular events leading to the establishment of chiasmata during meiosis I in human oocytes. Am J Hum Genet 76: 112–127

9. Yuan L, Liu JG, Zhao J, Brundell E, Daneholt B, Hoog C (2000) The murine SCP3 gene is required for synaptonemal complex assembly, chromosome synapsis, and male fertility. Mol Cell 5:73–83

10. Yuan L, Liu JG, Hoja MR, Wilbertz J, Nordqvist K, Hoog C (2002) Female germ cell aneuploidy and embryo death in mice lacking the meiosis-specific protein SCP3. Science 296:1115–1118

11. Hamer G, Novak I, Kouznetsova A, Hoog C (2008) Disruption of pairing and synapsis of chromosomes causes stage-specific apoptosis of male meiotic cells. Theriogenology 69:333–339

12. Pelttari J, Hoja MR, Yuan L, Liu JG, Brundell E, Moens P, Santucci-Darmanin S, Jessberger R, Barbero JL, Heyting C, Hoog C (2001) A meiotic chromosomal core consisting of cohesin complex proteins recruits DNA recombination

proteins and promotes synapsis in the absence of an axial element in mammalian meiotic cells. Mol Cell Biol 21:5667–5677

13. Borum K (1961) Oogenesis in the mouse. Exp Cell Res 24:495–507

14. Evans CW, Robb DI, Tuckett F, Chanlloner S (1982) Regulation of meiosis in the foetal mouse gonad. J Embryol Exp Morphol 68: 59–67

15. Kan R, Sun X, Kolas NK, Avdievich E, Kneitz B, Edelmann W, Cohen PE (2008) Comparative analysis of meiotic progression in female mice bearing mutations in genes of the DNA mismatch repair pathway. Biol Reprod 78: 462–471

16. Kolas NK, Svetlanov A, Lenzi ML, Macaluso FP, Lipkin SM, Liskay RM, Greally J, Edelmann W, Cohen PE (2005) Localization of MMR proteins on meiotic chromosomes in mice indicates distinct functions during prophase I. J Cell Biol 171:447–458

17. Wojtasz L, Daniel K, Roig I, Bolcun-Filas E, Xu H, Boonsanay V, Eckmann CR, Cooke HJ, Jasin M, Keeney S, McKay MJ, Toth A (2009) Mouse HORMAD1 and HORMAD2, two conserved meiotic chromosomal proteins, are depleted from synapsed chromosome axes with the help of TRIP13 AAA-ATPase. PLoS Genet 5:e1000702

18. Bolcun-Filas E, Bannister LA, Barash A, Schimenti KJ, Hartford SA, Eppig JJ, Handel MA, Shen L, Schimenti JC (2011) A-MYB (MYBL1) transcription factor is a master regulator of male meiosis. Development 138:3319–3330

19. Williams DL, Lafferty DA, Webb SL (1970) An air drying method for the preparation of dictyotene chromosomes from ovaries of Chinese hamsters. Stain Technol 45:133–135

20. Tarkowski AK (1966) An air-drying method for chromosome preparations from mouse eggs. Cytogenetics 5:394–400

21. Woods LM, Hodges CA, Baart E, Baker SM, Liskay RM, Hunt PA (1999) Chromosomal influence on meiotic spindle assembly: abnormal meiosis I in female Mlh1 mutant mice. J Cell Biol 145:1395–1406

Chapter 2

Analysis of Recombinants in Female Mouse Meiosis

Esther de Boer, Maria Jasin, and Scott Keeney

Abstract

During meiosis, homologous chromosomes (homologs) undergo recombinational interactions, resulting in the formation of crossovers (COs) or noncrossovers (NCOs). Both COs and NCOs are initiated by the same event: programmed double-strand DNA breaks (DSBs), which occur preferentially at hotspots throughout the genome. COs contribute to the genetic diversity of gametes and are needed to promote proper meiotic chromosome segregation. Accordingly, their formation is tightly controlled. In the mouse, the sites of preferred CO formation differ between male and female chromosomes, both on a regional level and on the level of individual hotspots. Sperm typing using (half-sided) allele-specific PCR has proven a powerful technique to characterize COs and all detectable NCOs at hotspots on male human and mouse chromosomes. In contrast, very little is known about the properties of hotspots in female meiosis. This chapter describes an adaptation of sperm typing to analyze recombinants in a hotspot, using DNA isolated from an ovary cell suspension enriched for oocytes.

Key words: Meiosis, Recombination, Crossover, Noncrossovers, Hotspot, Allele-specific PCR, Oocyte

1. Introduction

During meiosis, homologous chromosomes undergo recombinational interactions, which can yield crossovers (COs) or noncrossovers (NCOs). Studies in yeast have shown that both COs and NCOs initiate from the same event: the formation of programmed double-strand breaks (DSBs) (1). Subsequent repair of the DSBs through diverging pathways leads to the formation of the two different end products (2, 3). Whereas CO recombination leads to a reciprocal exchange of large chromosomal segments, NCO recombination results in a highly localized unidirectional transfer of genetic information.

Meiotic recombination events are not distributed evenly across the genome but tend to cluster in regions called hotspots (4). Analysis in sperm of human and mouse has shown that CO

Hayden A. Homer (ed.), *Mammalian Oocyte Regulation: Methods and Protocols*, Methods in Molecular Biology, vol. 957, DOI 10.1007/978-1-62703-191-2_2, © Springer Science+Business Media, LLC 2013

frequencies can vary considerably between hotspots, from 0.0004 to 2% (5, 6). Mammalian CO hotspots typically span 1–2 kb, often with the peak of COs and NCOs in the center (7, 8). The relative proportions of detectable COs and NCOs can vary widely from hotspot to hotspot, from a more than tenfold excess of COs to almost tenfold fewer COs than NCOs (9, 10).

The greater part of current knowledge of the relation between COs and NCOs in mammalian hotspots is derived from experiments with sperm from mice or men. However, large-scale CO analysis has shown that mouse chromosomes show sex-specific differences in CO recombination, both on a regional level and on the level of individual hotspots (11–13). What underlies these sex-specific differences remains elusive. Fully understanding the mechanisms and regulation of meiotic recombination thus requires parallel analysis of COs and NCOs in both male and female meiosis.

This chapter describes a strategy to analyze CO and NCO recombination in hotspots in female meiosis and assumes the hotspot of choice is already identified. For the identification of hotspots in the mouse see refs. 14, 15. The strategy we use to analyze COs and NCOs is based on allele-specific PCR, which has been very powerful in the analysis of mammalian CO hotspots in sperm (16, 17). We adapted this technique to allow parallel detection of COs and NCOs in mouse oocytes. Allele-specific PCR makes use of the presence of sequence polymorphisms (usually single nucleotide polymorphisms (SNPs), but also short insertion/deletions (indels)) between the parental haplotypes across the hotspot, both for the amplification of recombinants and for the mapping of COs and NCOs (Fig. 1). Inbred mouse strains with sufficient polymorphisms across the hotspot are crossed and recombination is analyzed in small pools of DNA purified from ovaries of the F1 hybrid mice. COs are selectively amplified by PCR with allele-specific forward and reverse primers (Fig. 1b), located outside the hotspot region. A half-sided allele-specific PCR with a pair of nested allele-specific primers on one side and nested non-allele-specific (universal) primers on the other allows parallel recovery of both COs and NCOs (Fig. 1c). The remaining internal polymorphisms are subsequently typed by hybridizing the PCR products with allele-specific oligos (ASOs) to map the CO and NCO breakpoints. The advantage of this technique is that recombination activity at all polymorphisms across a hotspot can be analyzed within the same experiment. In mouse, conversion tracts associated with NCOs are very short, typically 100 bp or less (10), so the detection of NCOs depends highly on the polymorphism density in the hotspot. This dependence may partially explain the above-mentioned variety in observed CO to NCO ratios between different hotspots in male meiosis. Additional information on allele-specific PCR and breakpoint mapping, but in sperm, is provided in refs. 18, 19. Additional information on the parallel

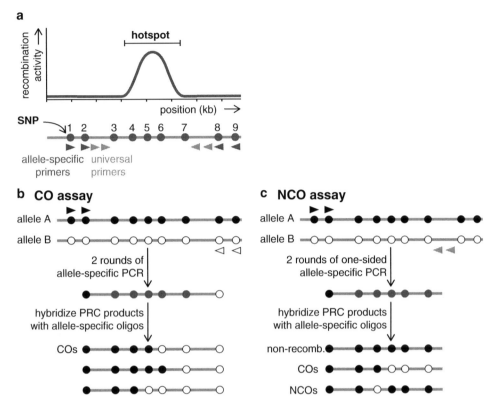

Fig. 1. Allele-specific PCR to amplify COs and NCOs across a hotspot. (**a**) Positions of PCR primers in relation to the hotspot. Depicted is a schematic representation of the sequence spanning a hotspot. Polymorphisms (e.g., SNPs) between the parental haplotypes are indicated as *closed circles* and *numbered*. In this example, allele-specific primers (*dark gray arrowheads*) are directed against SNPs 1, 2, 8, and 9. The remaining polymorphisms (SNPs 3–7) will later be used to map the COs and NCOs. Universal primers (*light gray arrowheads*) are located inside the positions of the allele-specific primers but still outside the hotspot region. (**b**) CO assay. Two rounds of nested allele-specific PCR are performed on pools of ovary DNA from the F1 of a cross between two mouse strains that have multiple SNPs across the hotspot. *Black and white circles* represent SNPs of the A and B parental haplotypes, respectively. Allele-specific primers are shown as *black or white arrowheads*. Only CO molecules will be amplified. To map the CO breakpoints, the remaining SNPs (*gray circles*) are typed by hybridizing the PCR products with allele-specific oligos (ASOs). (**c**) NCO assay. Two rounds of nested PCR are performed on small pools of F1 hybrid ovary DNA, with allele-specific primers on one side and universal primers on the other. This type of PCR will amplify non-recombinant molecules of one haplotype (the majority of PCR products) as well as both CO and NCO molecules. To map the CO and NCO breakpoints, the remaining SNPs (*gray circles*) are typed by hybridizing the PCR products with ASOs. Because the majority of the PCR products will be non-recombinants, only ASOs specific for the non-selected haplotype (in this example: allele B (*white*)) are used in SNP typing.

detection of recombinants in both male and female meiosis, focused on detection of NCOs involving a single polymorphism within a hotspot, is provided in ref. 20.

Compared to male meiosis, the analysis of COs and NCOs in female meiosis poses more of a challenge, mainly at the stage of sample collection and preparation. The number of oocytes that can be isolated from one female animal is about a 1,000-fold lower than the number of sperm cells that can be isolated from one male. A second problem is the high number of somatic cells in ovary samples. Sperm samples easily reach a purity of >99%, whereas the fraction of oocytes in an ovary cell suspension is usually between 10 and 15%. We adapted a technique used in in vitro fertilization research to isolate oocytes (21) to enrich the ovary cell suspension for oocytes, up to 60%.

2. Materials

To prevent contamination, keep reagents and materials "before PCR" and "after PCR" separate. Ideally, a separate set of pipettes is dedicated to the "before PCR" steps. In addition, equipment (pipettes, tube-racks, etc.) that is used "before PCR" should be regularly exposed to UV light to inhibit amplification of possible contaminant DNA.

2.1. Preparation of an Ovary Cell Suspension and Enrichment for Oocytes

1. F1 hybrid female newborn mouse (see Note 1).

2. Scissors and forceps for dissection. Forceps #5 and #5/45 are recommended for dissecting the ovaries.

3. Dissecting needles.

4. Small petri dishes (∅ 35 mm).

5. Small glass embryo dish.

6. Clean slides. Keep slides for at least 20 min in 100% ethanol, transfer to a slide rack, and submerge in boiling filtered ddH$_2$O for 25 min. Air-dry and store in a dust-free environment (see Note 2).

7. Tabletop microcentrifuge.

8. Phosphate buffered saline (PBS): 140 mM NaCl, 1.9 mM NaH$_2$PO$_4$, 8.9 mM Na$_2$HPO$_4$, pH 7.3–7.4. Filter-sterilize and store at room temperature.

9. Fetal calf serum (FCS). Make 1 ml aliquots and store at –20°C.

10. Collagenase: 20 mg/ml in PBS. Make fresh and keep on ice until use.

11. DNase 1: 400 µg/ml in PBS. Store at –20°C. Thaw and keep on ice until use.

12. 20× Saline-sodium citrate buffer (SSC): 3 M NaCl, 300 mM sodium citrate-dihydrate. Adjust pH to 7.0, sterilize by autoclaving, and store at room temperature. Dilute with ddH$_2$O to make 1× SSC. Filter-sterilize and store at room temperature.

2.2. Preparation of Slides and Quantification of Oocytes

1. 12-well microscope slides, clean (see item 6 in Subheading 2.1), and coverslips.

2. Humidified box: use a box that can be closed off completely, line the bottom with water-saturated filter paper or tissues.

Rest the slides on supports to avoid direct contact of the slides with the paper.

3. Heating plate at 37°C.

4. Tabletop microcentrifuge.

5. Coplin jars.

6. Nail polish (clear) or rubber cement.

7. Fluorescence microscope.

8. Filtered ddH$_2$O.

9. 100 mM sucrose, filter-sterilize and store at room temperature.

10. 2% (w/v) paraformaldehyde (PFA), pH 9.2, 0.15% Triton X-100. Prepare fresh. Add 0.5 g of PFA to 20 ml of ddH$_2$O, add 1–2 drops 1 N NaOH, and dissolve by stirring at 60°C. Cool to room temperature and adjust the pH to 9.2. Add ddH$_2$O to a final volume of 25 ml and filter through a 0.2 μm filter. Add 37.5 μl Triton X-100 and mix well. Use on the same day.

11. 0.4% (v/v) Photoflo (Kodak) in filtered ddH$_2$O. Store at room temperature.

12. PBS (see Subheading 2.1).

13. Blocking buffer: 5% (w/v) nonfat dry milk, 5% (v/v) goat serum, 0.01% (w/v) sodium azide (NaN$_3$) in PBS. Store at –20°C. Prior to use, add phenyl methyl sulfonyl fluoride (PMSF, make a 1 M stock in DMSO, store at –20°C) to a final concentration of 1 mM (see Note 3). Centrifuge at maximum speed in a tabletop microcentrifuge for 30 min and use the supernatant within 2 h.

14. A primary antibody against a meiosis- or oocyte-specific antigen (e.g., SYCP3).

15. The appropriate secondary antibody conjugated to a fluorescent label.

16. Mounting medium with DAPI (4′,6-diamidino-2-phenylindole): 10 μg/ml DAPI in Vectashield (Vector Laboratories Inc.). Store in the dark at 4°C.

2.3. DNA Extraction

1. Sterile petri dish.

2. Sterile razor blade or scalpel.

3. Size 80 μm mesh (Screens for CD-1, size 80 mesh; Sigma-Aldrich).

4. Tabletop microcentrifuge.

5. Heating block or water bath at 55°C.

6. Spectrophotometer.

7. 1× SSC (see Subheading 2.1).

8. β-Mercaptoethanol.

9. 10% (w/v) sodium dodecyl sulfate (SDS).

10. Proteinase K: 20 mg/ml in ddH$_2$O. Make fresh and keep on ice until use.

11. Phenol:Chloroform:Isoamyl alcohol 25:24:1 (v/v/v), saturated with 100 mM Tris–HCl pH 8.0.

12. Ethanol, 100% and 80%.

13. 3 M sodium acetate, pH 5.2.

14. 5 mM Tris–HCl, pH 7.5

15. Optional (see Note 4): Long polyacrylamide (LPA; e.g., GenElute™-LPA, Sigma-Aldrich).

16. Loading dye: 30% (v/v) glycerol in electrophoresis buffer of choice, add bromophenol blue to give blue color. Store at room temperature.

2.4. Allele-Specific PCR and Half-Sided Allele-Specific PCR

1. Thermal cycler.

2. 11.1× PCR buffer: 495 mM Tris–HCl pH 8, 121 mM ammonium sulfate, 49.5 mM MgCl$_2$, 73.7 mM β-mercaptoethanol, 48.4 μM EDTA pH 8, 11 mM dATP, 11 mM dTTP, 11 mM dGTP, 11 mM dCTP, 1.24 mg/ml bovine serum albumin (non-acetylated ultrapure BSA, Ambion). Store in 100–500 μl aliquots at –20°C (see Note 5).

3. 2 M Tris base (Tris(hydroxymethyl)aminomethane, Ultra grade for molecular biology, Fluka Chemie).

4. Taq polymerase.

5. Turbo *pfu* polymerase (Stratagene).

6. Genomic DNA of both parental genotypes. This can be prepared as described (see Subheading 3.3) or ordered from the Jackson Laboratory (http://www.jax.org/dnares/index).

7. S1 nuclease, diluted to 10 U/μl in S1 storage buffer (20 mM Tris–HCl pH 7.5, 50 mM NaCl, 0.1 mM zinc acetate, 50% (v/v) glycerol). Store at –20°C for up to 6 months.

8. S1 digestion mix: 20 mM sodium acetate pH 4.9, 1 mM zinc acetate, 100 mM NaCl. Aliquot and store at –20°C. Add 0.7 U/μl of S1 nuclease before use.

9. Dilution buffer: 10 mM Tris–HCl pH 7.5, 5 μg/ml sonicated salmon sperm DNA.

10. Loading dye (see Subheading 2.3).

2.5. Mapping CO and NCO Breakpoints by Hybridization with ASOs

1. 96-well dot blot manifold.

2. Hybridization oven with rotator.

3. Hybridization bottles.

4. Screw-cap microcentrifuge tubes.

5. Phosphorimager screen.

6. Geiger counter.

7. Whatmann filter paper.

8. Hybond-XL nylon membrane (Amersham/GE Healthcare).

9. Hybridization mesh (optional).

10. Denaturation buffer: 0.5 M NaOH, 2 M NaCl, 25 mM EDTA. Store at room temperature.

11. 2× SSC and 3× SSC: make 20× SSC (see Subheading 2.1), dilute in ddH_2O to make 2× and 3× SSC.

12. 10 mCi/ml (γ^{32}-P) ATP.

13. T4 polynucleotide kinase.

14. 10× kinase mix: 700 mM Tris–HCl pH 7.5, 100 mM $MgCl_2$, 50 mM spermidine trihydrochloride, 20 mM dithiothreitol (DTT). Aliquot and store at –20°C.

15. Sonicated salmon sperm DNA.

16. Tetramethylammonium chloride (TMAC) hybridization solution: 3 M TMAC, 0.6% (w/v) SDS, 10 mM sodium phosphate pH 6.8, 1 mM EDTA, 4 μg/ml yeast RNA, in 5× Denhardt's solution (see Note 6). Store at 4°C, prewarm before use.

17. TMAC wash solution: 3 M TMAC, 0.6% (w/v) SDS, 10 mM sodium phosphate pH 6.8, 1 mM EDTA (see Note 6). Store at 4°C, prewarm before use.

18. 0.1% (w/v) SDS.

3. Methods

3.1. Preparation of an Ovary Cell Suspension and Enrichment for Oocytes

1. Kill the newborn pups by decapitation.

2. Dissect the ovaries from the female newborns and store them in fresh PBS on ice until further use (see Note 7). Also collect a somatic tissue (e.g., liver or spleen) and store at –20°C. This will be used as a negative control for meiotic recombination.

3. Decapsulate each ovary on a clean slide with a droplet (25–50 μl) of PBS under the dissection microscope. Repeat for all collected ovaries.

4. Collect the ovaries in a small embryo dish containing 200 μl fresh PBS + 5% FCS at room temperature until all ovaries have been decapsulated.

5. Place the embryo dish under the dissection microscope and add 20 μl collagenase and 5 μl DNase 1 (see Note 8).

6. Using the dissecting needles, puncture and gently shake the ovaries at regular intervals for 1 h.

7. Remove as much debris as possible.

8. Collect the cell suspension in an eppendorf tube. Add PBS + 5% FCS to a final volume of 1.5 ml.

9. Centrifuge in a tabletop microcentrifuge for 2.5 min at 800 rpm ($60 \times g$). Discard the supernatant.

10. Gently resuspend the cell pellet in 1.5 ml PBS.

11. Centrifuge in a tabletop microcentrifuge for 2.5 min at 800 rpm ($60 \times g$). Discard the supernatant.

12. Resuspend the cells in 100–200 μl 1× SSC. Take a small aliquot (5–15 μl) of the cell suspension for slide preparation and quantification, store the remainder at –20°C until further use (see Note 9).

3.2. Preparation of Slides and Quantification of Oocytes

1. Place a clean 12-well microscope slide in a humidified box. Put a 5 μl drop of 100 mM sucrose in one or more wells.

2. Add 5 μl of the ovary cell suspension to the drop of 100 mM sucrose, put the lid on the humidified box and wait for 5 min.

3. Add 10 μl of 2% PFA, 0.15% Triton X-100 to the drop, and close the humidified box for ~1 h.

4. Remove the lid and leave for ~30 min.

5. Transfer the slide to a 37°C heating plate and allow the drop to dry to the consistency of toffee. This takes about 10~15 min.

6. Rinse the slide three times in filtered ddH$_2$O, dip once in 0.4% Photoflo, and finish with a rinse with 0.4% Photoflo, using a pipette.

7. Air-dry the slides at room temperature.

8. Either wrap the slides in aluminum foil and store at –80 °C until further use (see Note 10) or continue with immunocytological labeling (steps 9–25).

9. Wash the slide three times for 5 min in PBS.

10. Cover the slide with 500 μl blocking buffer.

11. Incubate for 30 min in a closed humidified box at room temperature.

12. Prepare 100 μl per slide of a meiosis- or oocyte-specific primary antibody, diluted in blocking solution (see Note 11).

13. Centrifuge the dilution for 30 min at maximum speed in a tabletop microcentrifuge at 4°C.

14. Pour the blocking buffer off of the slides and add 100 μl of the supernatant of centrifuged primary antibody dilution to the slide (see Note 12).

15. Cover the slide with a coverslip and incubate in a closed humidified box for ~2 h at 37°C (see Note 11).

16. Wash the slides three times for 5 min in PBS.

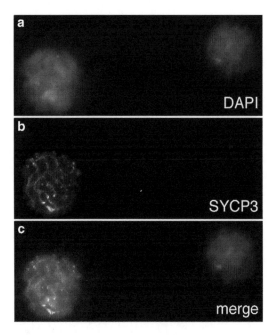

Fig. 2. Immunofluorescent labeling of SYCP3 and DAPI staining of ovary cells to identify oocytes. The nucleus on the left is from an oocyte and that on the right from a somatic cell. Both nuclei are stained with DAPI; however, only oocytes will show a positive SYCP3 signal.

17. Cover the slide with 500 µl blocking buffer and incubate for 30 min in a closed humidified box at room temperature.

18. Prepare 100 µl per slide of the secondary antibody, diluted in blocking solution, and centrifuge for 30 min at maximum speed in a tabletop microcentrifuge at 4°C.

19. Pour the blocking buffer off of the slides, add 100 µl of the supernatant of the centrifuged secondary antibody dilution to the slide, and cover with a coverslip.

20. Incubate the slide in a closed humidified box for 1 h at 37°C.

21. Wash the slide three times for 5 min in PBS.

22. Dip the slide three times in filtered ddH$_2$O (see Note 13).

23. Mount the slides in 10 µl Vectashield + DAPI and seal the coverslip with nail polish.

24. Using a fluorescence microscope, count the total number of nuclei (DAPI-positive) and the number of oocytes (positive for the meiosis- or oocyte-specific antibody) per well (see Fig. 2). The percentage of oocytes usually ranges between 40 and 60%.

25. Calculate the number of oocytes and total number of cells in the stored cell suspension.

3.3. DNA Extraction

Steps 1–10 describe the first steps for DNA extraction from somatic tissues. For DNA extraction from an ovary cell suspension, start from step 9. To avoid cross-contamination or mixing up the samples, it is preferable that DNA extractions from somatic tissue and ovaries are not performed side by side.

1. Prepare a chunk (2–3 mm^3) of somatic tissue (use fresh or stored at –20°C); put the tissue in a sterile petri dish.

2. Finely chop up the sample in both directions until mushy, add 1 ml 1× SSC.

3. Use a sterile transfer pipette to wash off the petri dish; collect the macerated tissue in 1× SSC at a corner of the dish.

4. Make a cone out of the mesh, place it on top of an eppendorf tube and pipette the sample up and down through the mesh.

5. Spin for 2 min in a tabletop centrifuge at 3,000 rpm (830 × g).

6. Remove the supernatant; spin again briefly and remove residual 1× SSC.

7. Resuspend the pellet in 1 ml 1× SSC (make sure there are no clumps, vortex if needed).

8. Repeat the 1× SSC washes at least two more times.

9. Resuspend the somatic cell pellet in 960 μl 1× SSC. For DNA extraction from ovaries, pool the ovary cell suspensions (see Note 14) and add 1× SSC to a final volume of 960 μl.

10. Add 120 μl β-mercaptoethanol, 100 μl 10% SDS, and 20 μl freshly made proteinase K (20 mg/ml).

11. Invert to mix; incubate 2–3 h at 55°C with occasional gentle mixing.

12. Split the contents of the tube into two tubes (600 μl each).

13. Add an equal volume of phenol:chloroform:isoamyl alcohol, mix well, and spin for 5 min in a tabletop microcentrifuge at maximum speed.

14. Use a 1 ml pipet tip with the tip cut off to transfer the aqueous layer (the top layer) to a clean eppendorf tube.

15. Re-extract residual DNA from the organic layer by adding 200 μl 1× SSC, 0.15% SDS, mix well, and spin for 5 min in a tabletop centrifuge at maximum speed. Pool the aqueous layer with aqueous solution from step 14.

16. Repeat steps 13–15.

17. Add 2 volumes of ice-cold absolute ethanol to aqueous solution, tap and invert tube until the DNA precipitates (see Note 4).

18. Spin for 5 min in a tabletop centrifuge at maximum speed, remove the supernatant, and wash the pellet with 1 ml 80% ethanol.

19. Spin for 5 min in a tabletop centrifuge at maximum speed and remove the supernatant.

20. Dissolve each pellet in 100 μl ddH$_2$O; pool the appropriate DNA solutions and reprecipitate by adding 1/10 volume of 3 M sodium acetate, pH 5.2, and 3 volumes of 100% ice-cold ethanol.

21. Repeat steps 18 and 19.

22. Air-dry the pellet and dissolve in 50–100 μl 5 mM Tris–HCl pH 7.5 (see Note 15).

23. Determine the concentration of the DNA by measuring the absorbance at 260 nm in a spectrophotometer. Add 5 mM Tris–HCl pH 7.5 to a working stock concentration of ~100 ng/μl. Store at –20°C.

24. Run a sample on a 0.8% agarose gel in the presence of a marker in various known amounts to verify the DNA concentration and to check the quality of the DNA. Good quality DNA should be high molecular weight and give a discrete band at the resolution limit of the gel.

3.4. Designing Allele-Specific and Universal PCR Primers

CO and NCO assays require two to three pairs of nested PCR primers that flank the region of the hotspot (see Fig. 1). Although most mammalian hotspots span a region of 1–2 kb, most PCR primer pairs need to be further apart because of the dependence on the presence of SNPs for the allele-specific PCR primers. Using the long-range PCR protocol described in this chapter, amplification of sequences up to 13 kb is feasible (see Note 16).

The behavior of PCR primers is difficult to predict so all primers have to be tested and optimized (described in Subheading 3.5). Below are some general guidelines for designing allele-specific and universal PCR primers:

1. Avoid repetitive regions and make sure the primer sequence is unique. On the Web, http://www.repeatmasker.org will show repeat regions in the sequence of interest and http://www.ncbi.nlm.nih.gov/blast allows determining whether the primer sequence is unique.

2. If possible, design primers with a GC content of at least 50% and a length between 18 and 22 nucleotides. For allele-specific primers, the polymorphic nucleotide that determines the allele specificity is always at the 3′ end of the primer. This means that the GC content of an allele-specific primer is dictated by the sequence context of the selected polymorphic nucleotide. If needed, the length of the allele-specific can be reduced down to 14 nucleotides (for high GC content) or increased up to 25 nucleotides (for low GC content) (see Notes 17 and 18).

3. The allele-specific primers should be located externally of and relatively close to the universal primers (see Fig. 1). Designing the allele-specific primers externally of the universal primers allows for the use of the same allele-specific primers in both the CO and NCO assay.

**3.5. Testing
and Optimizing
Allele-Specific
and Universal Primers**

1. Set up PCRs in a total volume of 8 μl with 0.72 μl 11.1× PCR buffer, supplemented with 12 mM Tris base, 0.03 U/μl Taq polymerase, 0.003 U/μl turbo Pfu polymerase, 0.2 μM of each primer, and 10 ng of genomic DNA per reaction.

2. Perform each PCR on genomic DNA from both parental strains that were used to generate the F1 hybrid that is going to be analyzed, and at various annealing temperatures (usually at 56, 59, 62, and 65°C).

3. PCRs with both forward and reverse universal primers serve as positive control and should amplify both parental strains equally efficiently and specifically at the optimal annealing temperature.

4. To test allele-specific primers and to determine their optimal annealing temperature, set up PCRs with an allele-specific primer in combination with a universal primer (e.g., allele-specific forward with universal reverse). The size of the resulting fragment should be as close as possible to the expected fragment size for the CO or NCO assay.

5. Cycle with the following conditions: 2 min 96°C (denaturation), followed by 35 cycles of amplification (20 s 96°C, 30 s at the annealing temperature and extension at 65°C for ~90 s/kb).

6. Analyze by adding 3 μl loading dye to each reaction and running 7.5 μl on a 0.8% agarose gel with 0.5 μg/ml ethidium bromide. Include longer exposure times when photographing the gel to enable visualization of weak bands.

7. The optimal annealing temperature for each allele-specific primer is the temperature at which the primer is both specific and efficient. Also note that the allele-specific primers that will be used together in a CO assay have to amplify both specifically and efficiently at the same annealing temperature (see Note 18).

**3.6. CO Assay:
Amplification of CO
Molecules by Allele-
Specific PCR**

In most cases it is advisable to set up a pilot experiment with varying amounts of input DNA to assess the activity of the hotspot before embarking on any large-scale experiments. If the activity of the hotspot is already known, proceed from step 9.

For both the pilot experiment and the subsequent large-scale CO assay, the DNA input has to be calculated. Since one haploid mouse genome contains approximately 3 pg of DNA, 6 pg of DNA will on average contain one molecule of each parental allele. From our experience, only about half of the molecules will actually generate PCR products, due to primer inefficiency and DNA damage. This means that on average one amplifiable molecule of each parental allele will be found in 12 pg of DNA. However, for both the CO and NCO, only the amplifiable molecules that are derived from oocytes should be considered. The DNA extracted from ovaries is a mix of DNA from somatic cells and oocytes and this mix varies between different preparations of an ovary cell suspension. Using the result from Subheading 3.2, step 25, calculate the final percentage of oocytes in the pooled ovary cell suspension that was used for the DNA extraction (Subheading 3.3, step 9). Given that oocytes have double the genomic content (4C vs. 2C) of most somatic cells, which are in G1/G0 phase, and assuming equal loss during the extraction for oocytes and somatic DNA, calculate the fraction of oocyte-derived DNA in the ovary DNA sample. If 12 pg of DNA contains on average one amplifiable molecule of each parental allele, then (12 pg/the fraction of oocyte DNA) will contain on average one oocyte-derived amplifiable molecule (see Note 19).

1. Set up PCRs with various amounts of ovary DNA, with somatic control DNA, and with no DNA (the latter two are negative controls). For instance, in a 96-well plate, set up 16 reactions each with 2,000, 1,000, 500, and 250 oocyte-derived amplifiable molecules per well. This pilot setup allows screening of 60,000 oocyte-derived amplifiable molecules at once, so even in very weak hotspots (recombination frequency $<1 \times 10^{-5}$), COs will be detected. As a negative control for contamination, set up eight reactions with no DNA. As a negative control for PCR artifacts, set up 24 reactions with 2,500 amplifiable molecules of somatic DNA. Both the total input and the maximum input per well of amplifiable molecules for somatic and ovary DNA should be in the same range (see Note 20).

2. Set up the first round of PCR with an outside forward allele-specific primer and the outside reverse primer that is specific for the other haplotype (Fig. 1b).

3. The total volume for each first PCR should be 8 μl with 0.72 μl 11.1× PCR buffer, supplemented with 12 mM Tris base, 0.03 U/μl Taq polymerase, 0.003 U/μl turbo Pfu polymerase, 0.2 μM of each allele-specific primer, and the determined amount of DNA. Start the PCR with 1 min denaturation at 96°C, followed by 25 cycles of amplification (20 s 96°C, 30 s at the optimized annealing temperature and extension at 65°C for ~90 s/kb).

4. As the first PCR is almost finished, prepare for S1 digestion by setting up a new 96-well plate with 5 μl of S1 digestion mix per well (see Note 21). Immediately upon completion of the first PCR, add 0.5 μl of each reaction to the corresponding well with the S1 digestion mix, and incubate at room temperature for 20 min. Dilute by adding 45 μl of S1 dilution buffer.

5. Set up the second round of PCR with the inside forward allele-specific primer and the inside reverse primer that is specific for the other haplotype. The direction and allele specificity of the nested primers has to be the same as in the first PCR (see Fig. 1b). The total volume for each second PCR should be 8 μl (0.72 μl 11.1× PCR buffer, supplemented with 12 mM

Tris base, 0.03 U/μl Taq polymerase, 0.003 U/μl turbo Pfu polymerase, and 0.2 μM of each allele-specific primer).

6. Seed each reaction with 1.6 μl of the S1 digested first PCR. Conditions for the second PCR are as described for the first PCR (see step 3), with 30–35 amplification cycles.

7. Add 3 μl loading dye to each reaction and run 5 μl of each on a 0.8% agarose gel with 0.5 μg/ml ethidium bromide. Increasing oocyte DNA input should correspond with an increase in the number of positive reactions and the negative controls should not show any PCR products.

8. Use the Poisson approximation to estimate the average number of COs per well (m) for each pool size: $m = -\ln$ (number of negative reactions/total number of reactions).

9. For large-scale CO assays, select an input DNA amount that yields an average of 0.4–0.6 COs per well. Using larger pools will make estimating the CO frequency less accurate because many of the positive PCRs will contain more than one CO molecule. On the other hand, smaller pools will decrease the number of positive reactions, increasing the number of PCRs required. About 100 COs are needed to accurately characterize the CO distribution in a hotspot.

10. Perform large-scale allele-specific PCRs with the selected amount of input oocyte DNA per well, as described above for the pilot experiment. Include negative controls with somatic DNA and no DNA. Do the same for the reciprocal CO orientation by switching the haplotypes targeted by the allele-specific primers (see Note 22). Adjust the annealing temperature in the reciprocal PCRs according to the optimization described in Subheading 3.5.

3.7. Mapping CO Breakpoints by Hybridization with ASOs

1. In a new 96-well plate, set up a third round of PCR with nested universal forward and reverse primers in a final volume of 22 μl for each reaction (1.98 μl 11.1× PCR buffer, supplemented with 12 mM Tris base, 0.03 U/μl Taq polymerase, 0.003 U/μl turbo Pfu polymerase, and 0.2 μM of each universal primer). Reserve the first column of wells for positive and negative controls and transfer 0.6 μl of each positive second PCR to one of the other wells. Keep a record of the origin (plate and well number) of each reaction. In the first column, seed two wells with 5 ng of genomic DNA from each parental strain that was used to generate the F1 hybrid. Seed the remaining six wells with 0.6 μl of randomly picked somatic controls from the second PCR.

2. The conditions for the third PCR are as described above (Subheading 3.6, step 3), with 35 amplification cycles.

3. Add 10 μl of loading dye to each reaction and run 2 μl of each on a 0.8% agarose gel with 0.5 μg/ml ethidium bromide. PCR amplification should be uniformly high.

4. Cut three Whatmann filter papers and ten pieces of nylon membrane to the appropriate size for the 96-well dot blot manifold, according to the manufacturer's instructions. Wet filter paper and a membrane with ddH$_2$O, assemble the dot blot manifold, and apply vacuum.

5. Add 300 µl of denaturation buffer to each well of the third PCR plate and pipet gently up and down to mix (see Note 23). Transfer 30 µl from each well to the corresponding well of the dot blot manifold. Make sure all the liquid is pulled through the membrane (see Note 24) and rinse each well with 150 µl 2× SSC. Remove the membrane, air-dry, and repeat for all replicate membranes. Ensure all membranes are completely dry before proceeding with the hybridization steps (see Note 25).

6. Prepare a working stock of 8 ng/µl of the ASO for each polymorphic site to be tested. ASOs are typically 18 nucleotides long with the SNP site located at the eighth base from the 5′ end (see Note 26). For each polymorphism, ASOs specific to both parental genotypes are used twice: once with one ASO to be labeled for hybridization and the ASO for the alternative allele as a competitor, followed by a hybridization in which the labeled and competitor ASO are switched. Store ASOs at a concentration of 800 ng/µl in ddH$_2$O at –20°C and dilute 1:100 to make the working stock.

7. For each dot blot, set up a 10 µl ASO labeling reaction in a screw-cap microcentrifuge tube for each polymorphism with 1 µl 10× kinase mix, 0.35 µl T4 polynucleotide kinase, 7.8 µl ddH$_2$O, 0.2 µl (γ-^{32}P)-ATP, and 1 µl of 8 ng/µl ASO. Incubate at 37°C for 45 min. Spin down briefly and inactivate the T4 polynucleotide kinase by incubating at 65°C for 20 min. Spin down briefly again and add 20 µl of 8 ng/µl unlabeled ASO of the other allele as competitor (see Note 27).

8. Soak the dot blots in 3× SSC and place in a small hybridization bottle with the DNA side facing inward. If blots from more than one experiment are to be hybridized with the same probe, they can be stacked with alternating layers of hybridization mesh. Stacks of up to eight blots have been successfully hybridized with the same probe. Keep in mind that the amount of probe and volumes of solutions have to be increased accordingly when hybridizing multiple blots. In addition, the length of time for all subsequent (pre-)hybridization and wash steps has to be increased to allow the solutions to fully penetrate the stack of blots.

9. For each blot, pre-hybridize with 3 ml of TMAC hybridization buffer at 58°C for 10–20 min in a rotator hybridization oven (see Note 28).

10. Discard the buffer and replace with 2.5 ml fresh TMAC hybridization buffer supplemented with 21 µg freshly denatured

sonicated salmon sperm DNA (boil for 5 min prior to use). Incubate in the rotator oven at 53°C for 5–10 min.

11. Add the probe solution containing the labeled ASO and the unlabeled competitor ASO and hybridize in the rotator oven for 1 h at 53°C.

12. Discard the hybridization solution. Wash three times with 3 ml pre-warmed TMAC wash buffer per blot at 56°C with rotation for 5–10 min per wash, followed by a 15 min wash with 4 ml pre-warmed TMAC wash buffer per blot.

13. Rinse the blots three times in the bottles with 2× SSC at room temperature, then transfer them from the bottle to a tray with 2× SSC and wash an additional two to three times. Blot off excess liquid, seal the dot blots in Saran wrap, and expose on a phosphorimager screen (see Note 29).

14. Strip the probes from the blots by repeated washes in boiling 0.1% SDS. Monitor probe removal using a Geiger counter and continue the washes until the signal is sufficiently reduced. For this step, all blots can be combined in a single tray with hybridization mesh between the layers of blots to increase the stripping efficiency.

15. Rinse the stripped blots several times in water and reprobe immediately or allow to air-dry and store free of dust at room temperature until further use (see Note 30). Continue probing and stripping the dot blots until all polymorphisms of interest have been hybridized with labeled ASOs from both parental genotypes. Ideally, include one or two SNPs on each side outside the hotspot region. To account for variation between the replicate dot blots, the same blot should be hybridized with labeled ASOs specific for both parental alleles for a given polymorphism. Blots can be stripped and reprobed up to ten times without significant loss of signal.

16. Score the positive signals on the dot blots (see Fig. 3), combine the data for both alleles in a table and order the COs according to their location (see example in Table 1).

17. Most PCR products will show a single CO breakpoint (e.g., Table 1, well 2A). However, some wells will contain mixed molecules, with a positive hybridization signal for both alleles at one or more polymorphisms (e.g., Table 1, wells 2C and 3C). This happens when two or more CO molecules are amplified in the same PCR. Only the two CO breakpoints that can be identified (both ends of the stretch of mixed SNPs) are scored. (In the example in Table 1, well 2C shows a CO breakpoint between SNP 3 and 4 and between SNP 6 and 7.) Note that it is possible that the PCR contained three or more CO molecules, but this assay does not allow for the identification of more than two. The occurrence of mixed molecules increases

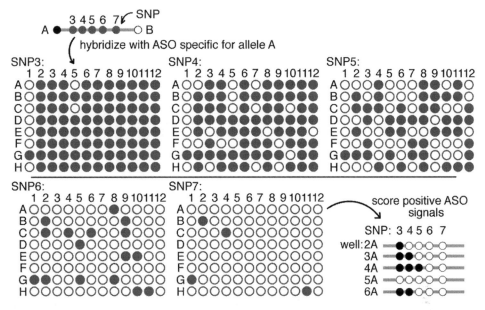

Fig. 3. CO breakpoint mapping. Allele-specific PCR will only amplify CO molecules, in this example molecules that are of the A genotype on the left side and of the B genotype on the right side (indicated by *black and white circles*, respectively). The remaining SNPs (*gray circles*) will be typed by ASO hybridization to determine the CO breakpoint for each CO molecule. The PCR products are transferred to *dot blots*, where every dot represents one PCR and ideally one (but possibly more) CO molecule. In this example, the first column in the plate contained 6 negative controls (wells 1A–1F) and the positive controls for alleles A and B (wells 1G and 1H, respectively). The *dot blots* are hybridized with ASOs that are specific for one of the parental haplotypes (in this example, the A alleles), for each subsequent polymorphism across the hotspot (SNPs 3–7 in this example, shown from *left to right*; positive hybridization signals are shown as *gray circles*, negative signals as *white*). The ASO hybridization signals are scored (in this example, results for wells 2A–6A, *black circles* represent positive hybridization signals, i.e., presence of the A alleles. *White circles* represent negative hybridization signals). For a CO assay the data for the A alleles are combined with the data for the B alleles (not shown).

with increasing amount of input DNA, which is why the initial pool size should not be too big (see Subheading 3.6, step 9) (see Note 31).

18. Sometimes one observes no switch from one haplotype to the other for any of the analyzed SNPs (see Table 1, well 2B). This either means that the CO breakpoint is located between one of the terminal SNPs and the inside allele-specific PCR primer or that the allele-specific PCR is not specific enough and is amplifying non-recombinant DNA. In either case, these wells are discarded from the analysis. In the case of mixed molecules only the single CO breakpoint that can be identified will be scored (see Table 1, well 3F, the CO breakpoint between SNP 5 and 6). Typing more SNPs outside the hotspot region may help avoid the problem of the apparent absence of a CO breakpoint. On the other hand, if the problem is due to nonspecific amplification, indicated by a high frequency of positives in the PCRs on somatic DNA, re-optimizing the allele-specific primers may help (see Subheading 3.5).

Table 1
Calculations for the CO assay

Well		Pool size	SNP				
Column	Row		3	4	5	6	7
2	A	300	A	B	B	B	B
2	F	300	A	B	B	B	B
3	E	300	A	B	B	B	B
3	H	300	A	B	B	B	B
2	C	300	A	M	M	M	B
2	D	300	A	A	B	B	B
2	H	300	A	A	B	B	B
3	A	300	A	A	B	B	B
3	B	300	A	A	B	B	B
3	C	300	A	A	M	B	B
2	E	300	A	A	A	B	B
3	D	300	A	A	A	B	B
3	F	300	M	M	M	B	B
3	G	300	A	A	A	B	B
2	G	300	A	A	A	A	B
2	B	300	A	A	A	A	A

	SNP interval	3–4	4–5	5–6	6–7
	Number of COs	5	5	5	2
	Analyzable pools	14	13	14	15
	Negative pools	9	8	9	13
	$P(0)$[a]	0.643	0.615	0.643	0.867
	m[b]	0.442	0.486	0.442	0.143
	Poisson-corrected number of COs[c]	6.186	6.312	6.186	2.147
	Number of analyzed molecules[d]	4,200	3,900	4,200	4,500
	Length SNP interval (bp)	500	200	300	600
	Recombination fraction (cM)[e]	0.147	0.162	0.147	0.048
	Recombination activity (cM/Mb)[f]	295	809	491	80

CO molecules that contained only the A allele are light gray ("A," data taken from columns 2 and 3 from the dot blots in Fig. 3), those containing only the B allele are dark gray ("B"), and those amplification products containing both parental alleles are black ("M"). The CO breakpoints are sorted according to their position. The number of COs per inter-polymorphism interval is corrected using Poisson approximation and the CO activity in cM/Mb is calculated for each interval

[a]$P(0) = $(number of negative pools/number of analyzable pools)
[b]$m = -\ln(P(0))$
[c]The Poisson-corrected number of COs $= m \times$ number of analyzable pools
[d]Number of analyzed molecules $=$ (number of analyzable pools $\times$ pool size)
[e]Recombination fraction $= 100 \times$ (Poisson-corrected number of COs/number of analyzed molecules) in cM
[f]Recombination activity $=$ (Recombination fraction/length of SNP interval in bp) $\times 10^6$

19. For each interval, count the number of COs. To account for the presence of possible hidden COs (i.e., two CO or more molecules with the same breakpoint), apply a Poisson correction. For this, count the number of analyzable pools per SNP interval. This number includes the pools that were included in the first and second PCR of the CO assay but were negative. For the example in Table 1, we assume 31 pools were initially analyzed for the retrieval of the 16 COs in columns 2 and 3 of the dotblot in Fig. 3. There are five COs in the SNP 4–5 interval (wells 2D, 2H and 3A–3C) and the number of analyzable pools is 28 (a total of 31 pools were initially analyzed, with 16 positive in the second PCR. Of the 16 COs, well 2B is discarded for all SNP intervals because it shows no switch between alleles across the analyzed SNPs. Wells 2C and 3F are considered not analyzable for the SNP 4–5 interval because the interval falls in a stretch of mixed molecules, which does not allow identification of any breakpoint in this interval). Because the fraction of negative pools is known: $P(0) =$ (number of negative pools/number of analyzable pools), the Poisson approximation can be used to calculate the average number of COs (m) per well: $m = -\ln P(0)$.

20. With m known, the expected (Poisson-corrected) total number of COs for each SNP interval can be calculated: Poisson-corrected number of COs $= m \times$ number of analyzable pools (see Table 1).

21. Calculate the recombination fraction in centimorgans (cM) for each SNP interval: recombination fraction $=$ (Poisson-corrected number of COs/total number of analyzed molecules) $\times 100$.

22. Calculate the recombination activity in cM/Mb for each interval: recombination activity $=$ (recombination fraction/length of the interval in bp) $\times 10^6$ (see Table 1).

23. An example graph of the recombination activity across a hotspot is depicted in Fig. 4.

24. Calculate the overall Poisson-corrected CO frequency across the hotspot by calculating m for the second PCRs as described in Subheading 3.6, step 8. The overall CO frequency $=$ (m/the number of amplifiable molecules per well).

3.8. NCO Assay: Amplification of CO and NCO Molecules by Half-Sided Allele-Specific PCR

The setup for the NCO assay (see Fig. 1c) is the same as for the CO assay (see Subheading 3.6) with the following exceptions:

1. Instead of the combination of forward allele-specific primers with reverse allele-specific primers that are specific to the opposite parental haplotype, the NCO assay is performed with allele-specific primers on only one side. Primers in the opposite direction will be universal primers. This "half-sided" PCR will

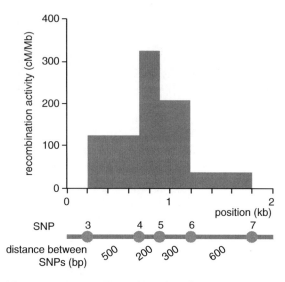

Fig. 4. CO activity across a hotspot. The position of the SNPs is indicated by *gray circles*. Data are from the example in Table 1.

amplify molecules that are CO, NCO, and non-recombinant, with the majority of products being of the latter type (see Fig. 1c).

2. Because most PCR products will be non-recombinant molecules, the pool size of input DNA needs to be much smaller than for the CO assay, usually about 15 amplifiable molecules per well. Having too many non-recombinant molecules in the PCR interferes with the detection of recombinants in the later ASO hybridization steps. As the input pool size is fixed, there is no need for a pilot experiment with different input pool sizes.

3. S1-nuclease digestion is usually not used in NCO assays (see Note 32). Immediately upon completion of the first PCR, add 35 μl dilution buffer to each well. Seed the second PCR with 0.6 μl of the diluted first PCR.

4. The second PCR for the NCO assay is set up in a larger reaction volume, usually 22 μl. The bigger reaction volume is needed because there is no re-amplification (or third PCR) prior to transferring the PCR products to dot blots.

3.9. Mapping NCOs and CO Breakpoints by Hybridization with ASOs

The procedure for mapping NCO and CO breakpoints by ASO hybridization is similar to the procedure for COs (see Subheading 3.7), with the following exceptions:

1. The second PCR is not re-amplified, but used directly. Add 10 μl of loading dye to each well and check the quantity and

quality of the PCR products by running 2 μl on a 0.8% agarose gel with 0.5 μg/ml ethidium bromide.

2. Prepare positive and negative hybridization controls by separately setting up and performing first and second half-sided PCRs on 5 μg genomic DNA from each parental strain that was used to generate the F1 hybrid. The negative hybridization control will be of the same haplotype as was selected for in the half-sided allele-specific PCR. The positive hybridization control is of the same haplotype as the ASOs that will be hybridized. For each parental haplotype, set up about 16 reactions of 22 μl and pool the products for each haplotype separately after completion of the second PCR. Add loading dye and check the quantity and quality of the PCR products by running 2 μl on a 0.8% agarose gel with 0.5 μg/ml ethidium bromide.

3. Replace one well on the 96-well plate of the completed second half-sided allele-specific PCR with 30 μl of the positive hybridization control. Add positive hybridization control to 6 wells in the following dilutions: 1:10, 1:30, 1:100, 1:300, 1:1,000, and 1:3,000. Empty one well, wash three times with water, and add 30 μl of negative hybridization control. Add 300 μl of denaturing buffer to each well and continue as described in Subheading 3.7.

4. Hybridize the dot blots with only the ASOs specific for the non-selected haplotype from the half-sided allele-specific PCR.

5. Use the dilution series of the positive hybridization control as a tool in scoring the hybridization signals. In general, hybridization signals that are stronger than the signals of 1:100 diluted positive controls are considered positive, i.e., represent bona fide CO or NCO molecules against the background of amplified non-recombinant molecules.

6. Score the positive hybridization signals and transfer the data to a table (see Fig. 5 and Table 2). Order the COs and NCOs separately according to their positions. Calculate the CO activity for each SNP interval and across the hotspot as described in Subheading 3.7. The overall CO frequency across the hotspot found in this assay should be comparable to the frequency found in the CO-specific assay.

7. Count the number of NCOs at each SNP. Calculate the Poisson-corrected average NCO number per well (m) for each SNP: $m = -\ln(\text{number of negative pools}/\text{number of analyzable pools})$. Note that pools that contain a CO are not analyzable for NCOs for the SNPs that give a positive signal in the ASO hybridization.

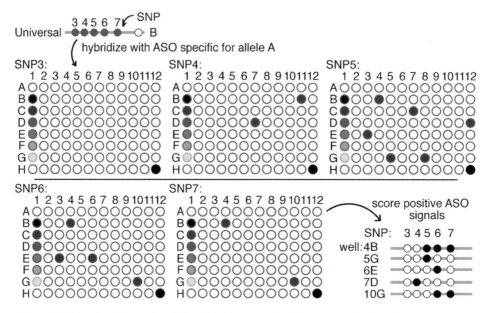

Fig. 5. CO and NCO breakpoint mapping. Half-sided allele-specific PCR will amplify CO, NCO, and non-recombinant molecules. In this example, amplified molecules are of the B genotype (indicated by *white circle*) on the right side and either genotype on the left side. The remaining SNPs (*gray circles*) will be typed to determine the CO or NCO breakpoint for each recombinant molecule. The PCR products are transferred to *dot blots*, where *every dot* represents one amplification product. In this example, the first column in each plate contained the controls with the negative hybridization sample (amplified from parental haplotype B) in well 1A and the dilution series of the positive hybridization control (amplified from parental haplotype A) in wells 1B–1G. Well 12H contains the undiluted positive hybridization control. The *dot blots* are hybridized with ASOs specific for the parental haplotype that was not selected for in the PCR (in this example, the A haplotype), for each SNP across the hotspot (SNPs 3–7 in this example, shown from left to right; positive hybridization signals are shown as *gray circles*, negative signals as *white*). The positive ASO hybridization signals are then scored. The example (*bottom right*) shows two COs (wells 4B and 10G) and three NCOs (wells 5G, 6E, and 7D), with *black circles* representing the presence of A alleles on the recombinant molecule (i.e., positive hybridization signals on the *dot blots*) and *white circles* representing inferred B alleles on the recombinants (i.e., negative hybridization signals).

8. Divide the Poisson-corrected NCO frequency per well by the pool size to give the NCO frequency per molecule for each SNP: NCO frequency = (m/pool size).

9. Calculate the overall NCO frequency across the hotspot as described for COs (see Subheading 3.6, step 8). Keep in mind that for the overall NCO frequency, each co-conversion (see Table 2, well 3E) should be counted as one event.

10. NCO activity across a hotspot can also be visualized by plotting either the relative or absolute NCO numbers for each SNP (see Fig. 6). Again, co-conversions are counted as one event, so each SNP involved will contribute an equal part of the event. For the example in Table 2, well 3A, this means that 0.5 of the recombinant involves SNP 5 and 0.5 of the recombinant involves SNP 6.

Table 2
Calculations for the NCO assay

Well		Pool size	SNP				
Column	Row		3	4	5	6	7
4	B	15			A	A	A
10	G	15				A	A
7	D	15		A			
11	B	15		A			
5	G	15			A		
7	C	15			A		
8	G	15			A		
12	D	15			A		
3	E	15			A	A	
6	E	15				A	

	SNP	3	4	5	6	7
	Observed number of NCOs	0	2	5	2	0
	Analyzable pools[a]	88	88	87	86	86
	Negative pools	88	86	82	84	86
	Poisson-corrected number of NCOs[b]	0.000	2.023	5.149	2.024	0.000
	Poisson-corrected NCO frequency[c]	0.000	0.023	0.059	0.024	0.000

SNPs that showed a positive hybridization signal in a well are shaded gray ("A"), SNPs that showed no positive hybridization signal are left blank. The CO and NCO breakpoints are sorted separately, according to their position. Shown is the calculation of the Poisson-corrected NCO frequency per SNP; for CO calculations see Table 1
[a]In the 96-well plate in Fig. 5, 8 wells were occupied by controls, leaving 88 wells available for analysis
[b]See Table 1 for the calculation of the Poisson-corrected number of recombinants
[c]The Poisson-corrected NCO frequency = (−ln(number of negative pools/number of analyzable pools))/ pool size

---

4. Notes

1. Most of the oocytes in newborns are in the dictyate stage and have completed recombination. Although the number of oocytes in an ovary is the highest shortly before birth and rapidly decreases following birth (22), it is more convenient to use newborns rather than sacrificing dams to recover embryos.

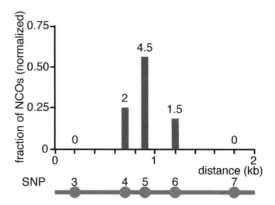

Fig. 6. NCOs across a hotspot. The fraction of Poisson-corrected NCOs per SNP is plotted against the distance in kb. The position of the SNPs is indicated by *gray circles*. The co-conversions were normalized (see Subheading 3.9 step 7), and the observed normalized number of NCOs is shown above each bar. Data are from the example in Table 2. Note that the polymorphism density in the hotspot as depicted is rather low, meaning that likely a great part of the NCOs across this hotspot remains undetected.

2. The objective is to remove all traces of grease and dust from the slides to reduce background in the immunofluorescence. Boiling the slides in a slide rack allows for free circulation between the slides and improves the removal of dust and grease.

3. Sodium azide and PMSF are added to inhibit bacterial growth and proteolytic breakdown, respectively, during the immune-incubation steps. PMSF is not stable in aqueous solutions and has to be added shortly before use. Note that both sodium azide and PMSF are highly toxic.

4. LPA can be added as a carrier to aid DNA precipitation.

5. Because the behavior of the primers can differ between different batches of 11.1× PCR buffer, it is advisable to make enough for at least one complete experiment (including optimization of the primers). Primers have to be re-optimized for every new batch of 11.1× PCR buffer.

6. TMAC allows the hybridization of oligos of different G/C content at the same temperature (23). Note that TMAC is highly toxic.

7. The ovaries can be kept on ice for some time but try to process them as quickly as possible. The most efficient way is to dissect out the ovaries of all the females in the litter, then continue with preparation of the cell suspension.

8. The treatment with collagenase and DNase I is required to remove part of the somatic cells during the subsequent wash steps and helps in making a clump-free cell suspension.

9. Additionally, 5–15 μl aliquots can be taken from the different preceding wash steps, included in the preparation of slides and quantified. This will give an indication of the change in the cellular composition of the cell suspension at each step.

10. Wrapped airtight, the slides can be kept at –80°C for several years.

11. The optimal dilution and incubation time depends on the antibody and should be tested beforehand.

12. The blocking buffer with the antibody is centrifuged to remove aggregates of denatured protein/antibody to help reduce background signals.

13. If the secondary antibody is conjugated to fluorescein isothiocyanate (FITC) or tetramethylrhodamine isothiocyanate (TRITC), adding a last dip in FITC buffer (150 mM NaCl, 50 mM sodium carbonate, pH 9.5, filter-sterilized) helps reduce fading of the fluorophore.

14. Avoid extracting DNA from too small amounts of ovary cell suspension; aim for a pooled suspension from at least 45 animals.

15. Do not over dry the pellet, as this will make the pellet harder to dissolve. If needed, dissolution can be aided by incubation at 50°C for up to 1 h.

16. Even though amplicon sizes up to 13 kb are feasible, the PCR efficiency is usually better for shorter ones.

17. Because the behavior of the allele-specific primers is unpredictable, it is advisable to test different lengths for the allele-specific primers simultaneously, e.g., an allele-specific primer of 20, 21, and 22 nucleotides long with the SNP at the 3′ end.

18. Only two nested allele-specific primers on each side of the hotspot are needed for the CO and NCO assays, however, the allele-specific primers that will be used together in a CO assay have to amplify both specifically and efficiently at the same annealing temperature. If possible, design and test allele-specific primers directed against more than two SNPs on either side of the hotspot.

19. Alternatively, set up the PCR assuming 100% oocyte DNA in the DNA sample and recalculate the number of oocyte-derived amplifiable molecules upon completion of the CO assay.

20. The somatic control is included to assess the level of nonspecific amplification, which tends to increase with higher input of negative control DNA. To aid the interpretation of the CO

assay, the amount of input DNA for the somatic control should not exceed the maximum input of ovary DNA.

21. S1 nuclease digestion improves the specificity of allele-specific primers and removes PCR artifacts by removing single-stranded DNA from the PCR products. PCR artifacts such as panhandle loop structures can interfere with the detection of COs (16). The plate with S1 digestion mix can be prepared several hours in advance and kept at 4°C until use.

22. Reciprocal allele-specific PCRs serve two purposes. First, they provide an internal control because the CO frequency should be comparable for both orientations. Second, differences in the initiation of recombination on the parental alleles will be revealed (10, 24).

23. Ensure the 96-well plate used for PCR is capable of holding this volume (~330 μl) beforehand.

24. Remove bubbles by poking a pipette tip in the well. Make sure to use a clean tip for every bubble to avoid cross-contamination. After the wash step with 2× SSC a single needle can be used for all bubbles.

25. Cross-linking the DNA to the membrane is not routinely performed for the Hybond-XL membranes but can be included. See the manufacturer's instructions.

26. If a particular ASO does not give a clear hybridization signal, washing at slightly different temperatures or trying hybridization with the reverse complement can help. The conditions of the ASO hybridization are optimized for a length of 18 nucleotides, so be prepared to try different washing conditions for ASOs with different length.

27. Optional: to check for the efficiency of the labeling reaction, load the sample on a spin column (e.g., Quick Spin Oligo Column, Roche) after T4 polynucleotide kinase inactivation. Spin the sample through the column and measure the incorporation of radioactive label using a Geiger counter.

28. Hybridization signals can be improved by increasing the pre-hybridization time. This is especially true for dot blots being hybridized for the first time.

29. Optimal exposure times differ between ASOs. We routinely expose overnight, followed by a separate 1 h exposure if needed.

30. Blots can be stored up to 2 years without significant loss of signal.

31. A method for further dissecting mixed molecules, applicable for both COs and NCOs, is described in ref. 10.

32. S1 digestion can be included if desired. Follow the steps as described for the CO assay (see Subheading 3.6).

Acknowledgements

We thank Liisa Kauppi and Francesca Cole for advice on allele-specific PCR. This work was supported by a Netherlands Organization for Scientific Research Rubicon Grant 825.07.006 (E.B.) and a National Institutes of Health Grant R01 HD53855 (S.K. and M.J.).

References

1. Keeney S (2007) Spo11 and the formation of DNA double-strand breaks in meiosis. In: Lankenau DH (ed) Recombination and meiosis. Springer, Heidelberg, Germany, pp 81–123

2. Allers T, Lichten M (2001) Differential timing and control of noncrossover and crossover recombination during meiosis. Cell 106: 47–57

3. Hunter N, Kleckner N (2001) The single-end invasion: an asymmetric intermediate at the double-strand break to double-holliday junction transition of meiotic recombination. Cell 106:59–70

4. Kauppi L, Jeffreys AJ, Keeney S (2004) Where the crossovers are: recombination distributions in mammals. Nat Rev Genet 5:413–424

5. Jeffreys AJ, Kauppi L, Neumann R (2001) Intensely punctate meiotic recombination in the class II region of the major histocompatibility complex. Nat Genet 29:217–222

6. Guillon H, de Massy B (2002) An initiation site for meiotic crossing-over and gene conversion in the mouse. Nat Genet 32:296–299

7. Jeffreys AJ, May CA (2004) Intense and highly localized gene conversion activity in human meiotic crossover hot spots. Nat Genet 36: 151–156

8. Guillon H, Baudat F, Grey C, Liskay RM, de Massy B (2005) Crossover and noncrossover pathways in mouse meiosis. Mol Cell 20: 563–573

9. Holloway K, Lawson VE, Jeffreys AJ (2006) Allelic recombination and de novo deletions in sperm in the human beta-globin gene region. Hum Mol Genet 15:1099–1111

10. Cole F, Keeney S, Jasin M (2010) Comprehensive, fine-scale dissection of homologous recombination outcomes at a hot spot in mouse meiosis. Mol Cell 39:700–710

11. Paigen K et al (2008) The recombinational anatomy of a mouse chromosome. PLoS Genet 4(7):e1000119

12. Billings T et al (2010) Patterns of recombination activity on mouse chromosome 11 revealed by high resolution mapping. PLoS ONE 5(12): e15340

13. de Boer E et al (2006) Two levels of interference in mouse meiotic recombination. Proc Natl Acad Sci U S A 103:9607–9612

14. Bois PR (2007) A highly polymorphic meiotic recombination mouse hot spot exhibits incomplete repair. Mol Cell Biol 27:7053–7062

15. Smagulova F et al (2011) Genome-wide analysis reveals novel molecular features of mouse recombination hotspots. Nature 472:375–378

16. Jeffreys AJ, Murray J, Neumann R (1998) High-resolution mapping of crossovers in human sperm defines a minisatellite-associated recombination hotspot. Mol Cell 2:267–273

17. Hubert R et al (1994) High resolution localization of recombination hot spots using sperm typing. Nat Genet 7:420–424

18. Kauppi L, May CA, Jeffreys AJ (2009) Analysis of meiotic recombination products from human sperm. Methods Mol Biol 557:323–355

19. Cole F, Jasin M (2011) Isolation of meiotic recombinants from mouse sperm. Methods Mol Biol 745:251–282

20. Baudat F, de Massy B (2009) Parallel detection of crossovers and non-crossovers in mouse germ cells. Methods Mol Biol 557:305–322

21. Eppig JJ, Schroeder AC (1989) Capacity of mouse oocytes from preantral follicles to undergo embryogenesis and development to live young after growth, maturation, and fertilization in vitro. Biol Reprod 41:268–276

22. McClellan KA, Gosden R, Taketo T (2003) Continuous loss of oocytes throughout meiotic prophase in the normal mouse ovary. Dev Biol 258:334–348

23. Wood WI et al (1985) Base composition-independent hybridization in tetramethylammonium chloride: a method for oligonucleotide screening of highly complex gene libraries. Proc Natl Acad Sci USA 82:1585–1588

24. Jeffreys AJ, Neumann R (2002) Reciprocal crossover asymmetry and meiotic drive in a human recombination hot spot. Nat Genet 31:267–271

Chapter 3

Studying Meiosis-Specific Cohesins in Mouse Embryonic Oocytes

Ji-hye Kim, Kei-ichiro Ishiguro, Nobuaki Kudo, and Yoshinori Watanabe

Abstract

Distinct meiotic cohesin complexes play fundamental roles in various meiosis-specific chromosomal events in spatiotemporally different manners during mammalian meiotic prophase. Immunostaining is one of the essential methods to study meiotic cohesin dynamics. For the study of cohesins in the meiotic prophase of oocytes, ovaries should be taken from the embryos during a very limited period before birth. Here we focus on some technical tips concerning the preparation of oocyte chromosome spreads for immunostaining. Further, we describe a method for chromosome fluorescence in situ hybridization (FISH) against immunostained oocytes.

Key words: Fetal oocyte, Embryonic ovary, Meiosis, Cohesin, Prophase, Axial element, Homolog synapsis, Synaptonemal complex, Chromosome spread, FISH

1. Introduction

Cohesin is essential for faithful chromosome segregation to establish cohesion between sister chromatids (1). The meiotic cohesin complex differs from that of mitosis since the mitotic RAD21/SCC1 subunit of the cohesin complex is largely replaced by meiotic counterparts, REC8 and RAD21L (2–5) in mammals. Also other meiosis-specific cohesin subunits, SA3 and SMC1β, are known to be expressed (6, 7). During meiotic prophase I, sister chromatids are organized into proteinaceous structures of axial elements (AEs) on which the synaptonemal complex (SC) is assembled to promote interhomolog recombination, a process yielding chiasmata between homologues (8, 9). The meiotic cohesin complexes, which interact with the SC components and localize along AEs, might act as a basis for SC assembly (10–14). Thus the cohesin complex is crucial

Hayden A. Homer (ed.), *Mammalian Oocyte Regulation: Methods and Protocols*, Methods in Molecular Biology, vol. 957, DOI 10.1007/978-1-62703-191-2_3, © Springer Science+Business Media, LLC 2013

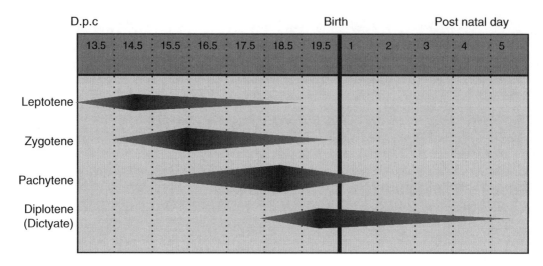

Fig. 1. Progression of meiotic prophase during female embryonic development. Shown is the embryonic age in days post corresponding with each specific meiotic prophase stage.

not only for sister chromatid cohesion but also for various meiosis-specific chromosomal events in meiotic prophase.

In contrast to male germ cells, which enter meiosis continuously and go straight into diakinesis after birth, female germ cells enter meiosis during fetal development and finally reach dictyate arrest that persists until puberty (Fig. 1) (15–17). Here we describe our technical knowledge concerning the study of cohesin in mouse fetal oocytes. Well-scheduled collection of embryos is required for oocyte preparation because meiotic prophase progression is chronologically coupled with the stage of embryogenesis. Unlike spermatocyte sampling from testis, special handling is required under a microscope due to the small size of the fetal ovary. Rapid preparation of single oocytes is achieved by physically crushing the fetal ovary followed by collagenase digestion with less physiological damage. The chromosome spread of oocytes based on a dry down technique (18) is made by hypotonic treatment and/or cytocentrifugation followed by fixation together with pre-extraction of free proteins providing a well-preserved chromosome axis structure. Fluorescence in situ hybridization (FISH) against immunostained chromosomes (19, 20) provides pairing/synapsis status of a pair of given homologues during meiotic prophase. These techniques are applicable not only to wild type but also to any strains of meiotic prophase deficient mouse.

2. Materials

2.1. Collection of Embryonic Ovaries

1. Fine-tipped tweezers.

2. Phosphate buffered saline (PBS): 1.47 mM KH_2PO_4, 8.1 mM Na_2HPO_4, 137 mM NaCl, 2.7 mM KCl.

3. M2 medium: containing 10% (v/v) fetal bovine serum, 1/100 (v/v) 200 mM L-glutamine, 10 U/ml penicillin, 10 μg/ml streptomycin. Store at 4°C.

4. Stereo microscope.

5. 1 mg/ml type I Collagenase in PBS. Store aliquots at –20°C.

6. 1 ml syringe with 18 G needle.

7. 0.05% trypsin

8. DMEM containing 10% (v/v) FBS

9. Hypotonic buffer: 17 mM trisodium citrate dihydrate, 50 mM sucrose, 5 mM EDTA, 0.5 mM DTT, 30 mM Tris–HCl, pH 8.2. Store aliquots at –20°C.

2.2. Processing of the Embryonic Oocyte Chromosome Spread by Dry Down

1. 2% (w/v) Paraformaldehyde (PFA), 0.2% (v/v) Triton X-100 in PBS, pH 9.2.

2. MAS-coated slide glass (MATSUNAMI).

3. PAP (hydrophobic barrier) pen (Vector laboratories, H-4000).

2.3. Embryonic Oocyte Spreading by Cytocentrifugation

1. 1% (w/v) PFA, 0.15% (v/v) Triton X-100 in PBS, pH 9.2.

2. Slide glass for cytocentrifuge (Thermo).

3. PAP (hydrophobic barrier) pen.

4. Disposable sample chamber (Thermo).

5. Cytocentrifuge machine (Thermo, Shandon cytospin 4).

2.4. Immuno-fluorescence

1. PBS.

2. 0.1% (v/v) Triton X-100 in PBS, pH 7.4.

3. Blocking solution: 5% (w/v) BSA in PBS. Store at 4°C.

4. Vectashield mounting medium containing 0.1 mg/ml of DAPI (Vector laboratories).

5. Coverslips.

6. Slide chamber.

7. Nail polish.

8. Anti-mouse-Alexa 555 (Cy3).

9. Anti-rabbit-Alexa 488 (FITC).

2.5. Chromosome Fluorescence In Situ Hybridization

1. 70, 80, 90, 100% ethanol.

2. 70% formamide/6× SSC.

3. 4% (w/v) PFA in PBS, pH 7.4.

4. 2× SSC.

5. 0.4× SSC/0.3% Tween-20.

6. PBS/0.1% Tween-20.

7. Mouse Chromosome FITC-labeled point probe (e.g., ID labs).

8. STARFISH hybridization buffer (ID labs) or 50% formamide/2× SSC.

9. Mouse Cot-1 DNA (Invitrogen).

10. Anti-rabbit-Alexa 488 (FITC).

11. Slide chamber.

12. 22×22 mm cover glass.

13. Nail polish.

14. Coplin jar.

15. Thermal cycler tube.

3. Methods

3.1. Collection of Embryonic Ovaries

1. Sacrifice the pregnant female at day 14.5–19.5 d.p.c or postnatal day 1–5 (see Note 1) by cervical dislocation.

2. Dissect out the uterine horns; remove embryos from the placenta; and transfer into a petri dish containing PBS.

3. Wash embryo with fresh PBS, cut away head, incise abdomen, and cut out side of legs (see Note 2) (Fig. 2).

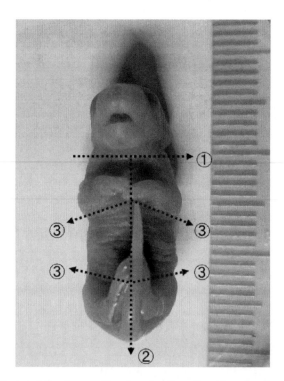

Fig. 2. Positioning of the d.p.c. 18.5 embryo for dissection. Dissect embryo by making incisions as indicated by the *dotted lines*.

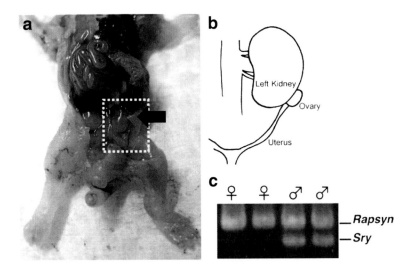

Fig. 3. (**a**) Exposed abdominal contents of a female embryo. The kidney is pseudocolored in *blue* and highlighted by a *black arrow*. (**b**) Shown is a schematic representation of some of the main structures located within the region demarcated by the *dashed box* in (**a**) illustrating that the ovary lies in close proximity to the infero-lateral surface of the kidney. (**c**) Sex is determined by PCR with male-specific *Sry* and autosome *Rapsyn* primers.

4. Open abdomen, displace loops of bowel upwards thereby exposing the ovaries located immediately inferior to the kidneys in the paravertebral gutters at the posterior wall of the peritoneal cavity (Fig. 3a, b) (see Note 3).

5. Remove both ovaries from each female fetus with fine-tipped tweezers under a stereo microscope and place in a 1.5 ml centrifuge tube containing 0.7 ml of M2 medium (see Note 4).

6. Crush ovaries into small pieces using an 18 G needle attached to a 1 ml syringe (see Note 5).

7. Spin down for 3 min at $400 \times g$.

8. Discard the supernatant (see Note 6) and add 1 ml of PBS.

9. Spin down for 3 min at $400 \times g$.

10. Discard the supernatant and add 0.5 ml of 1 mg/ml collagenase in PBS.

11. Incubate ovarian fragments with collagenase at 37°C for 15–30 min.

12. Gently pipette until large fragments of ovarian tissue are no longer visible.

13. Spin down for 3 min at $400 \times g$.

14. Discard the supernatant and add 0.5 ml of 0.05% trypsin (see Note 7).

15. Gently pipette and incubate at 37°C for 1 min.

16. Add 0.5 ml of DMEM containing 10% (v/v) FBS.

17. Gently pipette and spin down for 3 min at $400 \times g$.

18. Discard the supernatant and add 1 ml of PBS (see Note 8).

19. Resuspend cells by pipetting and spin down for 3 min at $400 \times g$.

20. Discard the supernatant and add hypotonic buffer (see Note 9).

21. Resuspend cells by pipetting and keep for 10 min or more at room temperature (see Note 10).

3.2. Chromosome Spread by Dry Down

1. Mark a circle approximately 1.0–1.5 cm in diameter on the slide with a hydrophobic barrier pen.

2. Put 10 μl of 2% (w/v) PFA containing 0.2% (v/v) Triton X-100 into each circle (see Note 11).

3. Add 10 μl of cell suspension in hypotonic buffer (prepared as in Subheading 3.1) to the PFA/Triton X-100 solution in each circle.

4. Slowly dry down in a humidity slide chamber for at least 2 h or overnight.

5. After drying, store the slides at –80°C.

3.3. Chromosome Spread by Cytocentrifuge

1. Mark a circle of approximately 1 cm in diameter on the slide using a hydrophobic barrier pen around the area where the cells will become attached following cytocentrifugation.

2. Install the slide equipped with the disposable sample chamber in the metal holder.

3. Add 100 μl of cell suspension in hypotonic buffer (prepared as in Subheading 3.1) to each chamber (see Note 12).

4. Install the holder equipped with slide and disposable sample chamber into the slot in the cytocentrifuge machine.

5. Cytocentrifuge for 5 min at 600–900 rpm.

6. Remove the disposable sample chamber from its slide.

7. Fix the attached cells on the slide in the marked area by adding 10–20 μl of 1% PFA containing 0.15% (v/v) Triton X-100.

8. Slowly dry down in a humidity slide chamber for a minimum of 2 h as in step 4 of Subheading 3.2.

9. After drying, store the slides at –80°C.

3.4. Immunostaining

1. Incubate frozen dry down or cytocentrifuge slides with 0.1% Triton X-100 in PBS for 10 min at room temperature.

2. Wash slides in PBS three times for 5 min.

3. Incubate slides with 100 μl of 5% BSA in PBS for 30 min to block nonspecific binding of the antibodies.

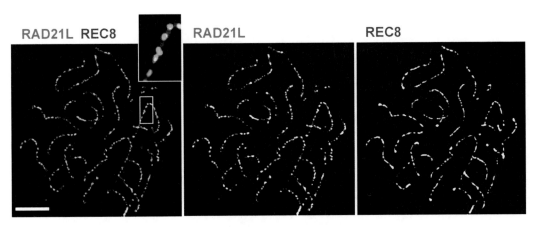

Fig. 4. Chromosome spread of pachytene oocyte obtained using the cytocentrifuge method immunostained for REC8 (*red*) and RAD21L (*green*). Shown inset is a magnified image of the *boxed* region. Scale bar, 5 μm.

4. Incubate slides with 50–100 μl of appropriately diluted mouse anti-REC8 and rabbit anti-RAD21L antibodies in PBS containing 5% BSA in a humidified chamber for 1 h at room temperature or overnight at 4°C.

5. Discard the solution and wash the slides three times in PBS, 5 min each.

6. Incubate slides with 50–100 μl of anti-mouse-Alexa 555 (Cy3) and anti-rabbit-Alexa 488 (FITC) secondary antibodies in PBS containing 5% BSA for 1 h at room temperature in the dark.

7. Discard the secondary antibody solution and wash three times with PBS for 5 min each in the dark.

8. Mount cover glass with a drop of Vectashield mounting medium containing DAPI and seal the edges of the cover glass using nail polish. Using this approach, both REC8 and RAD21L can be simultaneously imaged with fluorescence microscopy (Fig. 4).

3.5. Chromosome Fluorescence In Situ Hybridization and Immuostaining

1. Make chromosome spreads of oocytes as described in Subheading 3.2.

2. Stain the chromosome spreads with mouse anti-Sycp3 and rabbit anti-Sycp1 as described for REC8 and RAD21L in Subheading 3.4 (see Note 13).

3. Stain slides with the secondary antibodies anti-mouse-Alexa 555 (Cy3) and anti-rabbit-Alexa 647 (Cy5) as described in Subheading 3.4.

4. Wash with PBS and mount in Vectashield (see steps 7 and 8 in Subheading 3.4). Observe the slide under a microscope for confirmation of immunostaining.

5. Wash out mounting medium using PBS/0.1% Tween-20.

6. Fix antibodies on the slide glass with 4% PFA for 8 min at room temperature (see Note 14).

7. Wash three times with PBS for 5 min at room temperature.

8. Dehydrate immunostained samples by sequentially immersing in 70–80–90–100% ethanol for 5 min each at room temperature and air dry.

9. Denature double-stranded DNA by immersing immunostained slide samples in 70% formamide/6× SSC at 72°C for 10 min.

10. Dehydrate the samples by sequentially immersing in ice-cold 50–70–90–100% ethanol for 5 min each and air dry.

11. Mix 3 μl FITC-labeled chromosome point probe with 7 μl STARFISH hybridization buffer and 1 μl of Cot-1 DNA (final 0.1 μg/μl) in a thermal cycler tube.

12. Incubate 11 μl of the FISH probe/Cot-1 DNA/hybridization buffer mixture at 75°C for 5 min then chill immediately on ice.

13. Apply 11 μl of the probe mixture to the slide, cover with a 22×22 mm cover glass, and seal the cover with nail polish (see Note 15).

14. Place the slide in a humid chamber.

15. Hybridization: Incubate the chamber at 37°C for 12–16 h.

16. Carefully remove the nail polish from the slides and submerse the slides in a Coplin jar with 2× SSC at room temperature until the cover slips slide off (see Note 16).

17. Place the slides in Coplin jars with pre-warmed 0.4× SSC/0.3% Tween-20 solution at 73°C for 2 min (see Note 17).

18. Wash the slide in 2× SSC at room temperature for 1 min.

19. Mounting using Vectashield/DAPI (see step 8 in Subheading 3.4). Using this approach, chromosome-specific signals can be obtained whilst simultaneously labeling components of the SC (Fig. 5).

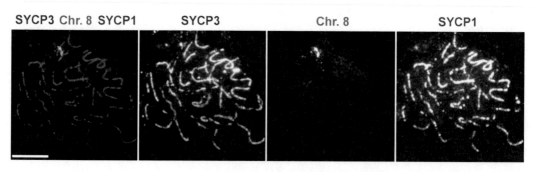

SYCP3 Chr. 8 SYCP1   SYCP3   Chr. 8   SYCP1

Fig. 5. FISH with Chromosome 8 A2 probe (*green*) against spread chromosomes immunostained with SYCP1 (*blue*) and SYCP3 (*red*). Scale bar, 5 μm.

4. Notes

1. The progression of meiosis in both sexes differs dramatically. In females, cells enter meiotic prophase during embryogenesis and arrest until the mouse matures after birth. After entering into meiosis in the embryonic ovary, oocytes pass through leptotene, zygotene, and pachytene stages before arresting in diplotene or dictyate stage, the last stage of meiotic prophase I. At leptotene, chromatin loops extend and condense, forming a filamentous meshwork called axial elements (AEs), which start undergoing synapsis or pairing at the zygotene stage. At the pachytene stage, the chromosome cores are completely synapsed and cells have 20 centromeres. The synaptonemal complex (SC) begins to disassemble at the diplotene stage, and homologous chromosomes move apart except at the chiasmata.

2. Making cuts deeply in the side of the legs is useful to fix the mouse firmly and to pick up the very small ovaries in the paravertebral gutters.

3. The early embryonic gonads are indistinguishable between the sexes; however, ovaries and testes become discernible from d.p.c 14.5 in terms of both morphology and location. The testes translocate to their final location at the anterior abdominal wall from d.p.c. 15.5.

4. If the sex of a given embryo is obscure, keep any part of the embryo body for genomic DNA extraction. The sex of each embryo from which gonads (fetal ovary) have been removed can be confirmed by PCR using primer sets for Y-chromosome-specific *Sry* and autosome control *Rapsyn* (Fig. 3c).

   Sry-F: 5′-CAGCCCTACAGCCACATGAT-3′

   Sry-R: 5′-GAGTACAGGTGTGCAGCTCTA-3′

   Rapsyn-F: 5′-AGGACTGGGTGGCTTCCAACTCCCAGAC AC-3′

   Rapsyn-R: 5′-AGCTTCTCATTGCTGCGCGCCAGGTTCA GG-3′

5. Using this approach, ovarian tissue can become retained and inadvertently lost within the syringe and needle. An alternative approach which minimizes such loss involves repeatedly pipetting ovaries through P1000 tips.

6. Supernatants should be discarded carefully using a pipetman because of the very small volume of the cell pellets.

7. Trypsin treatment is not mandatory. However, its use will improve the number of single cells.

8. Usually, a total of approximately 10^5 cells will be obtained per ovary after collagenase treatment. Meiotic prophase cells comprise approximately 5–10% of the total cells.

9. The volume of hypotonic buffer is adjusted in the next step of the method.

10. Longer hypotonic treatment gives a better chromosome spread that preserves intact axis associated proteins such as cohesins or SC components. For some other proteins, however, localization signals are decreased or even eliminated by prolonged hypotonic treatment.

11. Triton X-100 works for pre-extraction of free cytoplasmic/chromatin-unbound proteins during PFA fixation. The pre-extraction and fixation are simultaneously done until the nuclei are attached to the slide glass.

12. Cytocentrifugation aids in the rapid preparation of well-spread chromosomes. Cytocentrifugation induces cell nuclei to become physically attached to the slide glass prior to extraction of free proteins during the fixation stage. However, chromosome spreading is greatly affected by the input cell concentration. There is tendency for cytocentrifugation at a low cell concentration to give more widely spread chromosomes; however, nuclei with broken chromosome morphology are occasionally observed. Also significant numbers of cells can be lost through flow out from the slide glass. The cell concentration should be empirically determined for the next application.

13. The SYCP3 staining shows axial/lateral elements while SYCP1 staining shows central elements, both comprising the SC.

14. Fixation of antibodies following immunostaining is mandatory in order to prevent their dissociation from the chromosome during the subsequent heat denaturing step in 70% formamide/6× SSC. Poor fixation results in massive aggregation of dissociated antibodies, which results in high background noise. In canonical chromosome FISH, pepsin digestion of the spread chromosomes would usually be applied to enhance hybridization. However such protein digestion should be avoided when immunostaining is combined with FISH.

15. Rubber cement can be used for sealing the cover glass instead of nail polish.

16. Carefully remove the nail polish along the edge of the cover glass using a knife or forceps.

17. The appropriate conditions for the post-hybridization washing step should be determined empirically for each probe. Poor washing results in high background with a low signal:noise ratio, whereas extensive washing will reduce hybridization signals.

Acknowledgements

We thank Dr. Aya Sato for the technical advice on Chromosome FISH with co-immuostaining. This work was supported in part by a JSPS Research Fellowship (to J.K.), a Grant-in-Aid for Scientific Research on Priority Areas, a Grant-in-Aid for Young Scientists (to K.I.), the Global COE Program (Integrative Life Science Based on the Study of Biosignaling Mechanisms), a Grant-in-Aid for Specially Promoted Research (to Y.W.) from the Ministry of Education, Culture, Sports, Science and Technology of Japan.

References

1. Nasmyth K, Haering CH (2009) Cohesin: its roles and mechanisms. Annu Rev Genet 43: 525–558
2. Lee J, Hirano T (2011) RAD21L, a novel cohesin subunit implicated in linking homologous chromosomes in mammalian meiosis. J Cell Biol 192:263–276
3. Ishiguro K, Kim J, Fujiyama-Nakamura S et al (2011) A new meiosis-specific cohesin complex implicated in the cohesin code for homologous pairing. EMBO Rep 12:267–275
4. Gutierrez-Caballero C, Herran Y, Sanchez-Martin M et al (2011) Identification and molecular characterization of the mammalian alpha-kleisin RAD21L. Cell Cycle 10: 1477–1487
5. Lee J, Iwai T, Yokota T et al (2003) Temporally and spatially selective loss of Rec8 protein from meiotic chromosomes during mammalian meiosis. J Cell Sci 116:2781–2790
6. Prieto I, Suja JA, Pezzi N et al (2001) Mammalian STAG3 is a cohesin specific to sister chromatid arms in meiosis I. Nat Cell Biol 3:761–766
7. Revenkova E, Eijpe M, Heyting C et al (2001) Novel meiosis-specific isoform of mammalian SMC1. Mol Cell Biol 21:6984–6998
8. Page SL, Hawley RS (2004) The genetics and molecular biology of the synaptonemal complex. Annu Rev Cell Dev Biol 20:525–558
9. Kleckner N (2006) Chiasma formation: chromatin/axis interplay and the role(s) of the synaptonemal complex. Chromosoma 115: 175–194
10. Watanabe Y (2004) Modifying sister chromatid cohesion for meiosis. J Cell Sci 117: 4017–4023
11. Ishiguro K, Watanabe Y (2007) Chromosome cohesion in mitosis and meiosis. J Cell Sci 120:367–369
12. Pelttari J, Hoja MR, Yuan L et al (2001) A meiotic chromosomal core consisting of cohesin complex proteins recruits DNA recombination proteins and promotes synapsis in the absence of an axial element in mammalian meiotic cells. Mol Cell Biol 21:5667–5677
13. Prieto I, Tease C, Pezzi N et al (2004) Cohesin component dynamics during meiotic prophase I in mammalian oocytes. Chromosome Res 12:197–213
14. Eijpe M, Heyting C, Gross B et al (2000) Association of mammalian SMC1 and SMC3 proteins with meiotic chromosomes and synaptonemal complexes. J Cell Sci 113:673–682
15. Speed RM (1982) Meiosis in the foetal mouse ovary. Chromosoma 85:427–437
16. McClellan KA, Gosden R, Taketo T (2003) Continuous loss of oocytes throughout meiotic prophase in the normal mouse ovary. Dev Biol 258:334–348
17. Wilhelm D, Palmer S, Koopman P (2007) Sex determination and gonadal development in mammals. Physiol Rev 87:1–28
18. Peters AH, Plug AW, van Vugt MJ et al (1997) A drying-down technique for the spreading of mammalian meiocytes from the male and female germline. Chromosome Res 5:66–68
19. Kudo NR, Anger M, Peters AH et al (2009) Role of cleavage by separase of the Rec8 kleisin subunit of cohesin during mammalian meiosis I. J Cell Sci 122:2686–2698
20. Kauppi L, Barchi M, Baudat F et al (2011) Distinct properties of the XY pseudoautosomal region crucial for male meiosis. Science 331: 916–920

Chapter 4

Immunohistochemical Approaches to the Study of Human Fetal Ovarian Development

Jing He, Andrew J. Childs, Jieqian Zhou, and Richard A. Anderson

Abstract

The development of primordial germ cells into oocytes within primordial follicles involves a complex sequence of proliferation, developmental commitment, entry and arrest in meiosis, and association with surrounding somatic cells. These processes occur over the first few months of development in the human, with multiple stages of development present at any one time point. Immunohistochemistry has been hugely instructive in identifying the various key stages in ovarian development, by allowing simultaneous visualization of different stages of germ cell development, and their spatial arrangement. These studies allow comparison with other species and have identified key differences between human and murine ovarian development as well as giving a basis for functional studies. In this chapter we describe the main methodologies used in immunohistochemistry, using both chromogen and fluorescence approaches, and both single and double antigen detection.

Key words: Immunohistochemistry, Immunofluorescence, Germ cell, Detection systems, Antibodies, Antigen retrieval

1. Introduction

1.1. Germ Cell Development in Human Fetal Ovary

Immunohistochemistry is invaluable in furthering our understanding of the developing gonad by allowing precise localization of specific markers (usually proteins) in an organ characterized by complex and changing intercellular interactions. Two groups of cells contribute to the formation of the ovary; the germ cells, which are first detected in the epithelium of the yolk sac near the base of allantois at 3 weeks' gestation (called primordial germ cells), and the somatic cells, which appear as the genital ridge (1). After the primordial germ cells form, they proliferate and migrate to the genital ridge; this is complete by around 6 weeks' gestation (2). The gonads begin to differentiate into ovary or testis, with the gender morphologically identifiable at around 8 weeks' gestation

Hayden A. Homer (ed.), *Mammalian Oocyte Regulation: Methods and Protocols*, Methods in Molecular Biology, vol. 957, DOI 10.1007/978-1-62703-191-2_4, © Springer Science+Business Media, LLC 2013

with the formation of testicular cords in the male (3). In testis, the germ cells will not enter meiosis until puberty, while in the ovary, meiosis initiates at around 11 weeks' gestation (4), i.e., at the beginning of the second trimester, while mitotic proliferation continues in those germ cells that have not entered meiosis.

As development progresses germ cells form interconnected "nests" through incomplete cytokinesis (5). Interspersed within these nests are somatic cells that are destined to become pre-granulosa cells, and streams of mesenchymal somatic cells separate nest structures. From 18 weeks' gestation (i.e., in the late second trimester) breakdown of the germ cell nests and association of individual oocytes (arrested at diplotene stage in meiotic prophase I) with a single layer of surrounding pre-granulosa cells results in the formation of primordial follicles. Approximately two-thirds of germ cells are lost through apoptosis (and potentially other pathways) during the process of follicle formation (6). A summary of the key developmental events during human female fetal germ cell development is depicted in ref. 7.

1.2. Studying Human Fetal Ovary Development with Immunohistochemistry

Using immunohistochemistry to detect specific antigen expression is an important technology to study early ovarian development. The main advantage of immunohistochemistry for the detection of protein expression during development is in situ detection; revealing both the existence and distribution of the target antigen. The main principle of immunohistochemistry is the antigen–antibody reaction—using a primary antibody to recognize and bind to the target antigen, then using a secondary antibody to detect the primary antibody. The secondary antibody may carry a chromophore (and thus be directly visualized) but is more usually biotinylated or conjugated with peroxidase, requiring a further color reaction to stain the target antigen. This technology also allows detection of two or even more targets on the same section, making it a powerful tool to study potential interactions between biomolecules within identified specific cell types.

Immunohistochemistry has been particularly successful in revealing the asynchrony of human fetal germ cell development (8, 9)—in contrast to the situation in rodents in which germ cell differentiation occurs in a broadly synchronized wave (10, 11). Immunohistochemical detection of stage-specific markers of germ cell development, such as the primordial germ cell and pluripotency-associated transcription factor OCT3/4 and the cell surface marker c-KIT, and the RNA-binding proteins DAZL and VASA (which mark progressively more mature germ cells) has revealed the existence of a developmental gradient within the ovary; with less mature (OCT4-positive) germ cells around the edge of the ovary and with germ cells at increasing stages of differentiation found towards the interior of the ovary (8, 9). These markers can then act as an index of germ cell maturation, against which the

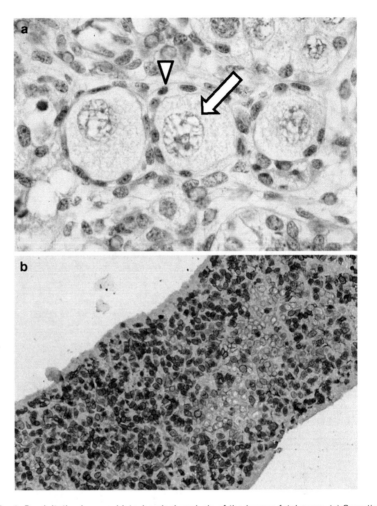

Fig. 1. Precipitative immunohistochemical analysis of the human fetal ovary. (a) Somatic cell-specific expression of the transcription factor FOXL2 in the 18-week human fetal ovary detected by diaminobenzidine (DAB) immunohistochemistry (*arrowhead*). Note the nuclei of the larger oocytes display no staining (*arrow*). For further details *see* ref. 12. (b) Double-precipitative immunohistochemistry showing co-expression of the cell surface marker KIT (*brown*; DAB) and the pluripotency-associated transcription factor OCT4 (blue nuclear staining; fast blue) in the germ cells of the 9-week human fetal ovary.

developmental stage at which other proteins of interest are expressed can be assessed. Alternatively, they can be used to demonstrate definitively whether proteins are expressed in germ or somatic cells by the presence or absence of co-localization, as we have demonstrated for FOXL2 and OCT4 (12). The distinct morphology of germ cells and somatic cells—the nuclei of germ cells are large compared to the smaller and more elliptical somatic cell nuclei—means these two populations can be readily discerned visually. Examples of the expression of FOXL2 and OCT4 in the human fetal ovary can be seen in Figs. 1a and 2b, respectively.

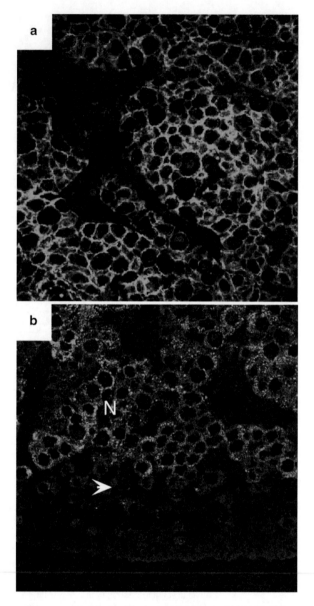

Fig. 2. Immunofluorescence analysis of the human fetal ovary. (**a**) Immunofluorescent detection of E-cadherin (*green*; propidium iodide nuclear counterstain, *red*) in the 14-week human fetal ovary outlining the cell membranes of germ cells in syncitial nests. Somatic cells in streams between germ cell clusters show no E-cadherin staining. For further details *see* ref. 19. (**b**) Double immunofluorescent detection of OCT4 (*red*) and DAZL (*green*) in the germ cells of the 14-week human fetal ovary revealing the existence of a cortical-medullary gradient of germ cell differentiation. *Arrowhead* denotes double-positive cell, *N* denotes germ cell nest. Blue nuclear counterstain: DAPI. Reproduced from ref. 8.

Immunohistochemical approaches have also yielded key insight into the expression—and directionality—of key signaling pathways regulating human fetal ovarian development. Immunolocalization of growth factors, their receptors, and downstream intracellular mediators/second messengers within the fetal ovary has provided considerable insight into how populations of cells in the fetal ovary converse with each other. We have used this in our recent analyses of TGF-beta superfamily growth factor signaling, which comprises the TGF-betas, activins, bone morphogenetic proteins (BMPs), and growth and differentiation factors (GDFs). Binding of these growth factors to cell surface receptors results in the phosphorylation and subsequent nuclear localization of SMAD proteins in target cells (for review see ref. 13). Specific SMAD proteins are utilized by particular groups of ligands; SMAD2 and 3 are phosphorylated in response to TGF-beta or activin signaling, whilst BMPs/GDFs activate SMAD1, 5, and 8. This specificity has enabled us to demonstrate that the somatic cells are the targets of activin signaling in the human fetal ovary (denoted by nuclear localization of SMAD2/3 in these cells (14)). Conversely, phosphorylated SMAD1/5 is detected only in germ cells, revealing these cells to be the only targets of BMP action during fetal ovarian development (15). Furthermore, the identification of specific cell types targeted by specific growth factors can inform hypotheses as to possible downstream gene targets. This is exemplified by our recent demonstration of activin-mediated regulation of brain-derived neurotrophic factor (BDNF) expression in the human fetal ovary, which arose from the observation that BDNF was expressed by human fetal ovarian somatic cells (16), which we had previously identified to be targets of activin (14). The demonstration that kit ligand is regulated by activin at a very specific stage of germ cell development also arose from careful observation of double-labeled immunohistochemical studies (14, 17).

1.3. Important Considerations

Each antibody requires optimization of fixative, tissue preparation, antibody concentration, and detection system. This chapter will introduce our methods for immunohistochemistry, and while not exhaustive will give the reader a basis for developing their own studies. We discuss peroxidase-based methods for standard and double-precipitative immunohistochemistry and immunofluorescence methods (particularly useful for double- or triple-labeling).

1.3.1. Choice of Antibody

Choosing the right antibody is extremely important. The ideal primary antibody is usually raised from a species which is far from the species to be investigated. Before buying the primary antibody, check the suppliers' Web sites and literature to see whether the antibody is suitable for immunohistochemistry. Antibody suppliers' Web sites will frequently list peer-reviewed publications in which the antibody has been used, and these can be a useful reference for determining whether the antibody is likely to be suitable for immunohistochemistry, appropriate dilutions, and to identify tissues in which the protein/antigen is known to be expressed, which can then act as positive controls. If the supplier provides images of immunohistochemical staining performed with the antibody, do not only pay attention to the positive results but also make sure the staining is in the right cell types (e.g., germ cells or somatic cells) and expected location (e.g., nucleus or cytoplasm).

Secondary antibodies are usually anti-primary species IgG. There are several kinds of secondary antibodies, classified according to the groups conjugated to the antibody. Simply, biotinylated antibodies are used in precipitative immunohistochemisty using diaminobenzidine (DAB)/Fast blue detection, fluorophore-conjugated antibodies are used in direct immunofluorescent detection, and peroxidase-conjugated antibodies are used for tyramide signal amplification (TSA)-enhanced fluorescence detection. Fluorophore-conjugated antibodies cannot be boiled in citrate buffer for double staining, because the fluorophores are attached to the primary antibody through secondary antibody, and will be removed with primary antibody after boiling. The other two staining methods are based on the chromogenic deposition near the antigen, thus do not depend on the continuing presence of the primary antibody.

For double- and triple-staining experiments, where the primary antibodies used have been raised in different species in one assay, try to use the secondary antibodies raised against these in the same species if available (e.g., if the primary antibodies are rabbit anti-human and mouse anti-human, then use goat anti-rabbit and goat anti-mouse secondary antibodies). If performing double staining with two primary antibodies raised in the same species, then Fab-fragment secondary antibodies can be used to minimize cross-reaction. The Fc (fragment crystalizable) parts are removed in the fab antibodies so this can minimize the nonspecific binding.

1.3.2. Precipitative Staining or Immunofluorescence?

Precipitative staining deposits a colored chromogen at the site of expression/antibody binding, which can be viewed using a brightfield microscope. DAB staining is perhaps most commonly used, although other stains (such as Fast Blue) are available. Immunoflourescent detection relies on the detection of photons emitted by the fluorophore or fluorescent chromogen in response to excitation at particular wavelengths of light. The choice of which approach to use depends on a variety of factors, including the quality of the primary antibody, whether single or double staining is planned, and local microscopy equipment availability. For single staining, DAB is often the first choice, because it is much cheaper than fluorescence. For double staining, however, the fluorescence methods are easier and more extensively used. If single DAB staining does not work or is very weak, then try fluorescence—it is generally more sensitive than DAB, especially with TSA staining (but which may introduce problematic background fluorescence). Alternatively, DAB staining can be enhanced using ABC (Avidin–Biotin Complex) kit instead of Streptavidin–HRP or a polymer-based detection system such as ImmPRESSTM (Vector Labs) instead of both secondary antibody and Streptavidin-HRP. Examples of single and double precipitative and fluorescence staining on the human fetal ovary can be seen in Figs. 1 and 2, respectively.

1.3.3. Optimizing Antibody Concentration

Optimizing the antibody to appropriate concentration may be the most difficult part of immunohistochemistry. If you are lucky, the manufacturer's literature may give the correct recommended concentration, but peer-reviewed publications which have used the antibody in similar procedures may be the best starting point. In this case, dilute the antibody to the recommended concentration and two higher/lower concentrations (e.g., recommended 1/200, then try 1/100 and 1/400). If no recommendation, the dilution ranges should be extended. We usually dilute the antibody to 1/100, 1/200, 1/500, 1/1,000, 1/5,000 and this will give us the rough ranges of dilution. If that range does not produce a result, the dilution can be changed to a higher/lower range.

If tissue is precious, try optimizing the antibody on fixed cell lines, if available. Where antibodies have been shown to be suitable for western blotting and the concentration for this application is known, usually a twofold higher concentration is suitable for DAB/fluorophore-conjugated staining (e.g., 1/200 in western, then try 1/100 or higher in immunohistochemistry). For tyramide-enhanced fluorescent immunostaining, the required concentration may be similar or lower than required for western blotting.

1.3.4. Controls in Immunohistochemistry

Controls are an essential part of an immunohistochemistry protocol to test for the specificity of the antibody and to ensure that the protocol as a whole has worked. Positive controls are fairly straightforward. They should be a tissue that is already known to express the protein of interest and are there to demonstrate that on the day the procedure worked.

Negative controls test for antibody specificity. Two types of negative controls are frequently used. The most robust involves pre-absorbing the primary antibody with the antigen (sometimes sold as a blocking peptide or immunogen against which the antibody was raised) prior to use. While this type of negative control is ideal it requires the availability of the purified antigen. The alternative method is to either omit the primary antibody all together, which is not a robust control as it then only really tests for the other aspects of the procedure and particularly the specificity of the secondary antibody. It is better to replace the specific primary antibody with normal serum of the same species as the primary antibody, or a nonspecific isotype-matched IgG used at a similar concentration to the primary antibody. Other aspects of controls for the methodology involve blocking by the use of normal serum pretreatment and removal of endogenous peroxidase activity by pretreatment with hydrogen peroxide. Additionally some tissues contain endogenous biotin. This is an issue if using the biotin–avidin detection system and blocking involves pretreatment of the section with avidin which is then saturated with biotin.

Autofluorescence can sometimes be a problem in fluorescence immunohistochemistry. The easiest way to check for this is to view the tissue section without any antibody incubation under the fluorescence microscope. There is no robust way of getting around this other than not to use fluorescent methodology.

1.3.5. Label the Slides!

Sometimes new users forget to label the slides, and this will cause a lot of confusion. There are several potential time points to label the slides: the very beginning, during blocking, or after adding primary antibodies. Labeling the slides before adding antibodies sounds ideal but sometimes accidents happen and tissue on important slides may be damaged. Labeling one by one after adding the antibodies is flexible and convenient but very easy to forget, especially when you are getting more practiced at the methods. For new users, labeling all the slides before adding the antibodies is recommended. Pencil may be the best option for labeling, because other pens may not work on wet slides or be erased during a lot of washing and wiping, and dipping in organic solvents.

2. Materials

2.1. For All Protocols

1. Xylene.
2. Graded alcohols (100%, 95%, and 70% prepared with dH_2O).
3. Citrate buffer (0.01 M): (diluted from 0.1 M stock, prepared by diluting 0.1 M citric acid in deionized water (dH_2O), and adjusted to pH 6.0 with 5 M sodium hydroxide solution. Store at 4°C).
4. Fume hood.
5. 3% hydrogen peroxide (H_2O_2): dilute 30% hydrogen peroxide stock 1/10 in methanol; mix well in a fume hood as the mixing liberates lots of gas.
6. Pertex glue (CellPath Ltd, Newtown Powys, UK) or Permafluor aqueous mounting medium (Thermo Scientific).
7. Glass staining dishes.
8. Rocking platform.
9. Standard pressure cooker.
10. Humidified chamber: made by lining a slide box with paper towel soaked in Tris-buffered saline (TBS).
11. Scott's tap water: 20 g $NaHCO_3$, 3.5 g $MgSO_4$ in 1 L H_2O. Mix thoroughly to dissolve.

2.2. Precipitative Immuno-histochemistry

1. Tris-buffered saline (TBS): 50 mM Tris-base, 150 mM NaCl, pH 7.4. Prepare 10× stock from 121.1 g Tris-base, 170 g NaCl, 2 L dH$_2$O, adjust the pH to 7.4 using 36% HCl.

2. Serum block (Serum/TBS/BSA): prepared by mixing normal serum (use serum from the same species as that which the secondary antibody is raised in) with TBS and bovine serum albumin (BSA) at a ratio of 4:16:1. This can be aliquoted and stored at –20°C for long-term use. Once thawed, working aliquots are good for 3 days and should be stored at 4°C.

3. Avidin/Biotin blocking kit (Vector Laboratories, Peterborough, UK).

4. Primary antibody: diluted to different concentrations as required in Serum/TBS/BSA.

5. Secondary antibody: anti-primary species-biotinylated, diluted 1/500 (up to 1/200) in Serum/TBS/BSA; store on ice or in 4°C before using.

6. Streptavidin–HRP (Vector Laboratories): diluted 1 in 1,000 in TBS.

7. DAB substrate (DAKO, Cambridgeshire, UK): 1 drop DAB diluted in 1 mL of DAB diluent.

8. ABC-AP (Avidin–Biotin Complex-Alkaline Phosphatase) kit (Vector Laboratories, for double-DAB staining only): dilute 10 μL Reagent A and 10 μL Reagent B in 1 mL TBS.

9. 0.1 M Tris, pH 8.2

10. Fast blue buffer (100 mL, for double-DAB staining only): 20 mg naphthol AS-MX phosphate disodium salt (N5000-500MG, Sigma) diluted in 2 mL dimethylformamide (Sigma) in a glass tube and then add 98 mL 0.1 M Tris pH 8.2. This can be stored at 4°C for up to 2 months.

11. 0.22 μm filters.

12. Fast blue working solution (for double-precipitative staining only, prepare fresh before use): 1 mg Fast blue BB salt (Sigma) diluted in 1 mL Fast blue buffer. Filter through a 0.22 μm filter.

2.3. Immuno-fluorescence Staining

1. 0.01 M citrate buffer (prepared as in **item 3** in Subheading 2.1).

2. PBS (phosphate buffered saline): 140 mM NaCl, 1.9 mM NaH$_2$PO$_4$, 8.9 mM Na$_2$HPO$_4$, pH 7.3–7.4. Prepare 10× stock solution by dissolving 100 PBS tablets in 2 L dH$_2$O. Store at room temperature.

3. 0.05% PBS-Tween: prepare by diluting Tween-20 (Sigma) 1:2,000 in 1× PBS.

4. 3% hydrogen peroxide (prepared as in **item 5** in Subheading 2.1).

5. Blocker (Serum/PBS/BSA): prepare as in item 2 in Subheading 2.2, substituting 1× PBS for TBS.

6. Primary antibody: diluted to different concentrations as required in Serum/PBS/BSA.

7. Secondary antibody: anti-primary species, either fluorophore or peroxidase-conjugated, diluted 1/500 (up to 1/200) in Serum/PBS/BSA; store on ice or 4°C before using.

8. TSA fluorescein (Perkin-Elmer, Waltham, USA): add 1 μL fluorescein to 50 μL amplification diluter per reaction (for one slide) (see Note 1).

9. DAPI (4′,6-diamidino-2-phenylindole) or PI (propidium iodide): diluted 1 in 1,000 in PBS.

3. Methods

The methods detailed here assume the reader has access to fixed tissue which has been sectioned using a microtome and sections mounted on electrostatic slides. Information on how to fix, wax-embed, and section tissues can be found in ref. 18.

3.1. Precipitative (DAB/Fast Blue) Immuno-histochemistry

Day 1 (for both single- and double-staining protocols)

1. Dewax the slides in xylene (2 × 5 min).

2. Rehydrate through graded alcohols (absolute ethanol: 2 × 20 s, 95% ethanol: 20 s; 70% ethanol: 20 s).

3. Rinse in tap water.

4. Retrieve antigens by pressure cooking slides. Bring 2 L 0.01 M citrate buffer to the boil in a standard pressure cooker, add slides, and seal. Release pressure 5 min after the appearance of steam, then remove from heat and allow to cool for 20 min.

5. Briefly wash the slides in tap water and rinse three times in dH$_2$O.

6. Incubate the slides in 3% hydrogen peroxide in methanol (to block endogenous peroxidase action) for 30 min at room temperature in glass staining dish on a rocking platform.

7. Rinse slides briefly in dH$_2$O three times and wash with TBS (2 × 5 min).

8. Take the slides one by one from the TBS and wipe away excess liquid from around the tissue section without touching the tissue.

9. Place the slides in the humidified chamber.

10. Apply 100 μL serum block (or enough to completely cover the section) per slide.

11. Incubate at room temperature for at least 30 min.

12. Drain the block from slides by standing the slides on end and touching the edge against a clean piece of paper towel.

13. Wash in TBS 2×5 min.

14. Take the slides from TBS and wipe as described in step 8.

15. Apply one drop avidin solution (from the avidin–biotin blocking kit) per slide (or sufficient to cover the tissue section) and incubate at room temperature for 15 min in the humidified slide tray.

16. Wash in TBS 2×5 min.

17. Wipe away excess TBS and apply one drop of biotin solution per slide and incubate at room temperature for 15 min.

18. Wash in TBS 2×5 min.

19. Dilute the primary antibody in serum block to the required concentration and place on ice until required.

20. Apply ~100 µL diluted antibody per slide and cover the tissue section with a small piece of parafilm to spread the antibody and prevent drying.

21. Incubate at 4°C in a humidified slide tray overnight. Include negative controls as required (see Subheading 1.3.2).

For single-staining protocols, proceed with step 22. For double staining proceed to step 38.

Day 2 (single-staining protocol)

22. Wash the slides in TBS 2×5 min. Do not attempt to remove the parafilm as this may damage the tissue; allow to lift off during the washes.

23. Dilute the secondary antibody (biotinylated and directed against the species in which the primary antibody was raised in) in serum block as required. Apply ~100 µL antibody per slide and incubate in humidified slide tray at room temperature for 30 min.

24. Wash the slides in TBS 2×5 min.

25. Dilute Streptavidin–HRP 1:1,000 in 1× TBS. Apply 50–100 µL per slide (or sufficient to cover the section) and incubate at room temperature in a humidified slide tray for 30 min.

26. Wash in TBS 2×5 min.

27. Dilute 1 drop of DAB chromogen in 1 mL DAB diluent to make DAB substrate.

28. One by one, apply 50–100 µL DAB substrate to each slide and assess color development (brown staining) under a light microscope.

29. Once the staining is appropriate, put the slide into TBS to stop reaction. For the negative control, wait for ~5 min after adding the DAB substrate then stop it in TBS.

30. Wash the slides in dH$_2$O for 5 min.

31. Stain the slides in hematoxylin for 5 min, then rinse in tap water.

32. Dry off excess liquid and transfer into acid-alcohol and shake quickly to decolor (no more than 2–3 s).

33. Rinse again in tap water, dry, and place slides into Scott's tap water for 20 s.

34. Assess counterstain intensity. If suitable, proceed to step 35. If counterstaining is too weak, restain slides in hematoxylin and then decolor again (steps 31–33).

35. Dehydrate tissue sections through graded alcohols (70% ethanol, 20 s; 95% ethanol, 20 s; 100% ethanol, 2 × 20 s; xylene 2 × 5 min). Drain excess liquid from the slide rack with paper towel between each step to avoid cross-contamination between solutions.

36. Transfer the slides to fresh xylene.

37. Take a cover slip (sufficient to cover the tissue section) and place a small amount of Pertex glue on it. Take a slide from xylene, only wipe the reverse side place face down onto the cover slip, until the glue is spread and the cover slip has stuck to the side. Using tweezers, squeeze out any bubbles and position the cover slip, then place the slides on a piece of clean paper towel dry.

Day 2 (double-staining protocol)

38. Perform steps 6–9, but use anti-primary species-biotinylated fab (fragment antigen binding) secondary antibody instead (see Note 2).

39. Microwave the slides in 400 mL boiling citrate buffer (pH 6.0) for 2.5 min to remove the primary antibody, then cool at room temperature for 30–60 min.

40. Rinse the slides in dH$_2$O, and then wash in TBS 2 × 5 min.

41. Repeat serum block (step 10), incubating slides in a humidified slide chamber for 30 min at room temperature. The serum used in the block at this stage should be from the same species that the secondary antibody used for detection has been raised in.

42. Wash the slides in TBS 2 × 5 min.

43. Repeat the Avidin/Biotin blocking (steps 14–17 above).

44. Wash the slides in TBS 2 × 5 min.

45. Prepare the second primary antibody in serum block and apply overnight as in steps 20 and 21 above.

Day 3 (double-staining protocol)

46. Wash the slides in TBS, 2 × 5 min.

47. Apply ~100 μL secondary antibody to each slide and incubate at room temperature for 30 min. The secondary antibody also should be directed against the second primary species-biotinylated. (And **fab** if both the primary antibodies are raised from same species.)

48. Wash the slides in TBS, 2 × 5 min.

49. Incubate the slides with ABC-AP kit diluted in TBS at room temperature for 30 min. Note that the ABC-AP solution needs to be made up 30 min prior to use.

50. Apply ~100 μL fast blue solution per slide and allow the color to develop for up to 30 min in the dark (see Note 3).

51. Check staining intensity using a light microscope.

52. Stop the reaction by putting slides in TBS.

53. Wash the slides in TBS, 2 × 5 min.

54. Mount the wet slides in Permafluor mounting medium (see Note 4).

3.2. Single-Tyramide/ Fluorophore-Conjugated Immunofluorescence

Day 1

1. Perform steps 1–7 of Subheading 3.1 substituting PBS for TBS.

2. Take slides one by one from PBS, drain the PBS from slides by touching the edge of the slide against the paper towel with one edge, then wipe around the tissue section (do not touch the tissue!).

3. Apply ~100 μL serum block (Serum/PBS/BSA) for at least 30 min at room temperature in a PBS-humidified slide tray.

4. Wash with PBS, 2 × 5 min.

5. Dilute primary antibody in serum block to required concentration. Apply ~100 μL primary antibody per slide, cover the sections with small pieces of parafilm, and incubate at 4°C overnight in a humidified slide tray. Include appropriate controls as required.

Day 2

6. Wash slides in PBS, 2 × 5 min, allowing parafilm cover slips to lift off themselves.

7. Apply ~100 μL secondary antibody (1/500–1/200 diluted in serum block) per slide and incubate in a humidified slide tray at room temperature for either 30 min (if using peroxidase-conjugated antibody for TSA detection) or 60 min (if using

fluorophore-conjugated secondary antibodies). Keep the slides in the dark from this step onwards.

8. Wash the slides in PBS, 2×5 min.

   If using fluorophore-conjugated antibody in step 7, proceed directly to step 11.

   If a peroxidase-conjugated antibody was used for TSA detection in step 7, then proceed first to step 9.

9. Apply ~50 μL TSA fluorescein per slide (up to ~100 μL per slide if not enough) and incubate for 10 min at room temperature (see Note 5).

10. Wash the slides in PBS, 2×5 min.

11. Counterstain the tissues with DAPI or PI diluted in PBS 1 in 1,000 for 10 min, applying ~100 μL per slide (see Note 6).

12. Wash in PBS, 2×5 min.

13. Place a drop of Permafluor mounting medium on a cover slip of appropriate size.

14. Remove the slides one by one from PBS (no need to wipe with tissue) and lower onto the cover slip. Gently touch the edge of the cover slip and move slowly until the Permafluor has spread evenly and the cover slip adheres. Wipe away the excess Permafluor and gently squeeze out bubbles with tweezers.

15. Leave the slides to dry at room temperature for ~4 h and then move to 4°C for storage, or put them straight to 4°C to dry (see Note 7).

3.3. Double-Tyramide Immunofluorescence

This protocol is only appropriate for TSA/TSA or TSA/fluorophore-conjugated secondary antibody staining.

Day 1

1. Perform steps 1–5 of Subheading 3.2, ensuring that the dilution of the primary antibody is adjusted to be suitable for TSA-enhanced detection on day 2.

Day2

2. Perform steps 6–10 of Subheading 3.2, but instead perform washes first with PBS-Tween for 5 min, then PBS for 5 min. DO NOT use a fluorophore-conjugated secondary antibody in step 7.

3. Microwave the slides in 400 mL boiled 0.01 M citrate buffer for 2.5 min to remove the primary antibody, cool for 30–60 min at room temperature.

4. Wash the slides in PBS, 2×5 min.

5. Apply serum block to sections and incubate at room temperature in a humidified slide chamber for 30–60 min. Note that

slides taken from warm citrate buffer will dry easily, so apply the block quickly.

6. Drain the serum block from the slides (optional: wash with PBS, 2×5 min).

7. Apply the second primary antibody, diluted to the appropriate concentration in serum block, cover with parafilm, and incubate overnight in a humidified slide chamber at 4°C.

Day 3

8. Proceed with steps 6–15 of Subheading 3.2 using PBS-Tween (5 min) then PBS (5 min) instead of 2×5 min PBS washes (except for washes after counterstaining, which remain PBS, 2×5 min).

4. Notes

1. TSA is a system developed by Perkin-Elmer to increase the sensitivity of detection. Tyramide is a substrate for HRP which converts the TSA reagents into a reactive-free radical that binds amino acids immediately adjacent to the target, i.e., primary antibody. The TSA reagent is labeled either directly with a fluorophore or for chromogenic detection. The increase in sensitivity can be very useful, but can also increase the background signal and is often not necessary.

2. Use of primary antibodies from the same species has traditionally been a major barrier to their simultaneous use. However the use of ImmPressive kit (ImmPress anti-primary species–HRP conjugate IgG or universal) can allow this. Incubate the tissue for 30 min at room temperature instead of both secondary antibody and Streptavidin–HRP incubation.

3. Fast blue solution is not stable, so always prepare fresh and keep away from direct light. Use it within 30 min of preparation.

4. The colors of fast blue and hematoxylin are quite similar so do not perform hematoxylin staining after double-precipitative immunohistochemistry. Fast blue is bleached by dehydration in ethanol and clearing with xylene, so do not mount the slides in Pertex. Permafluor is suitable for slides soaked in TBS as it is an aqueous mounting agent.

5. To avoid nonspecific background staining, avoid prolonged incubation in TSA. If staining many slides, try using two timers, starting the first and second timers when the TSA is added to the first and last slides, respectively.

6. The time length required for DAPI/PI incubation is variable. Tissues fixed in neutral buffered formalin (NBF) usually stain more strongly than those fixed in Bouin's fixative. Late gestation

human fetal ovaries usually show weaker staining than earlier ones (e.g., 18 week vs. 14 week). The staining may also become weaker when the reagent stock has been in use for a long time (especially freezed/thawed repeatedly). Prolonging the incubation period can often solve such problems; however, if the staining is still too weak then try sealing some DAPI/PI solutions onto the slides directly while mounting.

7. The slides will usually be dry after 4 h at room temperature and ready for microscopy. However if time is sufficient, leaving them to dry overnight at 4°C is recommended (as the slides may still not be completely dry after 4 h at room temperature, observation may introduce bubbles into the slides). The fluorescence will generally last for around 1 month, and up to 2 months sometimes, but will fade when stored for too long.

Acknowledgements

We are grateful to Anne Saunderson and the staff of the Bruntsfield Suite of the Royal Infirmary of Edinburgh for assistance with patient recruitment and specimen collection and to members of the Anderson Lab and Histology core facility at the MRC Centre for Reproductive Health for assistance in developing these protocols. This work was supported by the Medical Research Council (programme grant G1100357 to R.A.A.) and Medical Research Scotland (research grant 354 FRG to A.J.C.).

Funding

This work was supported by the Medical Research Council (programme grant G1100357 to R.A.A.) and Medical Research Scotland (research grant 354 FRG to A.J.C.).

References

1. Byskov AG (1986) Differentiation of mammalian embryonic gonad. Physiol Rev 66: 71–117

2. Wartenburg H (1981) Differentiation and development of the testes. In: Burger H, de Kretser DM (eds) The testis, 2nd edn. Raven, New York, pp 39–81

3. Hanley NA, Hagan DM, Clement-Jones M, Ball SG, Strachan T, Salas-Cortes L, McElreavey K, Lindsay S, Robson S, Bullen P, Ostrer H, Wilson DI (2000) SRY, SOX9, and DAX1 expression patterns during human sex determination and gonadal development. Mech Dev 91:403–407

4. Kurilo LF (1981) Oogenesis in antenatal development in man. Hum Genet 57:86–92

5. Pepling ME, Spradling AC (1998) Female mouse germ cells form synchronously dividing cysts. Development 125:3323–3328

6. Tingen C, Kim A, Woodruff TK (2009) The primordial pool of follicles and nest breakdown in mammalian ovaries. Mol Hum Reprod 15:795–803

7. Childs AJ, Anderson RA (2012) Experimental approaches to the study of human primordial germ cells. Methods Mol Biol 825: 199–210

8. Anderson RA, Fulton N, Cowan G, Coutts S, Saunders PT (2007) Conserved and divergent patterns of expression of DAZL, VASA and OCT4 in the germ cells of the human fetal ovary and testis. BMC Dev Biol 7:136

9. Stoop H, Honecker F, Cools M, de Krijger R, Bokemeyer C, Looijenga LH (2005) Differentiation and development of human female germ cells during prenatal gonadogenesis: an immunohistochemical study. Hum Reprod 20:1466–1476

10. Bullejos M, Koopman P (2004) Germ cells enter meiosis in a rostro-caudal wave during development of the mouse ovary. Mol Reprod Dev 68:422–428

11. Menke DB, Koubova J, Page DC (2003) Sexual differentiation of germ cells in XX mouse gonads occurs in an anterior-to-posterior wave. Dev Biol 262:303–312

12. Duffin K, Bayne RA, Childs AJ, Collins C, Anderson RA (2009) The forkhead transcription factor FOXL2 is expressed in somatic cells of the human ovary prior to follicle formation. Mol Hum Reprod 15:771–777

13. ten Dijke P, Hill CS (2004) New insights into TGF-beta-Smad signalling. Trends Biochem Sci 29:265–273

14. Coutts SM, Childs AJ, Fulton N, Collins C, Bayne RA, McNeilly AS, Anderson RA (2008) Activin signals via SMAD2/3 between germ and somatic cells in the human fetal ovary and regulates kit ligand expression. Dev Biol 314:189–199

15. Childs AJ, Kinnell HL, Collins CS, Hogg K, Bayne RA, Green SJ, McNeilly AS, Anderson RA (2010) BMP signaling in the human fetal ovary is developmentally regulated and promotes primordial germ cell apoptosis. Stem Cells 28:1368–1378

16. Childs AJ, Bayne RA, Murray AA, Martins Da Silva SJ, Collins CS, Spears N, Anderson RA (2010) Differential expression and regulation by activin of the neurotrophins BDNF and NT4 during human and mouse ovarian development. Dev Dyn 239:1211–1219

17. Childs AJ, Anderson RA (2009) Activin A selectively represses expression of the membrane-bound isoform of Kit ligand in human fetal ovary. Fertil Steril 92:1416–1419

18. Anderson G, Gordon K (1996) Tissue processing, microtomy and paraffin sections. In: Bankroft JD, Stevens A (eds) Theory and practice of histological techniques, 4th edn. Churchill Livingstone, New York

19. Smith SR, Fulton N, Collins CS, Welsh M, Bayne RA, Coutts SM, Childs AJ, Anderson RA (2010) N- and E-cadherin expression in human ovarian and urogenital duct development. Fertil Steril 93:2348–2353

Chapter 5

Protein Kinase Assays for Measuring MPF and MAPK Activities in Mouse and Rat Oocytes and Early Embryos

Jacek Z. Kubiak

Abstract

Protein phosphorylation plays a pivotal role in cell cycle regulation. MPF (M-phase Promoting Factor) and MAPK (Mitogen-activated protein kinase) are two major kinases driving oocyte maturation and early embryonic divisions. Their activities can be measured experimentally with kinase assays that use specific exogenous substrates. The activities of MPF and MAPK are measured using histone H1 kinase and MBP (Myelin Basic Protein) kinase assays, respectively. Here, we describe detailed procedures for measuring these two activities in mouse and rat oocytes and in early mouse embryos. The assays we describe can be performed using very small amounts of biological material and produce clearly discernible measurements of histone H1 and MBP kinase activities.

Key words: Cell cycle, CDK1/cyclin B, ERK 1/2, Histone H1 kinase assay, MAPK, MBP kinase assay, Meiosis, Mitosis, Mouse, MPF, Oocyte maturation, Rat

1. Introduction

Precise cell cycle regulation during oocyte maturation is required for the timely preparation of female gametes for fertilization and subsequent embryonic development (1). Protein phosphorylation is a major mechanism controlling the entry into M-phase upon induction of oocyte maturation at Germinal Vesicle (GV) Breakdown (GVBD), and the initiation of mitotic M-phase during embryonic development. In both cases the major M-phase serine/threonine protein kinase CDK1 (Cyclin-Dependent Kinase 1; the homologue of yeast *cdc2* or *cell division control 2* gene) is activated and triggers a cascade of phosphorylation events involving a number of mitotic substrates. CDK1-dependent phosphorylation is essential for Nuclear Envelope Breakdown (NEBD), chromatin condensation, bipolar spindle assembly, cessation of intracellular

Hayden A. Homer (ed.), *Mammalian Oocyte Regulation: Methods and Protocols*, Methods in Molecular Biology, vol. 957, DOI 10.1007/978-1-62703-191-2_5, © Springer Science+Business Media, LLC 2013

trafficking, cell rounding, and numerous other processes that accompany M-phase.

The active conformation of CDK1 is dependent upon its association with cyclin B and the dephosphorylation of CDK1 inhibitory sites. A network of other kinases and phosphatases (Myt1, Wee1, Cdc25, Plk1, PKA, PP1, PP2) (2) controls the timing of CDK1 activation. The activity required for initiating oocyte maturation and all meiotic and mitotic M-phases was first discovered several decades ago in frog (3) and mouse oocytes (4) and was initially named MPF (Maturation- and later M-phase Promoting Factor). Subsequently, MPF was identified as the CDK1 whose activity was modulated by cyclin B (5, 6) and which possessed a very specific histone H1 kinase activity (7–10). At that time the histone H1 kinase assay was established as an experimental tool for measuring MPF activity (11), a method we adapted to mouse oocytes (12).

At the beginning of the 1990s we started using histone H1 kinase assays for analyzing cell cycle regulation in mouse oocytes and embryos. We initially measured histone H1 kinase using pools of 50 oocytes (12–14). Very soon, however, we noted that lower numbers of mouse oocytes would suffice, and the method was adapted first for 40 oocytes (15) and subsequently for 30 oocytes (16, 17). Eventually, we managed to reduce the number of oocytes used for a single measurement to only three (18, 19), and at present, histone H1 kianse activity can be measured in a single oocyte.

In parallel with histone H1 kinase assays we developed the MBP kinase assay, which enabled us to measure the activity of two major oocyte MAPKs, ERK1 and ERK2. As confirmation that this assay is specific for these two MAPK activities we applied an in-gel-reaction and showed that in mouse oocytes the two main bands representing the kinases responsible for phosphorylating MBP migrate in PAGE at 46 and 44 kDa positions, corresponding to ERK1 and ERK2, respectively (17). In addition to the two major bands, three minor bands of approximately 52, 100 and 150 kDa were systematically detected (17). The identity of these bands has not yet been determined, but at least one of them, the 100 kDa band, may correspond to ERK5, an MAPK that is also known as BMK1 (for Big MAP kinase 1) (20). Significantly, no trace of either ERK1 or ERK2 activities was detected during the first embryonic mitotic division, showing that theses two MBP kinases are not activated during the first embryonic M-phase; notably however, other MAPK activities might be active in the early embryo as other higher-molecular weight (52, 100, and 150 kDa) MBP-phosphorylating bands were present in the in-gel-reaction experiments. Thus, the MBP kinase assay detects kinase activities corresponding in large part, albeit not exclusively, to ERK1 and ERK2. An important methodological advance was made when we noticed that the two assays, histone H1 and MBP phosphorylation, could be performed simultaneously in a single reaction (Fig. 1).

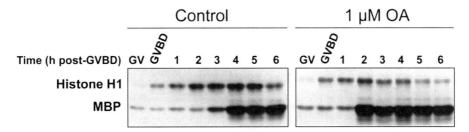

Fig. 1. Simultaneous assay for histone H1 and MBP kinase activities in rat oocytes during meiotic maturation. *Left panel:* Control; *right panel:* 1 µM okadaic acid (OA) treated oocytes. *GV* germinal vesicle stage, *GVBD* germinal vesicle breakdown, 1–6 h post-GVBD. Note that in the presence of OA histone H1 kinase is initially slightly higher (GVBD) than in controls, but then diminishes starting from 3 h post-GVBD to levels that are lower than that in controls. In controls, histone H1 kinase activity drops at 6 h post-GVBD concurrent with first polar body extrusion. In the presence of OA, MBP kinase activity is slightly higher between GV and 1 h post-GVBD. Thereafter, it is activated earlier after OA treatment (2 h post-GVBD in OA-treated oocytes versus 4 h in control oocytes) and to higher levels overall than in controls.

Exit from the M-phase follows inactivation of CDK1 and the dephosphorylation of its mitotic substrates (21). Protein dephosphorylation is mediated by a number of protein phosphatases (PP). Given that CDK1 substrates are phosphorylated on serine and threonine residues, serine/threonine phosphatases play a major role in the CDK1 inactivation that drives M-phase exit and the transition to interphase. PP1 and PP2 are the main players involved in these processes. These two major cell cycle-related phosphatases are inhibited by okadaic acid (OA). PP1 is more sensitive than PP2 and is totally inhibited by 100 nM OA, whereas PP2 requires 1 µM OA. As both PP1 and PP2 play a role in CDK1 and MAPK activation by removing inhibitory phosphorylation, OA results in rapid activation of these two kinases thereby triggering entry into a modified M-phase during which the major dephosphorylating enzymes remain inactive (22).

In mouse and rat oocytes, OA not only activates CDK1 and MAPKs, ERK1 and ERK2 (Fig. 1), but also induces characteristic changes in cell structure, e.g., in microtubule (MT) cytoskeleton organization (Figs. 2 and 3). Following GVBD and entry into M-phase, the MT cytoskeleton in OA-treated oocytes attempts to assemble a meiotic spindle. However, because of the disrupted equilibrium between protein phosphorylation and dephosphorylation in OA-treated oocytes, this tentative spindle collapses resulting in a number of spectacular and non-physiological MT cytoskeletal structures (Figs. 2 and 3).

Interestingly, the MTs of mouse and rat oocytes, behave differently when treated with OA. In the mouse, two sets of distinct MTs form: the first subpopulation of MTs radiates to the cytoplasm and associates with foci of pericentriolar material (PCM; in Fig. 2 detected by MPM-2 antibody), whilst the second morphologically distinct MT subpopulation encircles the condensing chromatin (Fig. 2). In contrast, in rat oocytes, OA treatment induces

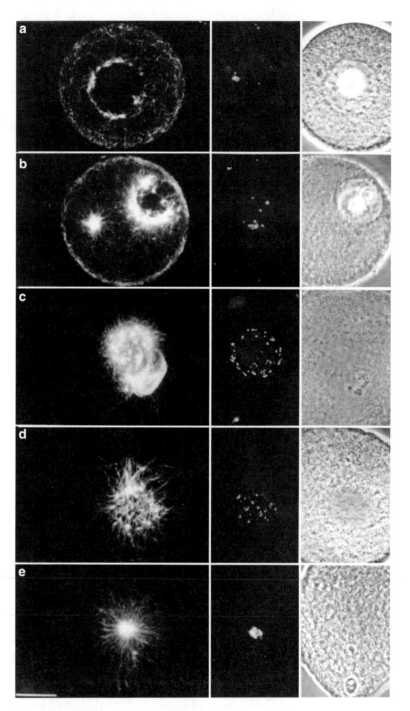

Fig. 2. Mouse oocyte maturation in the presence of 1 µM okadaic acid (OA). *Left panel* depicts microtubules (MTs) detected using anti-tubulin immunofluorescence (IF); *middle panel* depicts pericentriolar material detected using MPM2 antibody; images in the *right panel* show chromatin detected using phase contrast microscopy. Oocytes are shown at the GV stage (**a**), at GVBD (**b**), at 1 h post-GVBD (**c**), at 3 h post-GVBD, (**d**) and at 4 h post-GVBD (**e**). Note the presence of two distinct populations of MTs in (**c**), one surrounding the chromatin and the other surrounding PCM. MTs do not form a meiotic spindle but a central aster around hypercondensed chromatin (**e**).

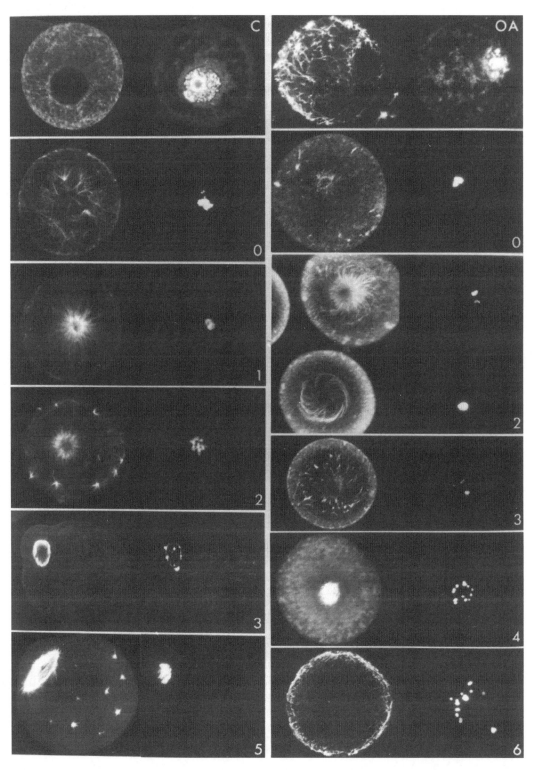

Fig. 3. Rat oocytes matured in control medium (C, *left panel*) and in the presence of 1 μM OA (OA, *right panel*). In each panel, images to the left depict MT immunofluorescence and images to the right show propidium iodide-stained DNA. Numbers to the bottom right corner of each panel represent time in h post-GVBD (0 refers to GVBD). Note that by 3 h post-GVBD in the presence of OA, a meiotic spindle is absent and MTs instead form long, curved asters.

the formation of a central aster of MTs so elongated that they curve along the oocyte membrane (Fig. 3).

This spectacular behavior of the MT cytoskeleton is certainly due to the specific non-physiological conditions triggered by OA treatment in which CDK1 is activated to a moderate level and the two pivotal MAPKs, ERK1 and ERK2, are activated to extremely high levels. This differential activation of the two kinases is clearly apparent in the results of histone H1 and MBP kinase assays (Fig. 1). This phenomenon also illustrates the usefulness of measuring histone H1 and MBP kinase activities for better understanding the regulation of distinct cellular processes by CDK1 and MAPKs, ERK1 and ERK2.

2. Materials

Prepare all solutions using ultrapure water and analytical grade reagents and store at room temperature (unless indicated otherwise).

2.1. Preparation of Oocytes and Embryos

1. M2 medium (Sigma) (see Note 1).
2. Laboratory stereo-microscope.
3. Watch slides or small Petri dishes (see Note 2).
4. 100–300 IU/ml hyaluronidase solution in PBS.
5. 20 mg/ml Bovine Serum Albumin (BSA) solution in molecular grade water.

2.2. Histone H1 Kinase Assay

1. Histone Kinase (HK) buffer stock solution without protease inhibitor cocktail: 80 mM β-glycerophosphate, 15 mM $MgCl_2$, 20 mM EGTA pH 7.3, 1 mM DTT in water. Buffer can be stored at 4°C for up to 1 month or at –20°C for longer periods of time.

2. Protease inhibitor cocktail for HK buffer: 1 mM AEBSF, 1 mg/ml leupeptin, 1 mg/ml pepstatin, 1 mg/ml aprotinin. For the complete HK buffer, add the protease inhibitor cocktail to the solution in item 1 above just before use.

   Prepare 250 μL of the complete HK buffer as follows: mix 40 μL of 80 mM β-glycerophosphate, 7.5 μL of 15 mM $MgCl_2$, 50 μL of 20 mM EGTA pH 7.3, 2.5 μL of 1 mM DTT, and 5 μL of each protease inhibitor: 0.1 M AEBSF, 1 mg/ml leupeptin, 1 mg/ml pepstatin, 1 mg/ml aprotinin. Adjust to a final volume 250 μL adding 130 μL of molecular grade water.

3. Radioactive mixture containing histone H1: 3.3 mg/ml histone H1, 1 mM ATP, 0.25 mCi/μL (^{32}P)-ATP.

Prepare 50 μL of the reaction solution by mixing 36.5 μL of water with 10 μL of 3.3 mg/ml histone H1 stock solution, 1 μL of 1 mM ATP, 2.5 μL of 10 mCi/ml (^{32}P)-ATP.

2.3. MBP Kinase Assay

1. HK buffer stock solution (see item 1 in Subheading 2.2).
2. Protease inhibitor cocktail (see item 2 in Subheading 2.2).
3. Radioactive mixture containing MBP: 6 mg/ml MBP, 1 mM ATP, 0.25 mCi/μL (^{32}P)-ATP.

The reaction solution for MBP kinase assay is prepared as follows: mix 13.5 μL of water with 34 μL of 6 mg/ml MBP stock solution, 1 μL of 1 mM ATP, 2.5 μL of 10 mCi/ml (^{32}P)-ATP.

2.4. Combined Histone H1 and MBP Kinase Assay

1. HK buffer stock solution (see item 1 in Subheading 2.2).
2. Protease inhibitor cocktail (see item 2 in Subheading 2.2).
3. Radioactive mixture containing both histone H1 and MBP: 3.3 mg/ml histone H1, 6 mg/ml MBP, 1 mM ATP, 0.25 mCi/μL (^{32}P)-ATP.

The reaction solution containing both H1 and MBP should be prepared as follows: mix 3.5 μL of water with 10 μL of 3.3 mg/ml histone H1 stock solution, 34 μL of 6 mg/ml MBP stock solution, 1 μL of 1 mM ATP, 2.5 μL of 10 mCi/ml (^{32}P)-ATP.

2.5. SDS-PAGE and Autoradiography

1. Standard SDS Polyacrylamide minigel electrophoresis (PAGE) components and equipment (see Note 3).
2. Laemmli electrophoresis buffer: 1 M Tris–HCl pH 7.5, 20% SDS, 0.5 M EDTA, 100% glycerol, Bromophenol blue, water.
3. Standard autoradiography procedure using X-ray films or a phosphorimager.

3. Methods

All procedures are carried out at room temperature unless specified otherwise.

3.1. Isolation of Oocytes and Embryos

1. Place the dissected ovaries in pre-warmed M2 containing BSA for obtaining GV-stage oocytes. Release the oocytes by puncturing the ovarian follicles (visible under the stereo-microscope as clear bulges at the ovarian surface) using a sharp needle (see Note 4).
2. Puncture the ampulla of the oviduct in pre-warmed hyaluronidase solution to obtain ovulated oocytes and one-cell embryos (see Note 5).

3. Maintain the oocytes/embryos in hyaluronidase solution for 3–6 min at 30°C by which time the surrounding follicular cells will usually become detached thereby making the oocytes/embryos easily visible.

4. Transfer cumulus-free oocytes/embryos to hyaluronidase-free M2 containing BSA using a mouth pipette (see Note 6).

3.2. Lysis of Oocytes and Embryos

1. Separate oocytes/embryos into groups of 3, 10, or 50 depending on the sample size to be used in the assay.

2. Calibrate a mouth pipette for 1 µL by gently aspirating 1 µL of fluid and marking the level of the meniscus using an indelible pen.

3. Using the calibrated mouth pipette, aspirate each sample of oocytes/embryos in 1 µL of water containing 20 mg/ml of BSA (see Note 7).

4. Transfer the 1 µL sample of lysed oocytes/embryos to an empty 1.5 ml Eppendorf tube. Ensure that all oocytes/embryos are drawn into the pipette. Blow the sample out with sufficient vigor to ensure that all of the oocytes/embryos are expelled into the tube (see Note 8).

5. Place the bottom of the tube on dry ice to snap-freeze the oocytes/embryos.

6. Repeat freezing/thawing twice and store the tube at –20°C.

3.3. Histone H1 Kinase Assay

1. Thaw Eppendorf tubes containing frozen oocytes or embryos (from Subheading 3.2) on crushed ice. Add 2 µL of HK buffer, to bring the total volume to 3 µL and gently spin down the sample (e.g., $100 \times g$ for 5 s) using a table centrifuge (see Note 9).

2. Place the samples back on crushed ice and add 1 µL of the radioactive reaction solution on to the inside wall of the Eppendorf without shaking so as to avoid immediate mixing of the radioactive solution with the 3 µL solution containing HK and the lysed cells. Every tube (containing a total volume of 4 µL) is now ready for the *in vitro* protein phosphorylation reaction (see Note 10).

3. Centrifuge all tubes once at low speed (e.g., $100 \times g$ for 5 s) using a table centrifuge (see Note 11).

4. Following centrifugation, transfer the tubes immediately to a 37°C incubation bath for 50 min.

5. Remove all tubes from the bath, place on crushed ice and immediately add 20 µL of Laemmli buffer to obtain a final volume of 24 µL (see Note 12).

6. Heat the tubes for 5 min at 90°C and then cool on crushed ice (see Note 13).

7. Freeze the tubes at −20°C before electrophoresis (see Note 14).

3.4. MBP Kinase Assay

1. Lyse oocytes/embryos in a 1 μL sample as described in Subheading 3.2.

2. Add 2 μL of HK buffer to make a total volume of 3 μL as described in step 1 of Subheading 3.3.

3. Add 1 μL of the radioactive reaction solution containing MBP to the inside wall of the Eppendorf as described in step 2 of Subheading 3.3.

4. Complete the assay by following steps 3–7 outlined in Subheading 3.3.

3.5. Mixed MBP and Histone H1 Kinase Assay

1. Lyse oocytes/embryos in a 1 μL sample as described in Subheading 3.2.

2. Add 2 μL of HK buffer to make a total volume of 3 μL as described in step 1 of Subheading 3.3.

3. Add 1 μL of the radioactive reaction solution containing both H1 and MBP to the inside wall of the Eppendorf as described in step 2 of Subheading 3.3.

4. Complete the assay by following steps 3–7 outlined in Subheading 3.3.

3.6. SDS-PAGE and Autoradiography

1. Prepare 8%, 10%, or 12% SDS-PAGE mini-gels as appropriate (see Note 15).

2. Thaw samples in Eppendorf tubes at room temperature.

3. Load gels with equal volumes of each sample, e.g., 6 μL and commence the electrophoresis (suggested parameters: 16 mA, voltage 80 V for 10 min and then 160–200 V) (see Note 16).

4. Continue the electrophoresis until the Bromophenol blue-colored front completely runs off the base of the gel (see Note 17).

5. Remove the gels and fix using 70% ethanol and 10% acetic acid.

6. Stain gels with Coomassie Brilliant Blue solution in 70% ethanol and 10% acetic acid (see Note 18).

7. Dry the gel in a vacuum dryer.

8. Expose the gel to X-ray film in a dark room or on a phospho-imager screen.

9. Develop film (Fig. 1 shows an example of the results obtained from a combined histone H1 and MBP kinase assay; Fig. 4 shows an example of the results obtained from a histone H1 kinase assay).

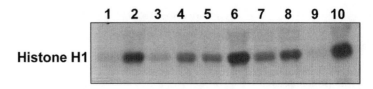

Fig. 4. Histone H1 kinase assay using the equivalent of three intact mouse one-cell embryos per sample. 1 and 9: three two-cell embryos at the G2 stage; 2 and 10: three two-cell embryos in mitosis; 3–6: 12 halves of two-cell embryo blastomeres, and more precisely, 3 and 5: anuclear blastomere halves progressing towards mitosis, 4 and 6: nuclear halves progressing towards mitosis; 7 and 8: six one-cell embryo halves, and more precisely, 7: anuclear halves and 8: nuclear halves. Note that based on histone H1 kinase assays, it can be seen that CDK1 activity is higher in nucleated than in anucleated oocyte halves.

4. Notes

1. GV-stage (or immature oocytes) are isolated from ovaries, whereas mature (ovulated) oocytes (arrested in meiosis II) and early-stage embryos (zygotes) are isolated from oviducts in M2 medium containing BSA. The pH of M2 medium is adjusted with HEPES, which permits open-air manipulation without significant pH variation. CO_2-equilibrated medium cannot be used for manipulations since their pH changes during exposure to air.

2. Glass slides are preferred since they are less easily scratched by forceps and needles thereby allowing for easier collection of oocytes.

3. For protein electrophoresis we use the Mini PROTEAN II™ system (Bio-Rad).

4. The heating provided by the electric bulb under the stage of the microscope is usually sufficient for sustaining a suitable temperature during oocyte and embryo manipulations. However, if a cold light is used, all solutions should be placed on a heating plate next to the microscope in order to maintain a temperature of ~30°C.

5. Ovulated oocytes and zygotes are surrounded by *cumulus oophorus*, which is composed of a mass of very sticky follicular cells. In the absence of hyaluronidase, cumulus-covered oocytes/embryos have a tendency to stick to the pipette, forceps, or needles and can be easily lost. It is therefore important that ovulated oocytes/embryos are liberated from the oviduct in the presence of hyaluronidase.

6. The *zona pellucida* of oocytes and embryos does not affect the kinase assays and need not be removed especially since

zona-free oocytes and embryos become relatively sticky and fragile and are easily lost during further manipulations.

7. The measurement of kinase activity is performed using an extract derived by lysing oocytes or embryos, which liberates CDK1 and ERK1 and ERK2 proteins. In order to compare the activities present in different groups of oocytes or at different stages of embryo development the same numbers of oocytes or embryos should be lysed in the same volume of BSA-treated water. We recommend lysing the oocytes in molecular grade water with 20 mg/ml BSA, which provides a very hypotonic environment that accelerates disruption of the oolemma. Importantly, the high BSA content stabilizes kinase activity following lysis. The volume of 1 µL does not have to be measured exactly; however, the same volume of molecular grade water with 20 mg/ml BSA (obtained with the same calibrated glass pipette) should be used for all samples in a single experiment to keep a constant volume of all samples.

8. It is normal for the sample of oocytes/embryos to disperse over the base of the tube. After the addition of all reagents, centrifuge the tube in order to concentrate the oocytes/embryos into a pellet at the bottom of the tube.

9. After centrifugation the lysed cells are mixed with HK buffer supplemented with protease inhibitors in order to prevent protein degradation. As it is better not to thaw the samples before adding HK buffer, it is advisable that not more than three tubes be thawed/handled at once.

10. Placing the 1 µL of radioactive solution on the inside wall of each tube resting on the crushed ice keeps it separate from the other 3 µL of the reaction mixture thereby preventing an uncontrolled start of the reaction.

11. The reaction begins when the tubes are centrifuged.

12. There should be minimal delay in between samples so as to avoid differences in the starting points of the reaction for the different samples. Placing tubes on crushed ice slows down the reaction and allows further manipulations to be done in a timely manner. Adding the Laemmli buffer stops the reaction. In order to minimize the delay between when the reaction is stopped in the first and the last tube, all tubes should be first opened on the surface of crushed ice and 20 µL of Laemmli buffer should be added as quickly as possible to each tube.

13. Heating the samples followed by cooling speeds up denaturation of proteins by SDS present in Laemmli buffer. Most importantly, it allows the complete inhibition of phosphatases and conserves the native state of phosphoproteins.

14. If electrophoresis is not performed soon after sample denaturation, samples can be stored at −20°C.

15. The lower percentage gels are better adapted for histone H1 kinase assays, whilst 12% gels are better suited to MBP kinase assays because MBP migrates at approximately 17 kDa and can easily run off the base of 8% or 10% gels and be lost.

16. For reliably comparing the levels of phosphorylation of histone H1 or MBP between samples, it is important that the volumes loaded on the gels are kept constant. If low numbers of cells are used in the reaction the final volume can be increased to 10 or 20 µL.

17. It is important to allow the Bromophenol blue colored front to completely run off the front of the gel because it contains unbound ^{32}P. If for any reason, the blue colored front remains in the gel it should be cut off before fixation as the unbound ^{32}P greatly augments the background noise.

18. Staining the gel with Coomassie Brilliant Blue serves to confirm equal sample loading for all lanes of the gel.

Acknowledgments

I thank Malgorzata Kloc for valuable discussions and critical reading of the manuscript. This work was supported by ARC grants to JZK.

References

1. Kubiak JZ, Ciemerych MA, Hupalowska A, Sikora-Polaczek M, Polanski Z (2008) On the transition from the meiotic to mitotic cell cycle during early mouse development. Int J Dev Biol 52:201–217

2. Marteil G, Richard-Parpaillon L, Kubiak JZ (2009) Role of oocyte quality in meiotic maturation and embryonic development. Reprod Biol 9:203–224

3. Masui Y, Markert CL (1971) Cytoplasmic control of nuclear behavior during meiotic maturation of frog oocytes. J Exp Zool 177:129–145

4. Balakier H, Czolowska R (1977) Cytoplasmic control of nuclear maturation in mouse oocytes. Exp Cell Res 110:466–469

5. Gautier J, Norbury C, Lohka M, Nurse P, Maller J (1988) Purified maturation-promoting factor contains the product of a Xenopus homolog of the fission yeast cell cycle control gene cdc2+. Cell 54:433–439

6. Gautier J, Minshull J, Lohka M, Glotzer M, Hunt T, Maller JL (1990) Cyclin is a component of maturation-promoting factor from Xenopus. Cell 60:487–494

7. Labbé JC, Picard A, Karsenti E, Dorée M (1988) An M-phase-specific protein kinase of Xenopus oocytes: partial purification and possible mechanism of its periodic activation. Dev Biol 127:157–169

8. Labbé JC, Picard A, Peaucellier G, Cavadore JC, Nurse P, Doree M (1989) Purification of MPF from starfish: identification as the H1 histone kinase p34cdc2 and a possible mechanism for its periodic activation. Cell 57:253–263

9. Arion D, Meijer L (1989) M-phase-specific protein kinase from mitotic sea urchin eggs: cyclic activation depends on protein synthesis and phosphorylation but does not require DNA or RNA synthesis. Exp Cell Res 183:361–375

10. Meijer L, Arion D, Golsteyn R, Pines J, Brizuela L, Hunt T, Beach D (1989) Cyclin is a component of the sea urchin egg M-phase specific histone H1 kinase. EMBO J 8:2275–2282

11. Felix MA, Pines J, Hunt T, Karsenti E (1989) A post-ribosomal supernatant from activated Xenopus eggs that displays post-translationally regulated oscillation of its cdc2+ mitotic kinase activity. EMBO J 8:3059–3069

12. Kubiak J, Paldi A, Weber M, Maro B (1991) Genetically identical parthenogenetic mouse embryos produced by inhibition of the first meiotic cleavage with cytochalasin D. Development 111:763–769

13. Weber M, Kubiak JZ, Arlinghaus RB, Pines J, Maro B (1991) c-mos proto-oncogene product is partly degraded after release from meiotic arrest and persists during interphase in mouse zygotes. Dev Biol 148:393–397

14. Kubiak JZ, Weber M, Géraud G, Maro B (1992) Cell cycle modification during the transitions between meiotic M-phases in mouse oocytes. J Cell Sci 102:457–467

15. Szöllösi MS, Kubiak JZ, Debey P, de Pennart H, Szöllösi D, Maro B (1993) Inhibition of protein kinases by 6-dimethylaminopurine accelerates the transition to interphase in activated mouse oocytes. J Cell Sci 104:861–872

16. Kubiak JZ, Weber M, de Pennart H, Winston NJ, Maro B (1994) The metaphase II arrest in mouse oocytes is controlled through microtubule-dependent destruction of cyclin B in the presence of CSF. EMBO J 12:3773–3778

17. Verlhac MH, Kubiak JZ, Clarke HJ, Maro B (1994) Microtubule and chromatin behavior follow MAP kinase activity but not MPF activity during meiosis in mouse oocytes. Development 120:1017–1025

18. Ciemerych MA, Tarkowski AK, Kubiak JZ (1998) Autonomous activation of histone H1 kinase, cortical activity and microtubule organization in one- and two-cell mouse embryos. Biol Cell 90:557–564

19. Ciemerych MA, Maro B, Kubiak JZ (1999) Control of duration of the first two mitoses in a mouse embryo. Zygote 7:293–300

20. Maciejewska Z, Pascal A, Kubiak JZ, Ciemerych MA (in press) Phosphorylated ERK5/BMK1 transiently accumulates within division spindles in mouse oocytes and preimplantation embryos. Folia Histochem Cytobiol 49(3):528–534

21. Skoufias DA, Indorato RL, Lacroix F, Panopoulos A, Margolis RL (2007) Mitosis persists in the absence of Cdk1 activity when proteolysis or protein phosphatase activity is suppressed. J Cell Biol 179:671–685

22. Zernicka-Goetz M, Verlhac M-H, Géraud G, Kubiak JZ (1997) Protein phosphatases control MAP kinase activation and microtubule organization during rat oocyte maturation. Eur J Cell Biol 72:30–38

Chapter 6

Time-Lapse Epifluorescence Imaging of Expressed cRNA to Cyclin B1 for Studying Meiosis I in Mouse Oocytes

Janet E. Holt, Simon I.R. Lane, and Keith T. Jones

Abstract

The first meiotic division of mammalian oocytes physiologically occurs in the ovary in the hours preceding ovulation. Fortunately, oocytes removed from their follicular environment will readily undergo this process in culture. Their large size, optical transparency, and efficiency in translating exogenous cRNA make mouse oocytes very amenable to study this process in detail using fluorescence imaging-based techniques. Here we describe the process of microinjecting cRNA to proteins of interest that have been coupled to a fluorescent protein using cyclin B1 as an example.

Key words: RNA, Epifluorescence, Fluorescent protein, Oocyte, Imaging, Meiosis

1. Introduction

Epifluorescence imaging is a well-established technique that relies on the capture of fluoresced light, often from a fluorochrome located within a cell, to an electronic image capture device, usually a CCD (charge coupled device) camera. Epifluorescence is best set up using an inverted microscope, and the cell of interest is positioned on its stage.

Epifluorescence can rarely match the quality of image generated by a confocal laser scanning microscope (CLSM), although deconvolution software can reconstruct very good quality images with minimal out of focus image blurring. Furthermore, cheaper setup and running costs, and its ease of use make it the preferred method of imaging for many subcellular applications. Historically we, and others, have used epifluorescence microscopy on mouse oocytes to image the process of meiosis I, commonly known as oocyte maturation (1, 2). This process begins with the breakdown

Hayden A. Homer (ed.), *Mammalian Oocyte Regulation: Methods and Protocols*, Methods in Molecular Biology, vol. 957,
DOI 10.1007/978-1-62703-191-2_6, © Springer Science+Business Media, LLC 2013

of the Germinal Vesicle (GV), equivalent to nuclear envelope breakdown, and finishes with first polar body extrusion following the segregation of homologous chromosomes.

For us the process of epifluorescence imaging is preferred because under ideal conditions we can track changes in both location and stability of expressed constructs with reasonable definition during the 12–14 h period of meiosis I in the mouse. It is primarily for us a process to be used on live cells, whereas CLSM with higher light energy is often performed on fixed material. As such for example, we can track processes such as the translocation of cyclin B1 from the cytoplasm to the nucleus at nuclear envelope breakdown (3), as well as the degradation of cyclin B1 during the latter stages of maturation (1).

Here we describe the methods for establishing epifluorescence-based imaging of GFP-labeled cyclin B1, as a model system for studying live oocytes during their maturation with some information on problem-solving common issues we have experienced in over 15 years of imaging oocytes. It is assumed that the user wants to image a fluorescently labeled protein, whose fluorescence is to be tracked during maturation. Oocytes are transcriptionally silent, and as such cRNA needs to be made and injected.

2. Materials

2.1. cRNA Manufacture

1. A suitable cloning vector such as pMDL or the Gateway System-based pSPE3 vectors (Life Technologies, Carlsbad, CA, USA) (see Note 1).

2. Transformation-competent bacteria such as DH5α bacterial cells (Life Technologies).

3. Luria Broth medium (MP Biomedicals, Solon, OH, USA).

4. Agar plates prepared from Luria Broth medium and Agar (MP Biomedicals).

5. Super Optimal Broth Medium (Sigma, St. Louis, MO, USA).

6. Ampicillin for bacterial selection (Sigma).

7. Wizard Plus SV Miniprep DNA Purification System (Promega, Madison, WI, USA).

8. Restriction enzymes dependent upon gene insert, for pMDL and pSPE3 vectors use *Kpn*I (Promega) and *Sfi*I (Takara, Shiga, Japan).

9. Nuclease-free water (Life Technologies).

10. Phenol:chloroform:iso-amlyalcohol (IAA) (25:24:1).

11. Chloroform (Sigma).

12. 100% Ethanol (Sigma).

13. 3 M sodium acetate (Sigma).

14. Pellet Paint Co-precipitant (Novagen, Madison, WI, USA).

15. Gel electrophoresis apparatus.

16. Agarose (Sigma).

17. DNA electrophoresis buffer (e.g., 1× Tris-acetate-EDTA) (Sigma).

18. T3 Message Machine High Yield Capped RNA Transcription Kit (Life Technologies) (see Note 2).

2.2. Culture Medium for Oocytes

1. M2 medium prepared from individual constituents (Table 1) (see Note 3).

2. 18 MΩ water (e.g., MilliQ H_2O, Millipore, Billerica, MA, USA).

3. Sterile plasticware and tissue culture-only volumetric flasks for preparation and storage of culture medium (see Note 4).

4. 0.22 μm Bottle top filter (Corning, Inc., Corning, NY, USA).

5. pH meter.

6. Osmometer (e.g., Osmometer Type 15, LÖser, Berlin, Germany).

Table 1
Constituents of M2 media

Component	Amount (g/L)	Concentration (mM)
1. HEPES	4.969	20.85
2. Penicillin G (K^+ salt)	0.060	0.16
3. Streptomycin sulfate	0.050	0.07
4. $CaCl_2 \cdot 2H_2O$	0.252	1.71
5. NaCl	5.533	94.66
6. KCl	0.356	4.78
7. KH_2PO_4	0.162	1.19
8. $MgSO_4 \cdot 7H_2O$	0.293	1.19
9. $NaHCO_3$	0.349	4.15
10. NaPyruvate	0.036	0.33
11. Glucose	1.000	5.56
12. Phenol Red	0.010	0.03
13. NaLactate	2.610	23.28
14. Bovine serum albumin	4.000	–

7. Minimum essential medium (MEM)α (Life Technologies) (see Note 5).

8. Penicillin/Streptomycin solution, which contains 5000 U penicillin/5 mg streptomycin per mL (Life Technologies).

2.3. Oocyte Collection

1. 3–4 Week-old F1 hybrid female mice (e.g., C57Bl6 females × CBA males) (see Note 6).

2. Pregnant Mare's Serum Gonadotropin (PMSG; Folligon, Intervet, Summit, NJ, USA) (see Note 7).

3. Miniforceps (World Precision Instruments, Sarasota, FL, USA).

4. Vannas scissors (World Precision Instruments).

5. Dissecting scissors (World Precision Instruments).

6. 35 mm round *Petri dishes* (BD Falcon, Franklin Lakes, NJ, USA).

7. Bunsen burner with adjustable pilot flame.

8. Flame-pulled glass Pasteur pipettes (150 mm flint glass, Kimble Chase, Vineland, NJ, USA) (see Note 8).

9. 1 cc/mL sterile tuberculin syringes with 0.01 cc graduations (Terumo, Somerset, NJ, USA).

10. Embryo culture-tested light mineral oil (Sigma) (see Note 9).

11. Milrinone (Sigma) (see Note 10).

12. 100% Ethanol (Sigma).

13. DMSO (Sigma).

14. Mouth pipette assembly (Sigma) (see Note 11).

2.4. Micropipette Fabrication

1. Borosilicate glass capillaries 1.5 mm outside diameter; 0.84 mm inside diameter, with internal filament (World Precision Instruments).

2. P-97 Flaming/Brown Pipette Puller (Sutter Instruments, Novato, CA) with 3 mm box filament (Sutter Instruments).

3. Kimtech (Kimberly Clarke Professional, Dallas, Texas, USA) or other commercial low-lint tissue.

4. 1.5 mm diameter micropipette storage jar (World Precision Instruments).

5. 1 cc/mL sterile tuberculin syringes with 0.01 cc graduations (Terumo) (see Note 12).

2.5. Microscopy

1. Zoom stereomicroscope with a total magnification of 80×, including objective and eyepiece together.

2. Inverted microscope, fitted with 4× (optional), 10×, 20×, and 40× objectives (see Note 13).

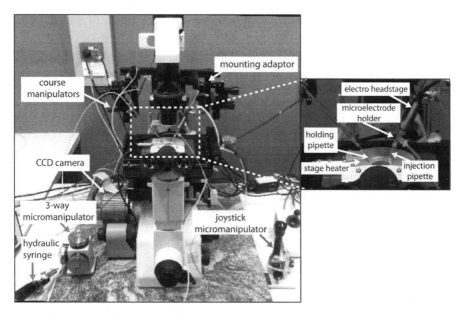

Fig. 1. Inverted microscope setup for microinjection. A Nikon TE300-inverted microscope fitted for epifluorescence, showing a front view with the associated microinjection and imaging equipment. *Inset right* is a magnified view of the heated stage and associated microinjection and holding pipettes.

3. An epifluorescence light source (Mercury, Metal-Halide, or Xenon lamp) (see Note 14).

4. Compound microscope, fitted with a 40× objective and 10× eyepiece.

2.6. Microinjection Apparatus and Assembly

1. Joystick-type or 3-way micromanipulators (MO-202U or MHW-3, Narishige, Tokyo, Japan) (see Note 15) (Fig. 1).

2. Coarse manipulators (MMN-1 Coarse Manipulators, Narishige) (Fig. 1).

3. Mounting adaptors (Microscope/Adaptor Combinations for Injection Systems, Narishige) (Fig. 1).

4. A hydraulic syringe (IM-5A Injector, Narishige) with a displacement of ~500 μm per turn of syringe (Fig. 1).

5. Injection Holder Set (Narishige) (see Note 16).

6. Prefabricated holding pipettes (e.g., G32801 Cook Medical, Bloomington, IN, USA) or may be fabricated with internal diameter 15 μm and external diameter 75 μm (Fig. 1).

7. Embryo culture-tested light mineral oil (Sigma).

8. Universal joint connector (UT-2 Universal Joint, Narishige).

9. A PV820 Pneumatic Picopump, (World Precision Instruments), capable of delivering compressed air at up to 50 psi.

10. An Electro 705 Electrometer (World Precision Instruments), possessing a "tickle" (negative capacitance) button.

11. A microelectrode holder (MEH2SF, World Precision Instruments).

12. A 37°C stage heater (Intracel, Royston, UK) (see Note 17) (Fig. 1).

13. Stage heater insert designed to hold a coverslip (see Note 18) (Fig. 1).

14. Round coverslips (22 mm, thickness No. 1, ProSciTech Thuringowa QLD, Australia).

15. Cell-Tak (BD Biosciences, Franklin lakes, NJ, USA).

16. Sterile-filtered buffer for microelectrode holder: 140 mM KCl, 1 M HEPES (pH 7.4) solution made from components (Sigma).

2.7. Imaging Equipment

1. CCD camera (e.g., Orca-R2, Hamamatsu Photonics K.K. Japan) capable of low intensity imaging (see Note 19) (Fig. 1).

2. Long-pass dichroic mirror suitable for imaging GFP (e.g., 505LP, Chroma Technology Corp., Bellows Falls, VT, USA) (see Note 20).

3. Emission filter suitable for imaging GFP (e.g. BA520, Nikon).

4. Lambda 10-3 control unit (Sutter, Novato, CA, USA).

5. Filter wheel (Sutter) (Coherent, Inc., Santa Clara, CA, USA).

6. Narrow bandpass excitation filter suitable for imaging GFP (e.g., 490 ± 7.3, Chroma).

7. PC running Metafluor/Metamorph software (Molecular Devices, Downingtown, PA) or Image J (National Institute of Health, Bethesda, MD, USA).

8. VS25 shutter (Uniblitz, Rochester, NY, USA) for controlling bright-field illumination.

3. Methods

3.1. cRNA Manufacture

1. Insert the coding sequence of cyclin B1 into the cloning vector using restriction enzyme cloning or recombination (see Note 1).

2. Transform competent bacteria using a 90-s heat-shock procedure followed by recovery in Super Optimal Broth Medium.

3. Plate out transformants on agar plates containing ampicillin (100 μg/mL) for selection.

4. Culture at least five clones in LB medium to obtain sufficient cells in log-phase growth for plasmid DNA preparation using the Wizard Plus SV Miniprep DNA Purification System.

5. Sequence plasmid DNA clones to ensure successful insertion and identity of the cyclin B1 construct.

6. Linearise 5 µg of plasmid downstream of the poly-A signal using an appropriate restriction enzyme (usually *Sfi*I).

7. Determine the efficiency of linearization prior to transcription by running 2 µL of the product on a 0.8% agarose gel.

8. Remove protein contamination by performing a standard phenol/chloroform extraction. Add an equal volume of choloroform:phenol:IAA (25:24:1) to reaction mix.

9. Vortex sample for 20 s and centrifuge at $10,000 \times g$ for 10 min.

10. Repeat steps 8 and 9 with the supernatant, using chloroform only.

11. Precipitate with 1/10 volume Na Acetate, $2.5 \times$ volume 100% ethanol and 2 µL pellet-paint to aid with visualization of the pellet. Wash pellet once with 70% ethanol.

12. Resuspend pellet in ~15 µL of nuclease-free H_2O.

13. Add 1 µg of the linearized, purified template per T3 Message Machine RNA Transcription reaction as per the manufacturer's guidelines and allow the reaction to proceed for 2 h.

14. Precipitate cRNA using lithium chloride (provided with T3 Message Machine kit). Wash pellet twice with 70% ethanol, and resuspend in nuclease-free H_2O at ~1 µg/µL.

15. Store RNA as single-use 1 µL aliquots at –80°C.

3.2. Fabrication of Micropipettes

1. Determine parameters of P-97 Flaming/Brown pipette puller empirically. The following parameters provide a starting point. Pressure = 500, Heat = Ramp + 10, Pull = 60, Velocity = 80, Time = 250.

2. Pull the borosilicate glass capillaries on the horizontal micropipette puller to a fine point with a shoulder to tip length of ~1 cm. The total length of the pulled pipettes should be ~4 cm.

3. Brush micropipette tip against the folded edge of lint-free tissue, at an angle of 45–90 °. After each brush inspect micropipette using a compound microscope. Repeat brushing until the tip is sufficiently broken to be visibly open by light microscopy. Discard tips with jagged edges. Store micropipettes in dust-free storage jars (see Note 21).

3.3. Oocyte Collection

1. Inject mice intraperitoneally with 10 IU PMSG 44–52 h prior to oocyte harvest.

2. Euthanize mice by cervical dislocation and remove ovaries carefully, minimizing residual fat and uterine tissue. Place the two ovaries immediately into a 35 mm Petri dish containing 1 mL of pre-warmed milrinone-treated M2 medium overlaid with mineral oil to maintain temperature and prevent evaporation of the medium (see Note 22).

3. Immobilize ovaries with mini-forceps in one hand and use a 30-gauge needle in the other hand to repeatedly puncture antral follicles. Discard unused ovarian tissue, leaving granulosa-enclosed oocytes in the droplet of medium.

4. Wash granulosa-enclosed oocytes into a clean droplet of medium and strip off granulosa cells mechanically by repeated pipetting using a glass pipette with an inner diameter approximately equal to that of the oocyte (see Note 23).

5. Wash oocytes through 3–4 oil-covered M2 droplets containing milrinone to remove all contaminating granulosa cells, and place in the dark at 37°C on a heat block (see Note 24).

3.4. Microinjection of cRNA

1. Assemble imaging chamber of the stage heater with 1 mL of M2 medium and overlay with mineral oil. Warm to 37°C. Focus the 10× objective on the coverslip of the stage heater insert, in the exact center of the imaging chamber.

2. Assemble the holding pipette on the left side of the inverted microscope. Move so that it rests gently on the coverslip in the center of the field of view.

3. Add oocytes into the chamber by mouth pipette. Depending on the speed of the operator, enough oocytes should be placed on the stage for 10 min of microinjection. Minimize room lighting and block illumination to the stage when not needed.

4. Backfill a pulled borosilicate micropipette (see Subheading 3.2) with a cRNA solution (~0.5 μL) using a narrowed elongated syringe (see step 5 in Subheading 2.4). Put aside temporarily whilst the microelectrode holder is prepared (see Note 25).

5. Load microelectrode holder with HEPES-buffered KCl solution and introduce the injection micropipette, where it should make a snug fit within the rubber O-ring. Firmly tighten seal. Attach the assembled holder to the Electro 705 head stage, which is held in place by the coarse micromanipulator on the right side of the inverted microscope (Fig. 1).

6. Lower the injection micropipette using the coarse micromanipulators into the stage-mounted imaging chamber. An angle of 70–80° above horizontal is ideal for effective microinjection.

7. Remove visible small air bubbles from the micropipette tip by increasing the pressure delivered by the PV820 Pneumatic Picopump to ~50 psi and using an extended injection interval of several seconds to push out the bubbles. Once the tip is cleared of air, reduce the pressure delivered by the Picopump to ~18 psi.

8. Raise tip of micropipette to just out of focus. Gently apply suction to the holding pipette using the hydraulic syringe to hold the oocyte in position. Lower the micropipette using the fine

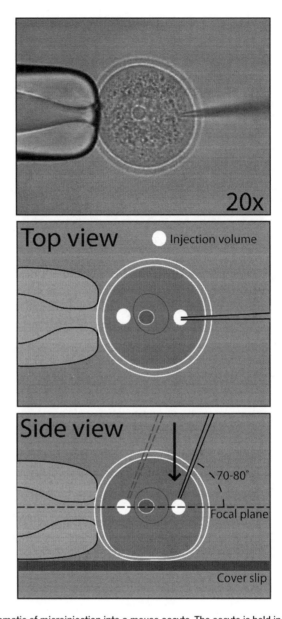

Fig. 2. Schematic of microinjection into a mouse oocyte. The oocyte is held in place by the holding pipette (*left*), whilst the microinjection pipette is lowered into the oocyte from above (*right*). The mRNA solution is delivered into a nucleolus-sized bolus in the plane of focus in either of the positions indicated. Upon injection a brief displacement of the cytoplasm reveals the size of the injection.

controls of the manipulator (Fig. 2) so that the micropipette tip advances across the zona pellucida; penetration through the zona pellucida is visible by its elastic recoil.

9. Penetrate the oocyte plasma membrane by activating the "tickle" button on the Electro 7605 so as to apply negative capacitance for a period of ~1 s (see Note 26).

10. Deliver a bolus of cRNA solution into the oocyte by applying a timed pressure-pulse using a setting of ~100 ms on the PV820 Pneumatic Picopump (see Note 27).

11. After each oocyte is injected, withdraw the injection micropipette by one turn of the joystick manipulator, and deposit the newly injected oocyte to a separate area of the coverslip by mechanically moving the stage (not the holding pipette) and displacing oocyte from the holding pipette.

3.5. Imaging of Oocytes

1. Following microinjection, maintain oocytes in the dark, in milrinone containing media under oil for >30 min so as to allow sufficient time for protein expression from the injected cRNA.

2. After allowing time for protein translation, place oocytes on the pre-heated stage of the inverted microscope.

3. Ensure the appropriate dichroic mirror and excitation filter required for GFP imaging are selected (see Subheading 2.7).

4. By means of the PC-controlled CCD camera, and a minimal level of brightfield illumination, adjust the focus and stage position in order to get the oocytes ideally positioned for imaging.

5. A suitable exposure time and binning value is established for brightfield and GFP by acquiring single images. An exposure time of ~100 ms for GFP may be used as a starting point. If the resulting image has low signal to noise ratio, increase exposure time. Conversely, if any part of the image is saturated, reduce exposure time (see Note 28).

6. Having determined an exposure time for GFP, the imaging software is then used to define time-lapse parameters including the interval between imaging cycles and the total number of cycles to be performed. Metafluor and Image J both have the ability to program more complex patterns of image acquisition, e.g., differing cycle lengths for different wavelengths or more intensive imaging at particular times during maturation, etc. depending on the requirements of the experiment (see Note 29).

3.6. Post-acquisition Analysis

1. Images are captured in TIFF (Tagged Image File Format) format, which is compatible with image processing software thereby allowing for post-acquisition analyses. Image J software is an open source software that is freely available and has the capability of displaying, editing, analyzing, processing, and printing 8-bit, 16-bit, and 32-bit images. Apart from TIFF, other formats compatible with Image J include PNG, GIF, and JPEG formats. Alternatively, commercially available software that are well suited to this purpose include Metamorph or MetaFluor.

2. In many cases, the objective of time-lapse fluorescence imaging is to monitor how the levels of a particular fluorescent protein change during progression through meiosis I. For an oocyte expressing a fluorescent protein of interest, any such changes will be represented by changes in total oocyte fluorescence. Thus, increases in oocyte fluorescence are indicative of net protein synthesis, whilst decreases in oocyte fluorescence represent protein destruction. In order to identify changes in oocyte fluorescence, the software is used to define a "Region of Interest" (ROI) that encompasses the entire oocyte; usually the ROI is obtained by outlining the oocyte's plasma membrane. The average pixel intensity for that ROI is then measured at each time-point for the duration of the experiment. Background subtraction may be necessary for each image in a series, and individually for each frame (see Note 30).

3. The intensity data is exported to an open Microsoft excel spreadsheet or other spreadsheet using an inbuilt dynamic data export function and used to plot a graph. The changes in average pixel intensity as depicted by the graph are representative of the changes in oocyte fluorescence and hence provides a readout of the changes in fluorescent protein levels during progression through meiosis I.

4. Notes

1. pMDL and pSPE3 were originally derived from the pRN3 vector (4, 5). Features of pRN3 conserved in these new generation vectors are the T3 RNA polymerase promoter site for in vitro RNA expression, a polyadenylation sequence located 3′ to a multiple cloning site and 5′ and 3′ UTRs for improved translation in vivo. pMDL has additional multiple cloning sites to allow insertion of a fluorescent tag, and pSPE3 has Gateway site-specific recombination sequences to allow high throughout cloning of vectors. We have adapted both pMDL and pSPE3 to contain a C-terminal GFP tag. Cyclin B1 can be inserted into pMDL-GFP using standard restriction cloning techniques. Insertion into pSPE3-GFP requires initial insertion into a donor vector (e.g., pDnr221) prior to transfer into the pSPE3 destination vector using recombination-mediated clonase reactions as per the manufacturer's protocols (Life Technologies). Both pMDL and pSPE3 vectors contain a rare-cutter restriction enzyme site (*Sfi*I) to aid in linearization of the vector prior to in vitro transcription.

2. The Message Machine High Yield transcription Kit (Life Technologies) adds a 7-methyl guanosine cap to the 5′ end of

the cRNA, aiding its stability and translation in oocytes. The kit is available for T3, T7, or Sp6 promoter-directed transcription (as separate kits). Both pMDL and pSPE3 vectors contain a T3 promoter.

3. M2 medium preparation: Dissolve HEPES in ~100 mL MilliQ H_2O and adjust to pH 7.4 with NaOH. Dissolve components 2–4 each (Table 1) in ~50 mL MilliQ H_2O. Add components 5–13 to ~500 mL MilliQ H_2O and allow to dissolve. Add to these components 1–4, followed by Na-lactate, and make up to 1 L with MilliQ H_2O. Test osmolarity of prepared M2 media. It should be 285–287 mOsmol/L. Discard any media outside this range and begin again. Add BSA to media of correct osmolarity, by sprinkling over media, allowing it to dissolve slowly prior to sterile filtration. Store media for up to 2 weeks at 4°C.

4. Oocytes are highly sensitive to detergent and heavy metal contamination found in tap water. Therefore all glassware used for preparing culture medium should be set-aside solely for this purpose and must be washed thoroughly in MilliQ H_2O. If a detergent is used, we recommend a 7× (MP Biomedicals). The greatest attention must be paid to extensive rinsing of glassware.

5. Maturation of oocytes is better in a CO_2 buffered medium such as MEM, which requires 5% CO_2 to maintain the correct pH. However, use of this medium is optional, given oocyte maturation will occur in a handling medium such as M2 which does not require additional CO_2. MEMα is purchased as powder and made up to volume with MilliQ H_2O and 1:200 pencillin: streptomycin stock solution.

6. In general, larger numbers of oocytes are obtained from F1 hybrid female mice compared with inbred females. Mouse age also has a significant impact on the numbers of GV oocytes that can be collected and it is best to use prepubescent mice (~18–21 days for C57Bl6/CBA F1 females). At this age females still lack the natural FSH/LH surges of the estrus cycle such that administration of PMSG results in greater yields of mature GV oocytes than can be obtained from reproductively mature female mice.

7. Prepare PMSG in sterile phosphate-buffered saline at 50 IU/mL. Store at –80°C in single use 0.5 mL aliquots in disposable syringes.

8. These are pulled horizontally over the pilot flame of a Bunsen burner by melting the tapered end approximately midway along its length. The glass is pulled in a rapid horizontal motion once the glass begins to melt. By adjusting the size of the pilot flame, the glass heating time, and the speed of glass pulling,

pipettes of the correct diameter can be pulled consistently. The tapered glass should be "hair-like" and flexible and should be kept intact until just prior to use, when the end can be broken to give a capillary length of around 2 cm.

9. Store mineral oil in the dark.

10. Prepare milrinone as a 10 mM stock in DMSO: absolute ethanol (1:1.37). Store for up to 3 months at 4°C.

11. Mouthpiece assembly consists of an aspirator mouthpiece, ~7 mm outer diameter soft plastic tubing, and with a glass pipette connected via a small segment of wider (~10 mm) tubing, filled with a cotton wool plug.

12. Plastic 1 mL syringe without plunger is melted in a flame and drawn out by hand, so as to produce a tapered end amenable for drawing up <1 μL volume solutions for delivery into the back of the injection pipette. The end must be just small enough to fit inside the microinjection pipette. A capillary that is too narrow makes filling the syringe difficult and one too large results in the solution easily being lost inside the syringe.

13. The 20× and 40× objectives should have a numerical aperture as large as possible, in order to increase the epifluorescent brightness of the fluorochrome. The 20× objective is usually air, and the 40× objective oil immersion.

14. A Xenon lamp is recommended because of its even light intensity across the electromagnetic spectrum.

15. Joystick manipulators are placed to left and right side of inverted microscope (Fig. 1). The left hand side micromanipulator moves a holding pipette, used to hold the oocyte during the microinjection procedure. The right hand side micromanipulator moves an injection micropipette, used to deliver cRNA into the oocyte. Both micromanipulators are attached to the stage of the microscope through connection with the coarse manipulators. They need to be attached to the microscope body through commercial mounting adaptors.

16. The Injection Holder Set is used to connect the syringe with the left-hand micromanipulator for holding the oocyte during injection. Fill syringe and tubing with mineral oil (Sigma) and when the prefabricated holding pipette is attached to the holder, allow oil to drip out of tip after turning the syringe to displace oil. Connect the hydraulic syringe to the coarse micromanipulators positioned on the left side of the microscope using a universal joint connector.

17. The microelectrode holder needs to (a) hold a glass micropipette of external diameter 1.5 mm (Subheading 2.4, item 1); (b) possess a male luer port, in order to connect with the air supply from the PV820 Pneumatic Picopump; and (c) possess

a female connector to attach to the Electro 705 head stage (Subheading 2.6, item 10) (Fig. 1, inset). Attach the microelectrode holder to the remote headstage, and the headstage to the coarse manipulators on the right side of the microscope stage through a universal joint.

18. The coverslip forms the base of a chamber in which the oocytes are placed for microinjection and imaging.

19. The CCD camera is attached to the C-mount of the inverted microscope. Such cameras are cooled so as to maximize signal:noise ratios.

20. The dichroic mirror is housed within the inverted microscope with the filter wheel placed between the dichroic and the illumination source. This controls the excitation light to the fluorochrome. The filter wheel and its shutter are controlled by the Lambda 10-3 control unit, which is itself attached to a PC running Metafluor/Metamorph software. The V235 shutter is placed between the bright-field illumination and the microscope condenser. This shutter is also connected to the Lambda 10-3 control unit and controlled by Metafluor/ Metamorph software.

21. With practice this is often achieved on the first or second attempt, and a consistent sized break can be routinely achieved. If the "natural break-point" of the pipettes leaves the tip too large or too small then subtle adjustments should be made to the parameters of the pipette puller in step 2. The settings of the micropipette puller will have to be changed empirically by observation of their usefulness at injection. Large micropipettes will be difficult for oocyte penetration and will cause excessive cell death; while small micropipettes will show excessive blocking.

22. Milrinone is a phosphodiesterase-3 inhibitor that maintains GV arrest at a concentration of 1–10 μM. Add to all media for use prior to, and during, microinjection.

23. Stripping of granulosa cells should take less than five repeated movements in and out of the glass pipette. Do not use overly small diameter pipettes that turn oocytes into a lozenge shape. Do not use overly large pipette that take >10 aspirations to remove all granulosa cells.

24. It is important to minimize the time of dissection and collection. Twenty minutes is allowed in total from the time of cervical dislocation until the granulosa cells are removed from all oocytes. Longer times compromise oocyte recovery rates following microinjection and oocyte maturation rates.

25. If cRNA dilution is desired, dilute with nuclease-free water, typically a 1:1 or 1:2 dilution. Careful flicking of the side of the

micropipette can force large air bubbles away from tip when held vertically.

26. The initial zona pellucida recoil may appear to the novice that the micropipette has penetrated the oocyte plasma membrane. Often this can happen but without further application of negative capacitance this is not guaranteed. Visually one can often believe injection is being made into the cytoplasm, when in reality it is being made into the perivitelline space. It is important always to inject along the line central to the axis of the holding pipette, otherwise the oocyte will roll to the side. In addition, care must be taken not to inject into the germinal vesicle. The microscope should be focused such that the tip of the injection pipette is always in focus in the center of the oocyte so that the injection size can be determined (Fig. 2).

27. A good size for microinjection is a bolus whose diameter is equal that of the nucleolus, however injection sizes up to that of the nucleus can be tolerated. The injection size is readily seen for a brief moment after injection, when the mRNA displaces the cytoplasm leaving a clear area around the tip of the pipette, in contrast to the more coarse appearance of the cytoplasm. The size of the injection can only be seen correctly if it is in the focal plane at the time of injection. After each injection the duration and pressure settings on the PV820 Pneumatic Picopump can be adjusted until the desired injection size is achieved.

28. Determining exposure times should always be done by acquiring single images and adjusting the exposure time according to the resulting image and not by using the "live-view" function, which leads to excessive imaging of the oocytes. If exposure times are very short (<10 ms) it may be better to introduce neutral density filters enabling longer exposures to be used, this gives better control of the exposure time (in a range 20–200 ms). When considering which binning to use it is best to consider the purpose of the experiment. If only whole-cell fluorescence intensities are required a higher binning such as 3 is recommended as this reduces the exposure time required for illumination thus preserving the oocytes health. However, if the nature of the experiment requires capturing detailed images, for example, lagging chromosomes, spatial resolution must be preserved and a binning of 1 is recommended. These images will be most suitable for publication.

29. Images should be captured for the entire duration of meiosis I which normally takes approximately 9 h, from germinal vesicle breakdown to extrusion of the first polar body. Consideration therefore needs to be given to the frequency of image capture to determine the optimal balance between acquisition settings that are frequent enough to capture the necessary detail but not too frequent to compromise oocyte health. Typically, image

acquisition of cyclin B1-GFP is performed every 15 min over a period of 10 h following washout from milrinone.

30. The ROI may need to be manually adjusted in each frame if the oocyte moves during the recording. This can happen if the temperature fluctuates, or oil is not used to cover the media. We recommend the use of Cell-Tak where possible to adhere oocytes to the coverslip. The background may change during the course of the experiment due to lamp fluctuations, an especially important consideration for mercury lamps and all lamps nearing the end of their life. Information about the ROI may be exported to a spreadsheet application (e.g., Microsoft Excel) for further analysis, and dependent on the experiment the user may decide to use average (i.e., mean) pixel intensity in the ROI or total pixel intensity count.

References

1. Reis A, Madgwick S, Chang HY, Nabti I, Levasseur M, Jones KT (2007) Nat Cell Biol 9:1192–1198

2. Homer HA, McDougall A, Levasseur M, Murdoch AP, Herbert M (2005) Reproduction 130:829–843

3. Holt JE, Weaver J, Jones KT (2010) Development 137:1297–1304

4. Lemaire P, Garrett N, Gurdon JB (1995) Cell 81:85–94

5. Roure A, Rothbacher U, Robin F, Kalmar E, Ferone G, Lamy C, Missero C, Mueller F, Lemaire P (2007) PLoS One 2:e916

Chapter 7

Using FRET to Study RanGTP Gradients in Live Mouse Oocytes

Julien Dumont and Marie-Hélène Verlhac

Abstract

Oocytes are extremely large cells that have to coordinate accurate chromosome segregation, asymmetric cytoplasm partitioning together with their own development as fertilizable gametes. For this, they undergo both global (cell cycle progression related) and local changes. It is therefore essential to be able to monitor local changes as they take place in live maturing oocytes. We describe here a method to follow RanGTP gradients using FRET technology in vivo.

Key words: RanGTPase, Ratiometric FRET, Live imaging, Mouse, Oocyte

1. Introduction

Spatial and temporal control of cell division is essential to maintain genomic integrity and to control daughter cell size and content. Mammalian oocytes are atypical cells that arise from extremely asymmetric divisions. Most of the cytoplasmic content is retained inside the large oocyte and two small nondividing polar bodies that contain one complement of the genome are generated. These two asymmetric divisions primarily rely on the functionality and proper positioning of the meiosis I and II microtubule-based spindles inside particularly large oocytes. In rodent oocytes, the meiosis I spindle assembles at or near the position of the nucleus, which is usually located at a near central position inside the oocyte. Asymmetry of the first and second meiotic divisions thus requires subsequent translocation of the spindle towards the oocyte cortex and anchoring of the cortically positioned spindle (1–5). At the onset of cytokinesis, ingression of the plasma membrane occurs in the cortical region directly surrounding meiotic chromosomes (for review see ref. 6). Cytokinetic furrow progression has to be restricted to this region in order to prevent symmetric cleavage of the oocyte.

Hayden A. Homer (ed.), *Mammalian Oocyte Regulation: Methods and Protocols*, Methods in Molecular Biology, vol. 957, DOI 10.1007/978-1-62703-191-2_7, © Springer Science+Business Media, LLC 2013

A local modification of the cortex surrounding chromosomes has been proposed to impose this restriction (4, 7).

In mouse oocytes, spatial and temporal coordination of cell division relies on signaling properties of meiotic chromosomes and their associated RanGTP activity (8). The small GTPase Ran whose active GTP-bound state is generated by the chromatin-localized exchange factor RCC1 (Regulator of Chromosome Condensation 1) has been primarily studied for its function in nucleo-cytoplasmic transport. During interphase, the high concentration of RanGTP in the nucleus controls the directionality of transport. In the cytoplasm, RanGAP and the 2 accessory proteins RanBP1/2 stimulate GTP hydrolysis and thus bring back Ran in the inactive GDP-bound state (9–14). After nuclear envelope breakdown (NEBD), a similar high chromatin-proximal RanGTP concentration participates in spindle assembly by locally activating a number of spindle assembly factors (SAFs) that have microtubule-directed activities (15–18). Ran also has actin-directed targets that participate in the control of asymmetric divisions (1, 19). In mouse oocytes, inhibiting RCC1 nucleotide-exchange activity prevents RanGTP accumulation around chromosomes, which affects the kinetics of spindle assembly, its proper positioning and anchoring at the oocyte cortex and the cortical differentiation essential to restricting cleavage furrow progression (1, 8). Proper coordination of these processes relies on constant monitoring of chromosome position inside the large oocyte through generation of a chromosome-centered gradient of active Ran.

Here, we describe a simple and efficient technique for visualizing the RanGTP gradient inside mouse oocytes using FRET (Fluorescence Resonance Energy Transfer) technology with the previously described Rango (*Ran*-regulated importin β car*go*) FRET probe that reports on Ran activity (20). FRET refers to the transfer of energy from an excited donor fluorophore (here CFP) to an acceptor fluorophore (here YFP) when in close proximity to one another (typically less than 10 nm) through non-radiative dipole–dipole coupling. The Rango probe corresponds to the importin β binding domain (IBB) of human snurportin 1 fused to CFP at its carboxy terminus and to YFP at its amino terminus. Rango behaves like an importin β cargo and is released from sequestration by importins in the presence of high RanGTP concentration (Fig. 1; 20). When outside of the high RanGTP-concentration area, Rango (20) is sequestered by importins thereby inducing a configuration in which the donor fluorophore is sufficiently distanced from the acceptor that the probe emits less FRET signal. In contrast, the FRET signal increases when Rango is liberated from importins as under such conditions, the donor is positioned close to the acceptor. The Rango probe therefore functions as an indirect reporter of RanGTP levels.

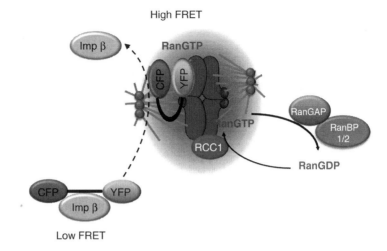

Fig. 1. Schematic of Rango probe functioning. When outside of the high RanGTP-concentration area, Rango (20) is sequestered by Importin β binding to the snurportin1-IBB domain resulting in low FRET signal. The FRET signal increases when Rango is liberated by high chromosome-proximal RanGTP concentration. Chromosomes are in *blue* with *purple* kinetochores; microtubules are in *green*; RanGTP gradient is *red*.

2. Materials

2.1. M2 + BSA Culture Medium Preparation

All solutions are made with embryo culture-tested water (see Note 1).

1. Solution A: NaCl, KCl, KH_2PO_4, $MgSO_4 \cdot 7H_2O$, 60% Sodium Lactate (solution kept at 4°C), D-glucose, 100× concentrated Penicillin G + Streptomycin sulfate (stock solution kept at −20°C). Prepare 10× concentrated stock solution A by dissolving 2.77 g NaCl, 0.18 g KCl, 0.08 g KH_2PO_4, 0.15 g $MgSO_4 \cdot 7H_2O$, 1.7 ml 60% Sodium Lactate (solution kept at 4°C), 0.5 g D-glucose, 0.5 ml 100× concentrated Penicillin G + Streptomycin sulfate (stock solution kept at −20°C) in water in a final volume of 50 ml.

2. 250 mM $NaHCO_3$ (solution B).

3. 250 mM HEPES, pH 7.4 (solution C). Adjust pH to 7.4 with 7 N NaOH.

4. 33 mM sodium pyruvate (solution D). Keep at 4°C.

5. 171 mM $CaCl_2 \cdot 2H_2O$ (solution E).

6. Bovine serum albumin Fraction V.

7. Precision scale, pH meter.

8. 0.22 μm filter cartridges.

9. M2 + BSA medium: mix five concentrated stock solutions (A, B, C, D, E). For 50 ml of M2 + BSA, mix 5 ml of stock A, 0.8 ml of stock B, 4.2 ml of stock C, 0.5 ml of stock D, and 0.5 ml of stock E. Add 200 mg of BSA and bring the volume to 50 ml with embryo-culture tested water. Adjust the pH to 7.4 by adding a few drops of 0.1 M NaOH. Filter sterilize using a 0.22 μm filter. Make 4 ml aliquots in 15 ml sterile Falcon tubes and store at 4°C.

2.2. Oocyte Harvesting and Culture

1. Dissection scissors (Fine Science Tools) and tweezers (Dumostar, N°5 Biologie).

2. Watch-glass dish, 60 × 15 mm.

3. Polystyrene Petri dishes (BD Falcon, 353004).

4. Glass filter pipettes.

5. Aspirator tube assembly (Sigma, A5177; see Note 2).

6. Mineral oil (Sigma, M8410; see Note 3).

7. Razor blades.

8. 25G needle (Terumo, NN-2526R) mounted on a 1 ml syringe.

9. Dissection scope with a diascopic stand and a heated stage (Tokai Hit).

10. 11-Week-old OF1 female mice (Charles River, see Note 4).

11. M2 + BSA medium.

12. Milrinone to maintain oocytes arrested in prophase I of meiosis (Sigma, M4659; see Note 5). Prepare stock solution by dissolving in DMSO to a final concentration of 10 mM. 20 μl aliquots are stored protected from light at –20°C.

13. 37°C incubator.

2.3. cRNA Preparation

1. pRN3 plasmids containing sequences encoding the Rango FRET probe, histone H2B-RFP (see Note 6), Ran-WT, Ran-T24N, and Ran-Q69L point mutants, which are dominant negative and constitutively active Ran mutants, respectively (see Notes 7 and 8).

2. SfiI restriction enzyme for plasmid linearization (NEB, R0123).

3. Qiaquick PCR purification kit (Qiagen, 28104).

4. T3 mMessage mMachine kit for in vitro transcription (Ambion, AM1348M).

5. RNeasy Mini Kit for cRNA purification (Qiagen, 74104).

6. Water bath set at 37°C.

7. TAE (Tris-acetate-EDTA) buffer: 40 mM Tris base, pH 8.0, 20 mM glacial acetic acid, 1 mM EDTA

8. 1% Agarose in 0.5× TAE buffer.

9. Electrophoresis apparatus and power supply.

2.4. cRNA Microinjection into Mouse Oocytes

1. 1-Well glass depression slide.

2. Autoclaved microloaders (Eppendorf, 5242 956.003).

3. 1.0 mm outer diameter × 0.78 mm internal diameter borosilicate glass capillaries model: GC1000TF-10 (Harvard Apparatus, 30-0038).

4. Holding capillaries, 100 μm outer diameter × 15 μm inner diameter VacuTip with a 35 ° angle (Eppendorf, 5175 108.000).

5. Horizontal capillary puller (Narishige, PN3).

6. Microinjection setup with 2 TransferMan NK2 (Eppendorf, 5188000012) mounted on an inverted microscope stand equipped with phase contrast, a 10× and a 40× phase lenses. Pressure for injection is supplied by a FemtoJet microinjector (Eppendorf, 5247000013). Oocytes are held using a CellTram Air pressure regulator (Eppendorf, 5176000017; see Note 9).

2.5. Live Imaging of Injected Mouse Oocytes and Image Analysis

1. Ludin imaging chamber (Life Imaging Services).

2. A wide-field fluorescence inverted microscope equipped with a 20×/0.7 NA objective.

3. Software for automated image acquisition and image analyses (e.g., Metamorph 6.0 software; Molecular Devices).

4. A cooled charge-coupled device (CCD) camera (CoolsnapHQ2, Photometrics) mounted on the microscope for image acquisition.

5. A Thermostatic chamber (Life Imaging Services) mounted on the microscope stage for maintaining ambient conditions suitable for oocyte maturation.

6. TetraSpeck microspheres of 4 μm diameter (Invitrogen, T-7283).

3. Methods

Analyzing the RanGTP gradient during meiotic maturation of the mouse oocyte will typically take 2 days (1 day for mouse oocyte collection and microinjection, followed by overnight time-lapse and half a day of image analysis). cRNAs have to be prepared in advance and stored at −80°C.

3.1. Preparation of cRNAs

pRN3-Rango probe, pRN3-histone H2B-RFP, pRN3-Ran-WT, pRN3-Ran-T24N, and pRN3-Ran-Q69L have been described

(8, 20, 21) and are available upon request from a number of experimenters (see Note 8).

1. Linearize a total amount of 5 μg of plasmid using *Sfi*I restriction enzyme for 3 h at 37°C in a final volume of 50 μl (see Note 10).

2. Purify the linearized product using a Qiaquick PCR purification kit following standard procedure. Elute the linearized plasmid in 40 μl of DEPC-treated water and stored at –20°C for months.

3. Use 4 μl of linearized plasmid (about 0.5 μg DNA) for in vitro transcription using the T3 mMessage mMachine kit. To these 4 μl, add 2 μl of DEPC-treated water, 10 μl of dNTP-Cap (2×), 2 μl of Buffer (10×), and 2 μl of T3-RNA polymerase. The tube is sealed with parafilm and incubated in a water bath at 37°C for 2 h.

4. Add 1 μl of DNAse I (provided with the T3 mMessage mMachine kit) to the reaction and incubate for 20 min at 37°C to remove the DNA template.

5. Proceed to cRNA purification using the RNeasy mini kit to remove all traces of enzymes and free rNTPs. Elute cRNAs in 40 μl of DEPC-treated water (final concentration of cRNAs is about 0.5 μg/μl) and make 4 μl aliquots that will be stored at –20°C (see Note 11).

6. Check the quality and amount of cRNAs by running a 2 μl sample on a freshly prepared 1% agarose gel in 0.5× TAE (see Note 12). Use the RNase-free RNA sample buffer that is provided with the T3 mMessage mMachine kit.

3.2. Oocyte Isolation from Ovaries

1. Pre-warm an aliquot of M2 + BSA medium supplemented with 1 μM milrinone (22) in a 37°C waterbath for about 20 min.

2. Pull a glass filter pipette using an alcohol burner and connect to the aspiration tube. The diameter of the glass pipette has to be slightly bigger than the oocyte diameter (80 μm).

3. Prepare a Petri dish with 30 μl drops of M2 + BSA medium (supplemented with 1 μM milrinone) covered with mineral oil to prevent evaporation and place in a 37°C incubator.

4. Using scissors and forceps, dissect a female and rapidly collect both ovaries in 2 ml of pre-warmed M2 + BSA medium supplemented with 1 μM milrinone in a watch glass dish on a 37°C heating plate and transfer to the 37°C heated stage of the dissection scope.

5. Skew each ovary on a 25G needle mounted on a 1 ml syringe and chop rapidly using a razor blade.

6. Remove remaining ovarian fragments from the dish with a pair of tweezers.

7. Pipette the liquid in the dish up and down several times using a 1 ml tip mounted on a p1000 pipette in order to fully dissociate follicles and release oocytes.

8. Identify fully grown prophase I-arrested oocytes and sort them from this mixture under the dissection scope using the glass pipette mounted on the aspiration tube and mouth pipetting techniques. Prophase I-arrested oocytes are identifiable by the presence of an intact nucleus or germinal vesicle (GV).

9. Transfer oocytes to a drop of M2 + BSA medium supplemented with milrinone in the pre-warmed Petri dish using the mouth pipette.

10. Wash oocytes from ovarian debris and denude the surrounding follicular cells by repeatedly aspirating them in and out of the pulled glass pipette.

11. Transfer denuded oocytes to clean M2 + BSA (with milrinone) drops (see Note 13).

12. Allow oocytes to recover for about 1 h in the 37°C incubator before proceeding to cRNA injection.

3.3. Injection of cRNAs into Mouse Oocytes

1. Centrifuge the in vitro synthesized cRNA mix (4 µl of Rango probe cRNA + 0.5 µl of histone H2B-RFP cRNA) for 45 min at $20,000 \times g$ to pellet potentially precipitated cRNAs that would otherwise clog the injection capillary.

2. Put a small drop (5 µl) of medium (M2 + BSA supplemented with milrinone) and cover with mineral oil in the well of a depression glass slide and place on the stage of the injection microscope.

3. Place the holding capillary in the holder of the left TransferMan NK2 motor if you are right handed (Place it on the right side if you are left handed).

4. Immerse the tip of the holding capillary in the medium and equilibrate the pressure by aspirating some medium (see Note 14).

5. Pull an injection capillary using the horizontal capillary puller (see Note 15).

6. Using a microloader mounted on a p10 pipette, aspirate 1 µl of cRNAs mix (see Note 16) and fill the injection capillary.

7. Mount the filled injection capillary in the holder of the right TransferMan NK2 (on the left side if you are left handed). Rapidly immerse the tip of the injection capillary in the medium to avoid drying of the tip that can lead to clogging.

8. Looking at the drop of medium through a 10×/NA 0.25 objective, transfer ten oocytes in the well of the depression glass slide by mouth pipetting.

9. Connect the injection capillary holder tubing to the FemtoJet. Fine movement of both the holding and the microinjection capillaries are controlled by Eppendorf TransferMan NK2 joysticks.

10. Gently break the tip of the injection capillary by very gently bringing it into contact with the tip of the holding capillary. This will start the flow of the solution for microinjection, which is performed under continuous flow to prevent clogging of the pipet.

11. Looking at the oocytes through a 10×/NA 0.25 objective, use the holding capillary to push oocytes and group them together in the top half of the field of view.

12. Stabilize an oocyte at the tip of the holding capillary by applying gentle suction using the CellTram control screw (see Note 14).

13. Introduce the injection capillary inside the oocyte and inject the cRNA solution into the ooplasm (see Note 17). It is important to proceed rapidly to injection after stabilizing the oocytes so as to minimize damages that the holding suction can cause.

14. Following injection, maintain oocytes in M2 + BSA medium supplemented with milrinone for 2–3 h to allow time for expression of exogenous proteins from injected cRNA.

3.4. Observing Live Oocytes with a Videomicroscope

1. Pre-warm an aliquot of M2 + BSA medium without milrinone in a 37°C waterbath 20 min before starting to image.

2. Prepare a Petri dish with 30µl drops of M2 + BSA medium covered with mineral oil to prevent evaporation and place in a 37°C incubator.

3. Prepare a Ludin chamber with one 30 µl drop of M2 + BSA medium covered with mineral oil and place in a 37°C incubator.

4. Transfer injected oocytes from the milrinone supplemented M2 + BSA medium to one of the drops of pre-warmed milrinone-free medium by mouth pipetting. Thoroughly remove milrinone from contact with oocytes by sequentially transferring them through four drops of milrinone-free medium (see Note 18). Resumption of meiotic maturation marked by NEBD is triggered upon release of the oocytes into milrinone-free medium.

5. Transfer rinsed injected oocytes to the pre-warmed Ludin chamber by mouth pipetting and place the chamber on the stage of an inverted fluorescent microscope enclosed in a thermostatic chamber set to 37°C (see Note 19).

6. Obtain images from histone H2B-RFP using a 546 ± 6 nm excitation filter, a 565 dichroic mirror and a 620 ± 20 nm emission filter.

7. Obtain images from Rango using a 440AF21 excitation filter, a 455 dichroic mirror, and 480AF30 emission filter for the CFP signals and a 535AF26 emission filter for detecting FRET (indicated by emission at the YFP acceptor peak).

8. Set exposure times to 100 ms and visualize the emissions produced using a Plan APO 20×/0.7 NA objective and a cooled charge-coupled device (CCD) camera.

3.5. Image Analysis and Measurement of FRET Ratio

Here we describe image analysis performed using Metamorph 6.0 and ImageJ softwares.

1. Check the xy pixel alignment of the CFP, FRET, and RFP channels. For xy calibration, use TetraSpeck microspheres of 4 μm diameter (Invitrogen, T-7283) and image them in the three channels (CFP, FRET, RFP).

2. Check the x and y translational alignment on an overlay image and if necessary shift the image in x and/or y to have a perfect co-alignment of the three channels. It is crucial that the fluorescence image pair used to calculate FRET ratios are perfectly aligned prior to division.

3. Apply the calibrated x/y pixel shift to the biosensor images in all three channels to generate the registered images.

4. Calculate the background of each image by measuring the average intensity value within a region of interest outside the cell (see Note 20).

5. Subtract the calculated background from each corresponding whole image.

6. In order to limit the area of FRET ratio calculation to the oocyte volume, apply a binary mask to each registered image, setting areas outside of the cell uniformly to zero.

7. Segment the FRET channel registered image by intensity thresholding ("Treshold Image" command in Metamorph). Inclusive thresholding is manually done by trial and error, setting the region inside the oocyte to 1 and outside the oocyte to 0.

8. Use this segmented image to generate a binary mask and multiply it by the CFP and FRET channel registered images to produce the corresponding masked registered images.

9. Generate the ratio image by pixel-wise divisions of the FRET masked registered image by the CFP masked registered image (I_{FRET}/I_{CFP}).

10. Scale the resulting 32-bit ratio image and display as a pseudocolor image by applying the ImageJ "Fire" look-up table (LUT; Figs. 2 and 3).

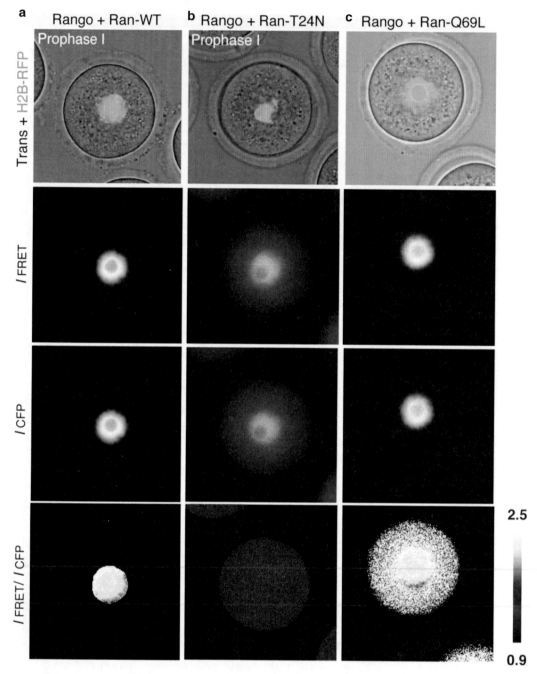

Fig. 2. Rango reports for RanGTP variations in live mouse oocytes. (**a–c**) Live prophase I-arrested mouse oocytes injected with Rango and histone H2B-RFP together with RanWT (**a**), RanT24N (**b**), or RanQ69L (**c**) encoding cRNAs. *Top panel*: Merge of transmitted light and chromosomes (in *green*); *Middle panel*: images of I_{FRET} and I_{CFP} emitted by the Rango probe. *Bottom*: Pseudo-color representation of the I_{FRET}/I_{CFP} ratio of the same oocyte expressing Rango. The highest I_{FRET}/I_{CFP} ratio values are observed in the chromosome region. Scale bar: 10 μm.

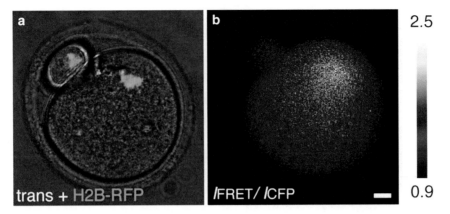

Fig. 3. Detection of the RanGTP gradient in live mouse oocytes. (**a**, **b**) Live MII arrested mouse oocyte injected with Rango and histone H2B-RFP encoding cRNAs. (**a**) Merge of transmitted light and chromosomes (in *green*); (**b**) Pseudo-color representation of the I_{FRET}/I_{CFP} ratio of the same oocyte expressing Rango. The highest I_{FRET}/I_{CFP} ratio values are observed in the chromosome region. Scale bar: 10 μm.

4. Notes

1. Stock solutions A, C, and E can be kept at 4°C for about 3 months. Stock B and D are only stable for 1 week at 4°C. M2 + BSA can be stored at 4°C for about 2 weeks.

2. This type of tube assembly is designed for using glass capillary tubes and requires specific adaptation for fitting larger diameter glass pipettes. We usually add a piece of silicon tubing, in which we plug the glass pipette, around the capillary holder.

3. We keep the stock of mineral oil protected from light at room temperature (light induces oil oxidation and renders it toxic for oocytes). A small amount of oil is also kept at 37°C for daily use.

4. OF1 mice can be purchased from Charles River. We usually order mice that are 28–32 g in weight, which are around 8 weeks old and let them grow for another 2–3 weeks before analysis.

5. Milrinone is a phosphodiesterase-3 inhibitor and thus maintains a high concentration of cyclic AMP in the oocyte, which prevents NEBD.

6. Histone H2B_RFP incorporates into the chromatin, thereby providing a reporter for the position of chromosomes in living oocytes, which is critical for setting the reference point for determining the Ran-GTP gradient.

7. pRN3 has been described (23) and contains *Xenopus* β-globin 5′ and 3′ untranslated regions (utr) and a 30 amino-acid long

poly A tail for stability of the exogenous cRNA in mouse oocytes. It also contains a bacteriophage T3 RNA polymerase promoter sequence for in vitro transcription.

8. Ran-WT is used as a control that should not perturb the distribution of RanGTP in the oocyte. Ran-T24N and Ran-Q69L mutants are used as negative and positive controls, respectively. Ran-T24N binds to RCC1 and blocks its capacity to exchange GDP thus resulting in RanGDP accumulation. Ran-Q69L is defective in its GTPase activity and leads to RanGTP accumulation. Both mutants should lead to Ran gradient flattening with either low or high Ran activity as evidenced by the Rango probe FRET efficiency (Fig. 2).

9. Alternatively CellTram Oil (Eppendorf, 5176000025) can be used. It is a hydraulic-based pressure regulator and provides smoother control of the holding capillary over air-based pressure regulators such as CellTram Air.

10. Linearization of the plasmids is critical to avoid longer than expected transcription products as a result of circular templates. To avoid problems, it can be useful to check for full linearization by running an agarose gel before proceeding to purification.

11. cRNAs can be stored for up to 3 months at −20°C. Longer storage will result in lower expression levels in mouse oocytes and can lead to clogging of the injection capillary.

12. Clean the gel pouring and running apparatus with 10% SDS and DEPC-treated water to avoid degradation of the cRNA that can arise from contaminating RNases.

13. Diameter of the pulled glass pipette is critical at this step. If too large, it makes the flux of liquid extremely hard to control and can lead to oocyte loss or to air bubble formation in the medium. Also, passing through a large pulled pipette won't clean oocytes. If too small, it can deform oocytes and lyse them.

14. The pressure inside the holding capillary has to be carefully equilibrated before starting the injection. This prevents uncontrolled movement of liquid inside the capillary that can lead either to oocytes not being properly immobilized during injection or to excessive suction of oocytes that can deform and lyse them.

15. The quality, size, and shape of the injection capillary will determine the survival rate of oocytes after injection. It can thus be useful to spend some time fine-tuning the capillary puller settings. We usually pull capillaries immediately before use to avoid accumulation of dust at the tip.

16. Avoid pipetting at the bottom of the tube as this is where precipitated cRNAs and dust accumulate.

17. Injection is performed with a continuous flow so that the amount of injected liquid only depends on the injection time. We usually inject until the oocyte just starts swelling. Injection of a large volume of liquid will result in excessive swelling and deformation of the plasma membrane that will lead to oocyte lyse.

18. Transfer as little medium as possible when putting oocytes in a new clean drop of M2 + BSA to avoid introducing milrinone that can interfere with NEBD.

19. We use a thermostatic chamber that fully encloses the microscope to avoid focus drift due to temperature changes inside the room. Alternatively, a smaller thermostatic chamber can be used providing that the microscope is equipped with an auto-focus system. However, temperature fluctuates more in these small chambers, especially upon loading of the Ludin chamber. This can be harmful for oocytes.

20. Use a region devoid of debris in the field of view for background subtraction.

Acknowledgments

This work was supported by grants from the *Ligue Nationale Contre le Cancer* (EL/2009/LNCC/MHV) and from the *Agence Nationale pour la Recherche* (ANR08-BLAN-0136-01) to MHV, and by grants from *Agence Nationale pour la Recherche* (ANR-RPDOC-005-01) and *Fondation pour la Recherche Médicale* (FRM "Amorçage Jeunes Equipes") to JD.

References

1. Deng M, Suraneni P, Schultz RM, Li R (2007) The RanGTPase mediates chromatin signaling to control cortical polarity during polar body extrusion in mouse oocytes. Dev Cell 12:301–308

2. Dumont J, Million K, Sunderland K, Rassinier P, Lim H, Leader B, Verlhac M-H (2007) Formin-2 is required for spindle migration and for late steps of cytokinesis in mouse oocytes. Dev Biol 301:254–265

3. Halet G, Carroll J (2007) Rac activity is polarized and regulates meiotic spindle stability and anchoring in mammalian oocytes. Dev Cell 12:309–317

4. Longo FJ, Chen D-Y (1985) Development of cortical polarity in mouse eggs: involvement of the meiotic apparatus. Dev Biol 107:382–394

5. Verlhac M-H, Lefebvre C, Guillaud P, Rassinier P, Maro B (2000) Asymmetric division in mouse oocytes: with or without Mos. Curr Biol 10:1303–1306

6. Azoury J, Verlhac M-H, Dumont J (2009) Actin filaments: key players in the control of asymmetric divisions in mouse oocytes. Biol Cell 101:69–78

7. Maro B, Johnson MH, Webb M, Flach G (1986) Mechanism of polar body formation in the mouse oocyte: an interaction between the chromosomes, the cytoskeleton and the plasma membrane. J Embryol Exp Morphol 92:11–32

8. Dumont J, Petri S, Pellegrin F, Terret M-E, Bohnsack MT, Rassinier P, Georget V, Kalab P, Gruss OJ, Verlhac M-H (2007) A centriole- and

RanGTP-independent spindle assembly pathway in meiosis I of vertebrate oocytes. J Cell Biol 176:295–305

9. Bischoff FR, Krebber H, Smirnova E, Dong W, Ponstingl H (1995) Co-activation of RanGTPase and inhibition of GTP dissociation by Ran-GTP binding protein RanBP1. EMBO J 14:705–715

10. Saitoh H, Dasso M (1995) The RCC1 protein interacts with Ran, RanBP1, hsc70, and a 340-kDa protein in Xenopus extracts. J Biol Chem 270:10658–10663

11. Wilken N, Senecal JL, Scheer U, Dabauvalle MC (1995) Localization of the Ran-GTP binding protein RanBP2 at the cytoplasmic side of the nuclear pore complex. Eur J Cell Biol 68:211–219

12. Lounsbury KM, Macara IG (1997) Ran-binding protein 1 (RanBP1) forms a ternary complex with Ran and karyopherin beta and reduces Ran GTPase-activating protein (RanGAP) inhibition by karyopherin beta. J Biol Chem 272:551–555

13. Pu RT, Dasso M (1997) The balance of RanBP1 and RCC1 is critical for nuclear assembly and nuclear transport. Mol Biol Cell 8:1955–1970

14. Plafker K, Macara IG (2000) Facilitated nucleocytoplasmic shuttling of the Ran binding protein RanBP1. Mol Cell Biol 20:3510–3521

15. Carazo-Salas RE, Guarguaglini G, Gruss OJ, Segref A, Karsenti E, Mattaj IW (1999) Generation of GTP-bound Ran by RCC1 is required for chromatin-induced mitotic spindle formation. Nature 400:178–181

16. Kalab P, Pu RT, Dasso M (1999) The Ran GTPase regulates mitotic spindle assembly. Curr Biol 9:481–484

17. Nachury MV, Maresca TJ, Salmon WC, Waterman-Storer CM, Heald R, Weis K (2001) Importin beta is a mitotic target of the small GTPase Ran in spindle assembly. Cell 104:95–106

18. Wiese C, Wilde A, Moore MS, Adam SA, Merdes A, Zheng Y (2001) Role of importin-beta in coupling Ran to downstream targets in microtubule assembly. Science 291:653–656

19. Yi K, Unruh JR, Deng M, Slaughter BD, Rubinstein B, Li R (2011) Dynamic maintenance of asymmetric meiotic spindle position through Arp2/3-complex-driven cytoplasmic streaming in mouse oocytes. Nat Cell Biol 13:1252–1258

20. Kalab P, Pralle A, Isacoff EY, Heald R, Weis K (2006) Analysis of a RanGTP-regulated gradient in mitotic somatic cells. Nature 440:697–701

21. Tsurumi C, Hoffmann S, Geley S, Graeser R, Polanski Z (2004) The spindle assembly checkpoint is not essential for CSF arrest of mouse oocytes. J Cell Biol 167:1037–1050

22. Reis A, Chang HY, Levasseur M, Jones KT (2006) APCcdh1 activity in mouse oocytes prevents entry into the first meiotic division. Nat Cell Biol 8:539–540

23. Lemaire P, Gurdon JB (1994) A role for cytoplasmic determinants in mesoderm patterning: cell-autonomous activation of the goosecoid and Xwnt-8 genes along the dorsoventral axis of early Xenopus embryos. Development 120:1191–1199

Chapter 8

Making cRNA for Microinjection and Expression of Fluorescently Tagged Proteins for Live-Cell Imaging in Oocytes

Mark Levasseur

Abstract

Fluorescently tagged proteins have become a crucial weapon in the armory of a successful cell biology laboratory. This chapter describes how to produce cRNA coding for a fluorescently tagged protein of choice, such that it is suitable for microinjection and subsequent expression studies in live oocytes.

Key words: Fluorescent proteins, In vitro transcription, cRNA, Oocytes, Live cell imaging

1. Introduction

Imaging proteins in live cells using fluorescent tags emerged as a hugely powerful technique following the discovery and cloning of green fluorescent protein (GFP) from jellyfish (1), since when a wide variety of GFP mutants and other natural fluorophores have been added to the color repertoire (2, 3) The conventional way of employing these fluorescently tagged proteins has been to express and purify them from bacteria. However due to a variety of problems such insolubility and lack of biological activity due to incorrect folding and modification, an elegant and attractive alternative to making the proteins in bacteria, and which avoids these pitfalls, is to express the fluorescently tagged protein of choice from cRNA (defined as synthetic RNA produced by transcription from a specific DNA single-stranded template) injected into the cell. This is especially viable in oocytes, which are large enough to be easily microinjected, and with few exceptions, despite being meiotically arrested (4), are nevertheless translationally active (5, 6). A further advantage offered by cRNA expression over purified protein is that the

Hayden A. Homer (ed.), *Mammalian Oocyte Regulation: Methods and Protocols*, Methods in Molecular Biology, vol. 957,
DOI 10.1007/978-1-62703-191-2_8, © Springer Science+Business Media, LLC 2013

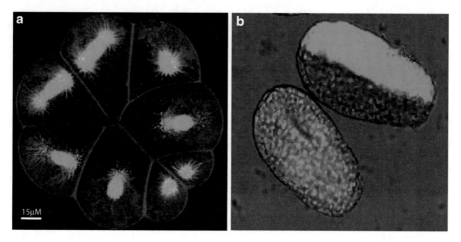

Fig. 1. GFP expression in ascidian embryos. (**a**) Shown is an 8 cell-stage embryo that was injected with MAP7::EGFP cRNA prior to fertilization. MAP7 can clearly be seen to localize to mitotic spindles (*green*). The plasma membrane is stained with FM4-64 (*red*). (**b**) Late stage embryos (just prior to tadpole formation) injected with mmGFP cRNA. Upper embryo was injected in one cell of a 2 cell stage embryo, lower was injected prior to fertilization and each allowed to develop. (Courtesy of Dr. McDougall obsevatoire oceanologique, Villefranche sur Mer, France).

protein continues to be expressed from the cRNA when the oocytes are fertilized and commence embryogenesis. This is beautifully illustrated in Fig. 1a where MAP7::EGFP decorates spindle microtubules (in green) at the eight cell stage of an ascidian embryo and further, in Fig. 1b where EGFP cRNA injected at the one or two cell stage zygote continues to be expressed in ascidian embryos nearing the tadpole stage. This also neatly demonstrates the non-toxicity of the fluorescent protein and that microinjection causes negligible disruption to normal development.

The following is an outline of how to produce fluorescently tagged cRNA suitable for microinjection and expression in living oocytes such as those of the primitive chordate the ascidian, or a mammal such as the mouse.

Essentially the protocol consists of three steps: (1) amplification of the target gene by PCR, (2) cloning of the amplified DNA into a suitable plasmid vector, (3) production of cRNA from linearized plasmid template.

1.1. Cloning Strategy

The vectors we use in our lab are derivatives of an in vitro transcription vector originally constructed by Patrick Lemaire, whilst working in John Gurdon's lab in Cambridge (7) called pRN3 (see Fig. 2a). The pRN3 vector consists of a pBluescript backbone, into which has been incorporated T3 and T7 polymerase promoters, three cloning sites (*Bgl*II, *Eco*RI, and *Not*I) as well as 5' and 3' globin untranslated regions (UTRs) which enhance biological activity of synthetic RNAs (8).

We have further modified pRN3 to produce a series of vectors each containing a gene for a different fluorescent protein, of which

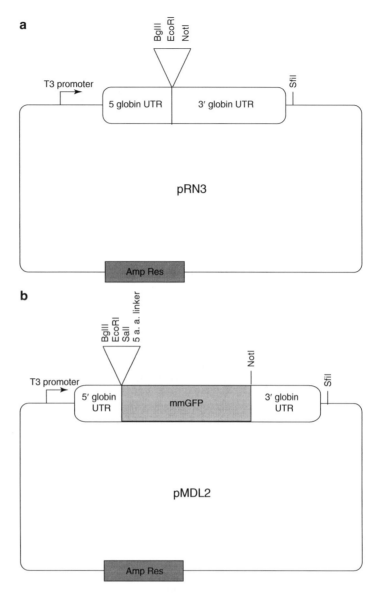

Fig. 2. In vitro transcription vectors pRN3 and pMDL2. (**a**) pRN3 (drawn from ref. (7)) showing key features needed for in vitro transcription (T3 promoter) and subsequent stability of transcripts (3′ and 5′ globin UTRs), restriction sites for cloning (*Bgl*II, *Eco*RI, *Not*I) and linearization (*Sfi*I) and antibiotic selection marker (Amp Res). (**b**) pMDL2 with inserted mmGFP, an extra *Sal*I cloning site, and a 5 amino acid linker for added structural flexibility in the resulting fusion protein.

pMDL2 is an example (see Fig. 2b), the construction of which has been previously described (9). As well as all the features of pRN3, pMDL2 contains mmGFP, a 50-fold brighter mutant of GFP, (10) as well as an extra *Sal*I cloning site. The *Sal*I site was added due to the paucity of existing sites, which were further reduced by the utilization of the *Not*I site for incorporation of the sequence encoding mmGFP, leaving only *Bgl*II and *Eco*RI for use in cloning upstream of mmGFP.

For experiments that utilize two or more fluorescently tagged proteins it is worth taking time to decide which combinations of fluorophores are most suited to the imaging capabilities of the lab and to first either obtain or construct pRN3 derivates which contain the desired genes. A common choice for dual fluorophore imaging is GFP and RFP (red fluorescent protein), which possess suitably non-overlapping excitation and emission spectra such that bleed-through (or signal cross-over) does not present a significant problem. There now exist a huge variety of other fluorescent proteins representing nearly all regions of the visible light spectrum should other colors be required. It is beyond the scope of this chapter to discuss these alternatives at length as they differ not only in their excitation and emission spectra but also in signal intensity. Useful information can be obtained from microscope manufacturer's websites, in particular Nikon; http://www.microscopyu.com/articles/livecellimaging/index.html.

Once the vector has been decided upon, the final piece of experimental planning is design of the primers to be used in PCR amplification (11) of the gene to be fluorescently tagged. This simply entails entering the sequence into a restriction site search program such as NEB cutter (http://tools.neb.com/NEBcutter2/index.php) and making sure the cloning sites in the vector to be used—for example, in the case of pMDL2, *Bgl*II, *Eco*RI, and *Sal*I—are not present. It is preferable to use directional cloning, i.e., a different site on the 5′ and 3′ primer such that insertion can only occur in the desired direction, but if only one site is available, this can be overcome later when screening the recombinants (see Note 1). The final considerations are to keep the sites in frame with the reading frame of the gene sequence, to incorporate the ATG start codon in the 5′ primer, and to omit the stop codon in the 3′ primer, otherwise the translated product will stop at the end of the target gene and not read through into the fluorescent tag (see Note 2 for an example set of primers). Should the fluorescent tag be required upstream in these vectors, the only site for cloning would be *Not*I, and the 5′ primer would not require an ATG start codon, but a stop codon would be necessary in the 3′ primer. Once designed, the primers can be synthesized by a commercial supplier of custom oligonucleotide synthesis (we use Invitrogen).

2. Materials

2.1. PCR-Based Sequence Amplification

1. Extensor Hi-fidelity PCR enzyme mix (supplied with 10× reaction buffer; Thermo Scientific).

2. dNTPs (Promega).

3. PCR tubes (Ambion).

4. Primers designed against gene of interest.

5. Target DNA (see Note 3).

6. Thermal cycler.

2.2. Assessment of Amplification

1. Agarose.

2. 50× TAE: 2 M Tris, 1 M acetic acid, 50 mM EDTA, pH 8.0.

3. Ethidium bromide solution (10 mg/ml, Sigma).

4. 6× gel loading dye: 30% glycerol, 0.25% Bromophenol blue, and 0.25% xylene cyanol.

5. DNA size ladders (Promega).

6. DNA (and RNA) mini subcell electrophoresis system with power supply (Bio Rad).

7. UV transilluminator.

2.3. Restriction Digestion and Ligation

1. Microspin 400 columns (GE Healthcare).

2. *Bgl*II, *Eco*RI, *Sal*I, *Not*I restriction enzymes (supplied with buffers and BSA from Promega).

3. T4 DNA ligase with LigaFast and standard ligation buffer (Promega).

2.4. Transformation, Selection, Plasmid Purification, and In Vitro Transcription

1. Competent strain of *E. coli* (e.g., NovaBlue competent cells, cloning efficiency (Novagen/Merck Bioscience)).

2. Polystyrene petri dishes (90 mm).

3. LB agar plates containing 100 μg/ml ampicillin. For making LB agar plates, dissolve 1 LB agar tablet (Sigma) per 50 ml of water, sterilize in an autoclave, cool to approximately 50°C, add ampicillin (Sigma) to 100 μg/ml, and pour 15–20 ml into petri dish. Allow to set with lids off under a lit Bunsen burner (to prevent contamination).

4. Ice.

5. Incubator (hot block) for Eppendorf tubes.

6. Platinum wire loop.

7. Wizard miniprep kit (Promega).

8. Orbital shaker/incubator.

9. Microcentrifuge.

10. *Sfi*I restriction enzyme (supplied with buffer and BSA from NEB).

11. Proteinase K (Roche).

12. Phenol:chloroform:IAA 25:24:1, RNase-free water, 3M sodium acetate pH 5.2 and tubes (Ambion), Pellet Paint (Novagen/Merck Bioscience).

13. mMESSAGE mMACHINE T3 transcription reaction kit (Ambion).

14. 70% ethanol made with RNase-free water.

15. 10×MOPS buffer: 0.2 M MOPS, 50 mM sodium acetate, 1 mM EDTA, pH 7.0.

16. RNA gel. To make the gel, melt 0.5 g agarose in 44 ml water, add 5 ml of 10× MOPS, cool to 50°C, and add 1 ml formaldehyde, pour and set.

17. Sample buffer: 500 µl neat formamide, 170 µl 37% formaldehyde, 70 µl 10×MOPS.

18. RNA size marker (Promega).

3. Methods

Carry out all procedures at room temperature unless otherwise stated and wear surgical gloves at all times.

3.1. PCR-Based Sequence Amplification

1. In a thin-walled PCR tube, set up the PCR amplification reaction

1 µl	Diluted target DNA (approx. 10 ng)
2.5 µl	5′ primer (20 µM dissolved in water)
2.5 µl	3′ primer (20 µM dissolved in water)
8 µl	dNTPs (1.25 mM)
5 µl	10× Reaction buffer
0.25 µl	Taq polymerase (Extensor Hi-fidelity PCR enzyme mix)
30.75 µl	Nuclease-free water
50 µl	

2. Place tube in a thermal cycler programmed to perform the following set of thermal cycles

94°C 1 min	1 cycle of initial denaturation
94°C 30 s	30 amplification cycles of denaturation, annealing and extension
x°C* 30 s	
68°C** 1 min	
68°C 5 min	Full extension of any unfinished PCR products

*Annealing temperature—at least 3°C lower than lowest primer T_m (melting point, shown on the synthesis sheets with custom made primers).
**68°C for this enzyme, other types commonly use 72°C. Longer may be necessary for longer targets.

3.2. Assessment of Amplification

1. Dissolve 0.4 g agarose in 50 ml 1×TAE buffer by boiling in a microwave.

2. Cool and add 2.5 μl ethidium bromide.

3. Pour into gel mold containing well former and allow to set.

4. Add 10 μl water and 3 μl 6× gel loading dye to 5 μl of PCR product.

5. Remove well former and place gel in the running tank and submerge under 1×TAE buffer.

6. Load wells with all 18 μl of sample and load an aliquot of DNA size ladder in an adjacent well.

7. Connect to power supply and run at 50–100 V until Bromophenol blue dye front is approximately 1/3–1/2 way along the gel.

8. Visualize DNA bands on a UV transilluminator, where they will fluoresce due to presence of ethidium bromide (wear eye protection) and the size markers can be used to assess if the band is of the expected size.

3.3. Restriction Digestion and Ligation

1. Perform buffer exchange on a Microspin 400 column (GE Healthcare) according to the manufacturer's protocol. Buffer exchange maximizes restriction enzyme cutting efficiency.

2. Add 10 μl of PCR product to the following:

2 μl	Restriction enzyme (i.e., approximately 20 units to ensure complete digestion)
0.5 μl	BSA (10 mg/ml)
5 μl	10× enzyme reaction buffer
32.5 μl	Nuclease-free water
50 μl (large volume to aid digestion efficiency)	

3. Incubate for 3–4 h at 37°C.
   If the enzymes have compatible reaction buffers, digest with both simultaneously by adding 2 μl of the 2nd enzyme, if not, perform the first restriction, buffer exchange again with a Microspin 400 column, and then digest the entire eluate with the second restriction enzyme.

4. Digest 1–2 μg of the vector as described in steps 2 and 3.

5. Check an aliquot of both PCR product and vector digests (1/10th of each digest should be sufficient) on agarose gel (see Subheading 3.2).

6. Estimate the concentration of the PCR product and vector digests from known amounts of the size marker lane.

7. Use these estimates to decide the amount of PCR product and vector to use in the subsequent ligation step to achieve the desired ratio. A good starting point is three molecules of insert (i.e., PCR product): 1 molecule of vector, and this can be determined using the following calculation (50 ng vector will be sufficient)

$$\frac{\text{ng of vector} \times \text{insert size (in kb)}}{\text{vector size (in kb)}} \times \text{insert : vector ratio}$$

$$= \text{ng of insert required}$$

8. Perform ligation based on the above calculations as follows:

$x\mu l$	Insert DNA
$y\mu l$	Vector DNA (50 ng)
1 µl	T4 DNA ligase
5 µl	2× LigaFast buffer or 1 µl standard 10× ligase buffer
$z\mu l$	Nuclease-free water
10 µl	Total volume

9. Incubate at room temperature for 5 min if using LigaFast buffer, or overnight at 4°C if using standard ligation buffer.

3.4. Bacterial Transformation, Selection, and Plasmid Purification

1. Combine ligation mix with a thawed aliquot of cells in an Eppendorf using a maximum of 2 µl of ligation reaction per 50 µl of competent *E. coli* cells.

2. Cool on ice for 20 min.

3. Heat shock at 42°C for 90 s. To do this place Eppendorf in a hot-block set to 42°C.

4. Plate out the transformation on standard LB agar plates containing ampicillin by pipetting a maximum of 150 µl of transformation onto the plate and then spread evenly using a flame-sterilized glass spreader. This selects for transformed cells as pRN3 and derivative vectors all contain the Amp resistance gene, so only transformants will grow on ampicillin.

5. Make a reference plate by drawing a grid on the base of a fresh LB plate, for example, a 6×6 grid that will accommodate 36 streaked out colonies (see Note 4).

6. Streak individual colonies from the transformation plate into each space on the grid using the platinum wire loop.

7. Incubate the reference plate overnight at 37°C.

8. On the following day, make up a PCR master mix sufficient for the number of streaked colonies to be screened. For example, for 20 reactions (of 20 μl each) mix the following:

20 μl	Primer 1 (20 μM)
20 μl	Primer 2 (20 μM)
40 μl	10× Taq polymerase buffer
64 μl	dNTPs (1.25 mM)
2 μl	Extensor Hi-fidelity polymerase*
254 μl	Water
400 μl	

*This enzyme is potent enough to work at this dilution.

9. Label the appropriate number of PCR tubes. For instance, 20 labeled tubes will be required for screening 20 colonies.

10. Dispense 20 μl aliquots of PCR master mix to each of the labeled PCR tubes.

11. Touch a shortened pipette tip (such that they fit in a PCR tube with the lid closed) to each colony streak to be screened and drop into a separate labeled tube containing PCR master mix.

12. Set up a positive control consisting of 1 μl of diluted (1:100) PCR product (from step 2, Subheading 3.1) plus 19 μl of master mix.

13. Perform PCR as used for amplification of the target gene (see step 2, Subheading 3.1).

14. Run all of each reaction on agarose gel with a DNA size marker ladder (as detailed in Subheading 3.2).

15. Record positives (i.e., those samples giving a band the same size as that in the positive control) and so identify and label the corresponding recombinant plasmid-containing clones on the reference plate, which will be used for generating cRNA for microinjection. In order to do this we first need to propagate amounts of recombinant plasmid DNA sufficient for use in repeated in vitro transcription reactions.

16. Select a healthy positive streak from the reference plate, and use a sterilized platinum wire loop to inoculate a suitable volume of LB broth containing ampicillin. 10 ml is sufficient for a miniprep, which generates enough DNA for approximately 10 transcription reactions.

17. Grow overnight at 37°C in a shaking incubator.

18. Generate plasmid DNA of sufficient purity for use as template in in vitro transcription reactions. We use Promega's Wizard miniprep kit, which comes complete with full instructions.

In brief, the steps involve harvesting the bacteria by centrifugation, resuspension of the cell pellet, cell lysis, precipitation and removal of bacterial DNA and cell debris, and finally purification of plasmid DNA in the supernatant by binding to a silica matrix, washing with alcohol and eluting purified plasmid in water or aqueous buffer.

19. Check a small aliquot of the plasmid prep. after restriction digestion on agarose gel (as detailed in Subheading 3.2). This step is optional.

3.5. Preparation of Linear Template

In vitro transcription is best performed on a linear template which ends shortly after the region which will generate the poly A tail. Hence, the pRN3 vector was designed with an *Sfi*I site (see Note 5) just downstream of the poly T sequence, such that when the plasmid is digested with *Sfi*I, a linear template is generated which comprises of the upstream T3 RNA polymerase promoter site, followed by the target gene fused in frame to the fluorescent tag and finishing with a run of 30Ts which will be transcribed as the poly A tail.

1. Digest 10 μl of miniprepped plasmid DNA (approximately 1.5–2 μg) with *Sfi*I by mixing the following:

10 μl	Plasmid DNA
2 μl	*Sfi*I
0.5 μl	BSA (10 mg/ml)
5 μl	10× reaction buffer
x μl	*Sfi*I oligo duplex (see Note 6)
y μl	Nuclease-free water
50 μl	

2. Incubate for 5 h at 50°C. Digestion with *Sfi*I can be problematic so that digestion conditions may need to be optimized (see Note 6).

3. Remove traces of RNase left from the plasmid prep procedure by adding 1 μl of 10 mg/ml Proteinase K and 2.5 μl 10% SDS and incubating for a further hour at 50°C. Proteinase K and restriction enzyme are then removed by phenol extraction as described in the following steps.

4. Add 100 μl nuclease-free water followed by 150 μl phenol:chloroform:IAA, 25:24:1, pH 7.9.

5. Vortex mix thoroughly and centrifuge at full speed in a microcentrifuge for 1 min.

6. Carefully remove upper aqueous layer (do not disturb the meniscus where the protein lies) to a fresh RNase-free tube.

7. Extract trace phenol by adding 150 μl chloroform and mixing and spinning as in step 5.

8. Transfer the aqueous layer to a fresh RNase-free tube and add 15 μl RNase-free 3 M Sodium Acetate (pH 5.2), 2 μl Pellet Paint (which makes resulting pellet easily visible by coloring it pink but does not interfere with the subsequent transcription reaction) and 450 μl 100% ethanol.

9. Place at –80°C for at least 1 h.

10. Centrifuge at full speed for 20 min.

11. Discard the supernatant and wash the pellet with 500 μl 70% ethanol (made with RNase-free water).

12. Centrifuge to remove the ethanol, centrifuge again and remove final ethanol traces.

13. Air dry for 15 min to allow final traces of ethanol to evaporate (see Note 7).

3.6. In Vitro Transcription

For in vitro transcription, we use Ambion's mMESSAGE mMACHINE T3 kit. This kit is highly reliable (see Note 8) and has been specifically designed to produce high yields of stable RNA from linearized DNA template.

1. Resuspend the linear template DNA in 6 μl RNase-free water. It is now ready for use in the in vitro transcription reaction.

2. Follow the kit instructions for assembly of the reaction.

3. Set up a T3 control reaction with linear template supplied with the kit.

4. Incubate both reactions at 37°C for 2 h.

5. Remove template with DNase and lithium chloride precipitation as per kit instruction, for which the reagents are all included in the kit.

6. Wash the pellet and gently dry for 5 min at room temperature.

7. Thoroughly resuspend pellet in 20 μl RNase-free water. Check transcription product as described in the following steps.

8. Prepare samples by adding 1 μl ethidium bromide and 13.5 μl sample buffer to 1 μl of transcription product and 1 μl of the product from the T3 control reaction. Heat at 65°C and then add 4 μl standard agarose gel 6× loading dye.

9. Prepare RNA size marker in the same way as the samples in step 8.

10. Load samples and RNA size marker in an RNA gel and run in 1×MOPS as described in step 7 of Subheading 3.2 (see Note 9).

11. Once satisfied with the cRNA quality, dispense into 1 μl aliquots and store at –80°C. Concentration can be estimated spectrophotometrically if necessary, where the A_{260} value $\times 40 = \mu g/ml$ RNA (see Note 10).

4. Notes

1. If only a single restriction site is available for cloning, the gene can ligate into the vector in two ways, as the same site will be at the 5′ and 3′ ends. Recombinants identified by PCR screening can then be checked for correct insertion orientation by cutting with an enzyme that cuts once in the vector and once in the inserted gene (but not near the middle). The pattern of bands obtained when the products of the restriction digestion are run on gel will then reveal whether the insert is in 5′–3′ or 3′–5′ orientation in the vector.

2. Primer design; Each primer should consist of at least 15 bases homologous to the target region (shown in bold), an in frame restriction site, and a start region of four random bases to aid restriction digestion efficiency. Avoid GC-rich regions in the target as these elevate the T_m. An example is shown here for 5′ and 3′ primers to mouse Cdc20;

*Bgl*II site Start

5′ primer 5′-TGAT<u>AGATCT</u>**ATGGCGCAGTTCGTGTTC**-3′

Target sequence 5′-ATGGCGCAGTTCGTGTTCGAGAGC...-3′

*Sal*I site

3′ primer 5′-TGAT<u>GTCGAC</u>**ACGGATGCCTTGGTGGAT**-3′

Target sequence 3′-<u>A</u>G<u>T</u>TGCCTACGGAACCACCTA-5′

Stop

3. Gene target DNA will typically be a plasmid containing the gene of interest, either already in the laboratory but not in the transcription vector, or received from another lab or DNA repository.

4. Growth on ampicillin selects for cells containing both recombinant and nonrecombinant plasmid. In order to screen for recombinants, percentages of which vary from one cloning experiment to another, we have found the most efficient way is to use a PCR-based screen with the primers used to amplify the target gene. To do this a reference plate must first be made so that positive clones can easily be located when required for propagation of recombinant plasmid DNA.

5. *Sfi*I has an 8 bp recognition site (it is actually GGCCNNNNNGGCC, a discontinuous site, where N is any base) which makes it exceptionally rare, and thus will not cut the cloned gene when preparing linear template.

6. Optimizing *Sfi*I digestion; Efficient cleavage with *Sfi*I is essential for a successful transcription reaction. *Sfi*I cleaves DNA by a mechanism which differs from other restriction endonucleases

(12). These are usually dimeric and operate on a single site, whereas *Sfi*I is a tetramer and needs to interact with two copies of its site before cleavage can occur. In order to aid this in the reaction, a synthetic double-stranded oligo containing the *Sfi*I cleavage site can be added to the reaction. To do this, two oligonucleotides must be synthesized; 5′-ATGTGGCCAACA AGGCCTATTG-3′ and 3′-CAATAGGCCTTGTTGGCCAC AT-5′ (12). These are then dissolved, mixed and heated to 95°C and allowed to cool in a large external volume of water to facilitate annealing. Optimize digestion with the resulting duplex by titrating its concentration from high nanomolar to low picomolar final concentrations in the digests. We find that as little as 2.5 pM results in 100% efficiency. If necessary overnight digestion with *Sfi*I at 50°C can also be effective in improving efficiency.

7. Transcription reaction efficiency also appears highly dependent on removing absolutely *all* traces of ethanol after the template preparation step, and thus it is worth trying extended pellet drying times of up to 30 min prior to resuspension in RNase-free water.

8. The Ambion mMESSAGE mMACHINE kit, although excellent, is relatively expensive and only supplies components for 25 reactions. Since the kit has a shelf-life of 6–9 months (in our experience) if feasible, once reactions are working, it is worth doing as many further transcription reactions together as possible, since in our experience, the cRNA produced is perfectly stable for many years when aliquoted and stored at −80°C. Aliquoting is essential as repeated freeze thawing of cRNA leads to degradation.

9. We have found these RNA gels somewhat unreliable and prefer to use microinjection of the newly synthesized cRNA and subsequent fluorescence in the injected eggs as the gold standard for a successful transcription reaction.

10. Concentration is typically around 0.5–1 μg/μl when resuspended in 20 μl, which is fine as a starting point for microinjection. Dilution may sometimes be necessary for reactions with unusually high yields (the resulting cRNA being too viscous to inject easily) or if the fluorescent signal is too bright.

References

1. Prasher DC (1995) Using GFP to see the light. Trends Genet 11:320–323
2. Shaner NC, Steinbach PA, Tsien RY (2005) A guide to choosing fluorescent proteins. Nat Methods 2:905–909
3. Day RN, Davidson MW (2009) The fluorescent protein pallette: tools for cellular imaging. Chem Soc Rev 38:2887–2921
4. Whitaker M (1996) Control of meiotic arrest. Rev Reprod 1:127–135

5. Levasseur M, McDougall A (2000) Sperm-induced calcium oscillations at fertilisation in ascidians are controlled by cyclin B1-dependent kinase activity. Development 127:631–641

6. Herbert M, Levasseur M, Homer H, Yallop K, Murdoch A, McDougall A (2003) Homologue disjunction in mouse oocytes requires proteolysis of securin and cyclin B1. Nat Cell Biol 5:1023–1025

7. Lemaire P, Garret N, Gurdon JB (1995) Expression cloning of Siamois, a *Xenopus* homeobox gene expressed in dorsal-vegetal cells of blastulae and able to induce a complete secondary axis. Cell 81:85–94

8. Kreig PA, Melton DA (1984) Functional messengers are produced by SP6 in vitro transcription of cloned cDNAs. Nucleic Acids Res 12:7057–7070

9. Homer HA, McDougall A, Levasseur M, Yallop K, Murdoch AP, Herbert M (2005) Mad2 prevents aneuploidy and premature proteolysis of cyclin B and securin during meiosis I in mouse oocytes. Genes Dev 19:202–207

10. Zernicka-Goetz M, Pines J, MacLean Hunter S, Dixon JPC, Siemering KR, Hasselhof J, Evans MJ (1997) Following cell fate in the living mouse embryo. Development 124: 1133–1137

11. Mullis KB, Faloona F (1987) Specific synthesis of DNA in vitro via a polymerase catalyzed chain reaction. Methods Enzymol 155: 335–350

12. Wentzell LM, Nobbs TJ, Halford SE (1995) The *Sfi*I restriction endonuclease makes a four-strand DNA break at two copies of its recognition sequence. J Mol Biol 248:581–595

Chapter 9

RNAi-Based Methods for Gene Silencing in Mouse Oocytes

Paula Stein, Petr Svoboda, and Richard M. Schultz

Abstract

RNA interference (RNAi) is an evolutionary conserved gene-silencing pathway that can be efficiently utilized as a tool to study gene function. RNAi is initiated by long double-stranded RNAs (dsRNAs), which are processed into small duplexes called small-interfering RNAs (siRNAs). In turn, these duplexes target mRNAs for degradation in a sequence-specific manner. Mouse oocytes, unlike most mammalian cell types, lack an interferon response to long dsRNA. Moreover, they are a rare example of a mammalian cell type with a robust endogenous RNAi pathway. For these reasons microinjection of either long dsRNAs or siRNAs results in efficient, sequence-specific gene silencing. Here, we describe a protocol for preparation and microinjection of long dsRNA into mouse oocytes.

Key words: RNAi, dsRNA, Mouse oocyte, Microinjection, Knockdown

1. Introduction

Since the discovery of the RNA interference (RNAi) phenomenon more than a decade ago (1), this gene silencing mechanism has been characterized and extensively utilized as a tool to study gene function in a myriad of organisms (2). The trigger of this pathway is a molecule of long double-stranded RNA (dsRNA), which is recognized by the RNase III enzyme Dicer and processed into 21–24 nucleotide long duplexes named small-interfering RNAs (siRNAs). One of the two strands of the siRNA, the guide strand, associates with a multiprotein complex named RNA-induced silencing complex (RISC). The other strand, known as passenger strand, is degraded. The siRNA guides the RISC complex to its target through base complementarity and the endonuclease activity of Argonaute 2, one of the components of RISC, cleaves the target mRNA in the center of the duplex region between the guide strand and the target mRNA (3).

Hayden A. Homer (ed.), *Mammalian Oocyte Regulation: Methods and Protocols*, Methods in Molecular Biology, vol. 957, DOI 10.1007/978-1-62703-191-2_9, © Springer Science+Business Media, LLC 2013

In the early years of RNAi there was skepticism that dsRNA could trigger sequence-specific gene silencing in mammals. The reason for this disbelief is that mammals were known for decades to possess another pathway triggered by long dsRNA, the interferon response (4). This pathway is an anti-viral strategy by which dsRNA produced during viral infections is recognized by the cell as foreign, resulting in transcriptional activation of interferon genes, inhibition of protein synthesis, and ultimately cell death (5).

Nevertheless, we and others reported in 2000 that dsRNA triggers RNAi in mammals (6, 7). Both studies were conducted using mouse oocytes and preimplantation embryos. In most mammalian cell types, however, introduction of long dsRNA (>30 bp) does result in an interferon response; therefore siRNAs must be used to trigger RNAi in these cell types. Mouse oocytes and early embryos, on the other hand, lack an interferon response and respond very efficiently and sequence-specifically to long dsRNA (8). In fact, mouse oocytes are a rare example of a mammalian cell type with a robust endogenous RNAi pathway (9, 10). Studies in our laboratory demonstrated that dsRNA prepared either by annealing sense and antisense RNA strands or by transcribing an inverted repeat that generates a hairpin RNA are equally effective and that adding an overhang to the hairpin does not affect RNAi efficacy (11). We also showed that the hairpin can be expressed from a plasmid (11); in this case, the construct has to be microinjected into the germinal vesicle of meiotically incompetent oocytes, since the fully-grown oocyte is transcriptionally inactive. We went on to develop a transgenic RNAi approach, in which an inverted repeat is expressed under the control of an oocyte-specific promoter (12).

In summary, RNAi triggered by long dsRNA is an extremely powerful tool to study gene function in mouse oocytes. It is equally effective in the early preimplantation embryo. dsRNA can be introduced into oocytes by microinjection or using a transgenic approach. Here we describe a protocol for preparation and microinjection of dsRNA into mouse oocytes. Two different strategies are discussed to prepare dsRNA: using a DNA template generated by polymerase chain reaction (PCR) and preparing a template from a plasmid. The use of an inverted repeat that generates a hairpin RNA upon transcription is also described. Detailed protocols are also presented for collection, culture, and microinjection of fully-grown mouse oocytes, as well as for RNA isolation and quantitative reverse transcription (qRT)-PCR to assess the efficacy of RNAi.

2. Materials

2.1. Preparation of Long dsRNA Molecules

1. TAE (Tris-acetate-EDTA) buffer: 40 mM Tris base, pH 8.0, 20 mM glacial acetic acid, 1 mM EDTA.

2. 1.5% Agarose gel, (nondenaturing) in 1× TAE buffer containing 0.5 μg/ml ethidium bromide.

3. Chloroform.

4. DNA template (5–10 μg).

5. 100 mM DTT (provided with SP6 polymerase).

6. Ethanol (95%).

7. H_2O (DEPC-treated).

8. 25 mM nucleoside triphosphates (NTPs).

9. PCR purification kit (e.g., QIAquick PCR purification kit; Qiagen) (for preparation of template using PCR; optional).

10. Phenol (pH 4.5).

11. Phenol:chloroform (1:1 v/v).

12. Plasmids carrying the chosen target sequence (for preparation of template from a plasmid).

13. 3 M potassium acetate (pH 5.5).

14. Restriction enzymes (for preparation of template from a plasmid).

15. RNase T1 (1,000 units/μl; Fermentas) (optional).

16. RNasin (40 units/μl; Promega) or another RNAse inhibitor.

17. RQ1 DNase (Promega).

18. SP6 buffer.

19. SP6 RNA polymerase (20 units/μl).

20. TE (Tris-EDTA) buffer: 10 mM Tris, 1 mM EDTA, pH 8.0. Prepare a 10× solution and dilute to 1× before use.

21. Electrophoresis apparatus for ~10-cm gel length (e.g., Hoefer/Amersham mini submarine electrophoresis apparatus), with power supply.

22. Hybridization oven preset to 65°C (optional).

23. Lid locks for microcentrifuge tubes.

24. Microcentrifuge (Eppendorf 5424 or equivalent).

25. Spectrophotometer.

26. Thermoblock, dry (e.g., Eppendorf Thermomixer).

27. Tubes, microcentrifuge (1.5-ml).

28. QIAquick Gel Extraction Kit (Qiagen).

29. TOPO-TA Cloning kit (Invitrogen).

2.2. Microinjection of dsRNA into Oocytes

1. Arcturus PicoPure Frozen RNA Isolation Kit (Applied Biosystems).

2. CZB medium.

3. 10 mM dNTPs.

4. dsRNA solution (10^6 molecules in 10 pl).

5. GFP mRNA.

6. MEM/PVP+M (Minimum Essential Medium/polyvinylpyrrolidone+milrinone) medium: MEM, 3 mg/ml PVP, 2.5 μM milrinone. Whitten's/HEPES/PVA+M (polyvinyl alcohol+milrinone) medium (HEPES-buffered Whitten's medium, 0.01% polyvinyl alcohol, 2.5 μM milrinone) can be used as an alternative.

7. Mice: 6-week-old NSA (CF-1) females (Harlan).

8. Milrinone (2.5 mM stock solution in DMSO).

9. Nuclease-free H_2O.

10. Pregnant mare's serum gonadotropin (Calbiochem).

11. Random hexamers (0.5 μg/μl).

12. RNasin (or other RNAse inhibitor).

13. RNase H.

14. Superscript II reverse transcriptase (includes 5× first strand buffer and 100 mM DTT).

15. TaqMan assays (Applied Biosystems).

16. TaqMan Gene Expression Master Mix (Applied Biosystems).

17. Dishes, tissue culture (sterile plastic) (35- and 60-mm).

18. Forceps.

19. Incubator (humidified; 5% CO_2 in air at 37°C).

20. Microinjection setup: micromanipulators (e.g., TransferMan NK2 micromanipulator) and injector (e.g., Pico-Injector Microinjection Systems; Harvard Apparatus).

21. Microscope.

22. Needle (27-gauge).

23. Nunc Lab-Tek chamber slide.

24. Paraffin oil (light).

25. Pipettes, glass (mouth-operated).

26. Borosilicate-glass capillary tubing for making mouth-operated handling pipettes and holding and injection pipettes used for

microinjection. Injection pipettes are hand-made by pulling in a mechanical pipette puller. They can be prepared in advance or as microinjection proceeds. Holding pipettes are pulled the same way, but then they are cut to a diameter of 80–120 μm and the tip is melt using a microforge. They can be prepared in advance (for detailed instructions, see ref. 13). Holding pipettes are also commercially available from various companies (e.g., Eppendorf VacuTips).

27. MicroAmp optical adhesive film kit.

28. Microcentrifuge (Eppendorf 5424 or equivalent).

29. Real-Time PCR instrument (e.g., Applied Biosystems' 7000, 7300, or 7500 Real-Time PCR System).

30. Refrigerated centrifuge for 96-well plates (e.g., Eppendorf 5810R).

31. Scissors.

32. Syringe (1-ml).

33. Thermoblock, dry (e.g., Eppendorf Thermomixer).

34. Tubes, microcentrifuge (1.5-ml).

35. Watch glasses or glass staining blocks (VWR).

36. 96-well optical reaction plates (Applied Biosystems).

3. Methods

3.1. Preparation of Long dsRNA Templates for In Vitro Transcription

This protocol is based on the methods described by Fire et al. (1) and Kennerdell and Carthew (14). DNA templates can be prepared in two different ways. PCR can be used to generate sense and antisense DNA templates from the gene of interest. Alternatively, a plasmid carrying a fragment of the gene and a suitable promoter for in vitro transcription can be used.

3.1.1. Preparation of Template by PCR

1. Design forward and reverse primers to amplify the selected region by RT-PCR of total RNA (see Note 1).

2. Gel purify this primary PCR product using QIAquick Gel Extraction Kit (Qiagen).

3. Dilute the purified PCR product 1:500–1:1,000 in H_2O.

4. Use diluted and purified PCR product as a template for two separate second PCR reactions in which either the forward or the reverse primer carries an SP6 or T7 promoter sequence at its 5′ end (see Note 2). The products of these two separate

PCR reactions will be sense and antisense templates for in vitro transcription (Fig. 1).

5. Purify the fragments by gel extraction using QIAquick Gel Extraction Kit (Qiagen), as in step 2.

6. Estimate the amount and quality of the templates spectrophotometrically.

The advantages of this method are speed and the absence of foreign sequence in the dsRNA. Using plasmid templates, on the other hand, is less expensive and generates higher quality templates.

3.1.2. Preparation of Template from a Plasmid

There are three options for generating templates from suitable plasmids carrying the chosen target sequence:

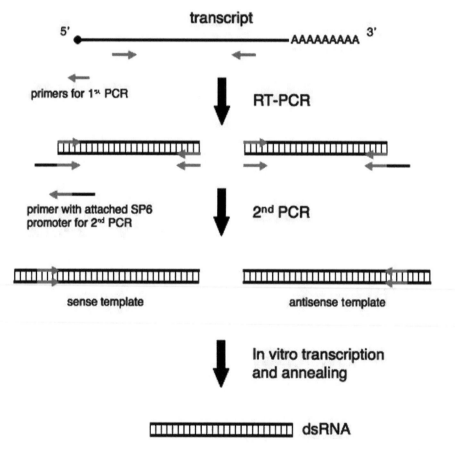

Fig. 1. Strategy for preparing dsRNA using sense and antisense fragments as template. RNA isolation and reverse transcription are performed, followed by a first PCR reaction that uses forward and reverse primers that define the targeted region. In the second PCR reaction, the primers are the same, except that either the forward primer (in the sense reaction) or the reverse primer (antisense reaction) contains SP6 or T7 promoter sequence at its 5′ end. In vitro transcription is then performed, followed by annealing of sense and antisense RNA strands to generate dsRNA.

Generate Two Plasmids, One with the Insert in the Sense Orientation and the Other One with the Insert in the Antisense Orientation.

This option is particularly useful if a plasmid with a single promoter is available or if only one RNA polymerase is available

1. Amplify the desired sequence by PCR.

2. Insert it into the plasmid using either TA or TOPO-TA Cloning kits (Invitrogen).

3. Pick positive colonies and start liquid cultures.

4. Perform DNA isolation, plasmid linearization, and purification using standard procedures (15).

Use One Plasmid that Has Appropriate Promoters Flanking the Multiple Cloning Site from Both Sites, Such as pCRII (Invitrogen)

1. Amplify the desired sequence by PCR.

2. Insert it into the plasmid using either TA or TOPO-TA Cloning kits (Invitrogen).

3. Pick positive colonies and start liquid cultures.

4. Perform DNA isolation, plasmid linearization, and purification using standard procedures (15).

A Very Convenient Method of dsRNA Preparation Uses a Plasmid Carrying an Inverted Repeat of the Target Sequence as Template for In Vitro Transcription

Although cloning an inverted repeat is not simple, the advantage of this method is that it generates dsRNA directly during in vitro transcription, without any need for estimating equimolar ratios of sense and antisense strands and annealing. The inverted repeat is made by ligating two almost identical PCR products. One of the PCR products is slightly longer than the other one at its 5′ end (Fig. 2).

1. To generate these two PCR products, design two different forward primers and one reverse primer according to the diagram depicted in Fig. 2a. The distance between the two forward primers should be ~20–50 bp. This primer design defines the targeted region (from F2 to R) and the size of the spacer in the middle of the inverted repeat (distance between forward primers). Add the same restriction site to both forward primers and a different restriction site to the reverse primer.

2. Amplify both PCR products (short and long).

3. Digest both PCR products with the restriction enzyme that recognizes the site present in both forward primers.

4. Ligate long and short fragments. Three possible products can result from this ligation reaction: short-short, long-long, and short-long (Fig. 2b). The first two ligation products are perfect inverted repeats without spacers. While they could in principle be good templates for in vitro transcription, they are often lost or recombined in bacteria; the long-short ligation product with a spacer (from the excess sequence in the longer PCR fragment) is usually more stable and cloned at much higher rate (see Note 3).

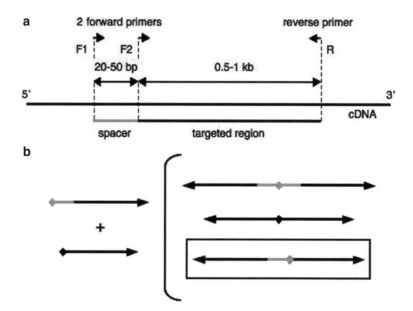

Fig. 2. Preparation of an inverted repeat. (**a**) Primer design to define the targeted region and the length of the spacer in the inverted repeat. The two forward primers should be 20–50 bp apart and the targeted region is usually 0.5–1 kb long. (**b**) Ligation of long and short PCR fragments can generate three possible products: long–long, short–short, and long–short. The targeted region is depicted in *black* and the spacer is depicted in *gray*.

5. After ligation, gel-purify the appropriate size band.

6. Add several hundred nanograms of this DNA to a PCR mixture with Taq polymerase and incubate for 10 min at 72°C (in this step, Taq polymerase adds A overhangs).

7. Process the product as a regular PCR product (as described in Subheadings "Generate Two Plasmids, One with the Insert in the Sense Orientation and the Other One with the Insert in the Antisense Orientation. This Option Is Particularly Useful if a Plasmid with a Single Promoter Is Available or if Only One RNA Polymerase Is Available" and "Use One Plasmid that Has Appropriate Promoters Flanking the Multiple Cloning Site from Both Sites, Such as pCRII (Invitrogen)").

8. Digest the plasmid completely downstream from the insert, preferentially with an enzyme that does not leave 3′ overhangs. If there is no alternative restriction site, blunt-end the 3′ overhang using the Klenow fragment prior to in vitro transcription.

9. Purify the digested plasmid by standard phenol extraction, precipitate it (for a more detailed protocol, see ref. 15), and resuspend it in the appropriate amount of TE. Keep the final DNA concentration >500 ng/μl.

The amount of plasmid needed for in vitro transcription is 5–10 μg. Therefore, a standard plasmid miniprep will provide enough template for a few in vitro transcription reactions and one in vitro transcription reaction usually generates enough dsRNA for many rounds of microinjection.

3.2. In Vitro Transcription

The following protocol describes in vitro transcription with SP6 polymerase. For other polymerases (T3, T7), simply use the desired RNA polymerase and suitable buffer.

1. Set up the reaction mixture:

5× SP6 buffer	20 μl
100 mM DTT	10 μl
40 units/μl RNAsin	5 μl
25 mM NTPs	20 μl
SP6 RNA polymerase (20 units/μl)	5 μl
DNA template	5–10 μg
H$_2$O (sterile, DEPC-treated)	up to 100 μl

2. Incubate the reaction for 2–4 h at 37°C.

3. Add 5 units of RQ1 DNase and incubate for 15 min at 37°C (for more than 5 μg of template, add ~6 min of incubation time for each additional microgram). If an inverted repeat was transcribed, also add 5 μl of RNase T1 and incubate the mixture with RQ1 DNase and RNase T1 for 30 min at 37°C.

4. Estimate the amount and quality of transcribed RNA by electrophoresis (see Note 4).

5. If sense and antisense templates were used, load 1, 2, and 4 μl of each reaction (from step 3) into a 1.5% nondenaturing agarose gel (in 1× TAE buffer, containing ethidium bromide). If an inverted repeat was utilized as template, load 4 μl to estimate the amount and quality of RNA and proceed to the purification step.

Usually more than one band appears on the gel, with the fastest-migrating band corresponding to half the size of the RNA compared to the DNA marker. The slowest migrating bands are dimers, trimers, etc., that form under nondenaturing conditions. The pattern is dependent on the RNA sequence. Loading different amounts of reaction product helps in estimating the ratio of sense to antisense reaction mixture volumes to be used for annealing.

3.3. Annealing

1. In a 1.5-ml microcentrifuge tube mix roughly equimolar amounts of sense and antisense RNAs (the ratio does not have to be exact; a slight excess of ssRNA will be removed during RNase T1 treatment after annealing and, in any case, unannealed ssRNA does not affect RNAi). Keep the concentration

of RNA at ~200–500 ng/ml in 100–200 µl of the annealing mixture. Adjust the volume with DEPC-treated sterile H_2O.

2. Secure the lid of the microcentrifuge tube with a lock to prevent tube opening and contamination with H_2O from the water bath.

3. Incubate the annealing mixture for 3 min in a 1-L beaker of boiling water. If the lid of the microcentrifuge tube is unsecured, boil the tube for only 1 min.

4. Remove the beaker from the heat, and leave the tube in the beaker of H_2O for 3 h at room temperature to allow it to cool down.

5. Check an aliquot (1–4 µl) of the reaction on a 1.5% nondenaturing agarose gel (in 1× TAE buffer, stained with ethidium bromide). If the annealing reaction is incomplete (i.e., a smear or a large amount of ssRNA can be seen), incubate the tube, still in the beaker, in a preheated 65°C hybridization oven for 1 h, followed by a gradual cooling down period of 2 h at room temperature.

6. (Optional) After annealing, add 5 µl of RNase T1 and incubate the mixture for 30 min at 37°C.

3.4. dsRNA Purification

1. Add 1 volume of phenol (pH 4.5) to the dsRNA, vortex, and centrifuge at maximum speed for 10 min at 4°C.

All centrifugations in this procedure are performed at 4°C.

2. Transfer the aqueous phase into a new microcentrifuge tube. Add 1 volume of phenol:chloroform (1:1 v/v) to the aqueous phase and extract as in step 1.

3. Repeat step 2, but with chloroform instead of phenol:chloroform.

4. Precipitate the dsRNA with 0.1 volume of 3 M potassium acetate and 3 volumes of 95% ethanol for >1 h at –20°C (it can be overnight).

5. Centrifuge at maximum speed for 20 min at 4°C to pellet the dsRNA.

6. Remove the supernatant and wash the pellet twice with 400 µl each of 75% ethanol. Air-dry the pellet. Do not dry the pellet under vacuum.

7. Resuspend the pellet in 20 µl of sterile H_2O (see Note 5), and estimate the quality and the amount of dsRNA.

8. Dilute the dsRNA to the desired concentration with sterile H_2O (200 ng/µl of 0.5 kb dsRNA is ~200,000 molecules/pl), aliquot it into 0.5-ml microcentrifuge tubes, and freeze the samples at –80°C (see Note 6).

3.5. Microinjection of dsRNA into Oocytes

The dsRNA is introduced into mouse oocytes by microinjection. The microinjected cells are cultured for the desired period of time until assayed for an RNAi effect. The first assay to conduct with microinjected oocytes should always be qRT-PCR to check that the cognate mRNA has been degraded, i.e., to prove the efficiency of RNAi (see Note 7). Specific biological assays depend on the experimental design and can range from assessing meiotic maturation, fertilization, and early embryo development to techniques such as Western blot, immunofluorescence, enzyme activity determination, and gene expression analysis using either microarray technology or RNA sequencing. A typical experimental outline of an RNAi experiment in mouse oocytes is shown in Fig. 3.

The duration of microinjection experiments depends on the phenotypic characteristic studied. Cell collection and microinjection of dsRNA are completed in 1 day, but in most experiments oocytes are cultured for 20–24 h, then either assayed for an RNAi effect or in vitro-matured and sometimes fertilized. In addition, female mice must be primed with gonadotropins 2 days prior to oocyte collection.

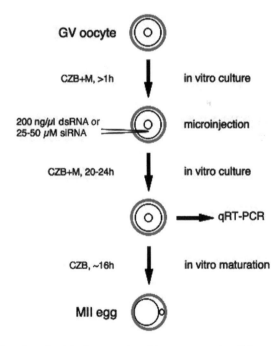

Fig. 3. Outline of a microinjection experiment in mouse oocytes. Fully-grown, GV-intact mouse oocytes are isolated and cultured for at least 1 h in CZB + M. Oocytes are microinjected with either long dsRNA or siRNA in MEM/PVP + M. After in vitro culture for 20–24 h, qRT-PCR can be performed to assess the extent of knockdown. The microinjected oocytes can also be matured to metaphase II for ~16 h in CZB medium. Metaphase II eggs can be assayed by immunofluorescence, Western blot, microarray technology, or RNA sequencing, among other techniques, to assess the phenotype resulting from knocking down the gene of interest.

1. Administer a single intra-peritoneal injection of 5 IU of pregnant mare's serum gonadotropin (PMSG) to 6-week-old NSA (CF-1) female mice 48 h prior to sacrifice. Hormonal treatment improves the yield of antral follicles containing fully-grown germinal vesicle (GV) intact oocytes.

2. Prepare culture medium: add 1 μl of 2.5 mM milrinone (see Note 8) to 1 ml of CZB medium (16).

3. Set up microdrop culture dishes: place several 50–100-μl drops of CZB containing 2.5 μM milrinone (CZB + M) on the bottom of a 60-mm sterile plastic tissue culture dish and cover the dish with light paraffin oil.

4. Place the dish in a humidified incubator containing 5% CO_2 in air at 37°C in advance to allow for temperature and CO_2 equilibration.

5. Sacrifice the females and remove the ovaries using forceps and scissors.

6. Place the ovaries in a watch glass with MEM/PVP containing 2.5 μM milrinone (MEM/PVP + M).

7. Release the antral follicles from the ovaries by puncturing them several times with a 27-gauge needle attached to a 1-ml syringe.

8. Use a mouth-operated glass pipette to collect the oocyte–cumulus cell complexes and transfer them to a clean watch glass containing ~2 ml of MEM/PVP + M.

9. Select large antral follicles containing cumulus-enclosed oocytes and discard the smaller pre-antral follicles or denuded oocytes.

10. Mechanically denude oocytes by pipetting the complexes up and down using a pipette with a tip diameter that is about the size of the oocytes.

11. Transfer the cumulus-free oocytes to the culture dish and place in the incubator (see Note 9).

1. Place a 5-μl drop of injection medium (MEM/PVP + M) on a Nunc Lab-Tek chamber slide (or a 100-mm Petri dish).

2. Place a 1-μl drop of dsRNA solution (see Note 10) as close as possible to the microinjection drop and then flood the dish with light paraffin oil.

3. When more than one dsRNA is injected, or if vehicle (usually H_2O) injection is used as a control, set up a set of drops (a 5 μl-drop of MEM/PVP + M and a 1 μl-drop of dsRNA or H_2O) for each substance to be microinjected (see Note 11).

4. Place the dish in the stage of the micromanipulator, position the injection and holding pipettes, and connect the gas supply.

5. Transfer a group of oocytes from the incubator to the microinjection drop and inject 5–10 pl of dsRNA into their cytoplasm.

6. Return microinjected oocytes to the incubator.

7. Repeat the procedure with another group until all oocytes are microinjected.

8. On completion of microinjection, check all microinjected cells under the microscope and remove those that did not survive.

9. Culture the oocytes in CZB + M for 20–24 h to allow time for RNAi effect.

10. Lyse oocytes for evaluating the efficacy of RNAi by RNA isolation and qRT-PCR.

3.5.3. RNA Isolation

RNA isolation is performed using Arcturus PicoPure Frozen RNA Isolation Kit (Applied Biosystems).

1. Prepare a master mix containing 10 μl of extraction solution and 1 μl of Gfp mRNA (~2 ng/μl) per sample (Extr + Gfp). Gfp mRNA serves as an internal control for the efficiency of RNA isolation and qRT-PCR.

2. Pipet oocytes into a microcentrifuge tube containing 11 μl of Extr + Gfp and freeze at −80°C (if RNA isolation is not going to be performed right away) or keep on dry ice until ready for RNA isolation. To achieve reproducible results, use at least 20 oocytes for RNA isolation.

3. Thaw frozen cells on ice and incubate at 42°C for 30 min.

4. Precondition RNA Purification Columns:

   (a) Pipette 250 μl Conditioning Buffer to purification column and incubate for 5 min at room temperature.

   (b) Centrifuge at $16,000 \times g$ for 1 min.

5. Pipette 10 μl of 70% EtOH into cell extract and mix well by pipetting (do not centrifuge).

6. Add cell extract/EtOH mixture to column and centrifuge at $100 \times g$ for 2 min and $16,000 \times g$ for 30 s.

7. Add 100 μl wash buffer 1 (W1) to column and centrifuge at $8,000 \times g$ for 1 min (see Note 12).

8. Add 100 μl wash buffer 2 (W2) to column and centrifuge at $8,000 \times g$ for 1 min.

9. Add 100 μl W2 to column and centrifuge at $16,000 \times g$ for 2 min.

10. Remove flow-through and centrifuge at $16,000 \times g$ for 1 min.

11. Transfer column to elution tube and pipet 11 μl of elution buffer *directly* onto membrane.

12. Incubate column for 5 min at room temperature; centrifuge for 1 min at $1,000 \times g$ and 1 min at $16,000 \times g$.

3.5.4. Reverse Transcription

1. Add 1 μl of random hexamers (0.5 μg/μl) and 1 μl of 10 mM dNTPs to the ~10 μl of isolated RNA and heat up to 65°C for 5 min; then chill on ice.

2. Add 4 μl of 5× first strand buffer, 2 μl of 0.1 M DTT, and 1 μl of RNAsin and heat up to 42°C for 2 min.

3. Add 1 μl of reverse transcriptase (Superscript II) and further incubate for 50 min at 42°C.

4. Inactivate the reaction by heating for 15 min at 70°C and let the samples cool down for a few minutes at room temperature.

5. Add 1 μl of RNase H and incubate at 37°C for 20 min.

6. Add enough nuclease-free H_2O to the cDNA to set the concentration at 1 oocyte equivalent per microliter (e.g., if there were 50 oocytes in the tube, then add 50-21 = 29 μl H_2O).

7. Store cDNA at –20°C.

3.5.5. Real-Time PCR

1. Set up the reaction mixture on ice and protecting TaqMan probes from light:

   10 μl TaqMan Gene Expression Master Mix

   1 μl 20× assay mix (TaqMan probe)

   1 μl cDNA

   8 μl nuclease-free H_2O

2. Transfer the volume of each reaction mixture to each well of a 96-well optical plate (see Note 13). Run the samples in triplicate.

3. Seal the plate using MicroAmp Optical Adhesive Film.

4. Centrifuge the plate briefly ($1,800 \times g$ for 3 min at 4°C).

5. Run the plate on a real-time PCR instrument (if using Applied Biosystems' 7000, 7300, or 7500 Real-Time PCR System, use default PCR conditions).

6. Calculate relative expression using the ΔΔCt method (ABI PRISM 7700 Sequence Detection System, User Bulletin #2, Applied Biosystems, 1997).

4. Notes

1. In our experience, dsRNA molecules longer than 0.5 kb work best. The upper limit for the length of the targeted region is

limited mainly by the efficiency of in vitro transcription and successful cloning of an inverted repeat. We would not recommend using fragments >1.5 kb and for most purposes we would recommend staying within the 0.5–1 kb range, since longer inverted repeats may be more difficult to clone and may be unstable due to possible secondary structures. With respect to selecting the region to target, choose any part of the coding sequence or even the 3′UTR and make sure it is unique. This is especially important when targeting a member of a family with high homology. If using TaqMan probes for real-time PCR you may want to find a region that does not overlap with the amplicon generated by the real-time PCR primers to avoid detecting the dsRNA in the PCR reaction.

2. When designing primers containing SP6 or T7 promoter sequence, add three to four additional nucleotides at the 5′ end of the promoter.

3. In our experience, inverted repeats containing spacers 20–200 bp-long work well both in term of cloning and efficiency of RNAi.

4. To avoid problems with RNA degradation, keep all material for RNA gels separate from reagents and equipment used for DNA gels, and always wear gloves.

5. Do not use DEPC-treated H_2O to resuspend the dsRNA after precipitation because it can be toxic to the oocytes. Use nuclease-free or sterile MilliQ H_2O.

6. The dsRNA aliquots can be stored at −20°C for a few weeks, but for longer periods of time −80°C is preferable.

7. We recommend performing qRT-PCR to assess the extent of knockdown. However, protein levels must also be determined, as a highly efficient knockdown at the mRNA level is not always accompanied by a reduction in the amount of protein. This is particularly true in oocytes, where the half-life of some proteins is extremely high.

8. Milrinone, a phosphodiesterase 3 inhibitor, must be added to all collection, culture, and microinjection media used for oocytes to prevent spontaneous meiotic maturation and keep the oocytes at the GV stage. Other phosphodiesterase inhibitors, such as IBMX, or cAMP analogs (e.g., dibutyryl cAMP) can be used instead.

9. After oocyte collection, allow at least 1 h before microinjection to let oocytes recover.

10. siRNAs can be used instead of long dsRNA. They can be purchased from a variety of vendors (e.g., Applied Biosystems, Dharmacon, Sigma-Aldrich, Qiagen, Santa Cruz Biotechnology, etc). The protocol for microinjection is exactly the same. We typically use them at a concentration of 25–50 μM.

11. The use of a control dsRNA (or siRNA) is highly recommended instead of using an uninjected control or a vehicle-injected control. We typically use Gfp dsRNA (prepared using the protocol described above). For siRNAs, usually a scrambled siRNA is used as control. Alternatively, several companies sell negative control siRNAs that do not target any known gene.

12. An optional DNase treatment can be performed during RNA isolation using the Picopure kit, between W1 and W2 washes. However, in our experience this is not necessary when using TaqMan probes that span intron–exon boundaries and thus do not recognize genomic DNA.

13. An endogenous control must be used in the real-time PCR reaction to normalize the results. We typically use Ubtf (upstream binding transcription factor) or Hist2h2aa1 (histone H2a). Many other transcripts can be used as endogenous controls, e.g., Gapdh, actin, etc. We also utilize a TaqMan probe against Gfp (custom-made) to analyze the efficiency of RNA isolation and real-time PCR.

Acknowledgments

The research in RMS's laboratory was supported by a grant from the NIH (HD022681 to RMS). The research in P. Svoboda's laboratory was supported by the Czech Science Foundation (grant GACR 204/09/0085).

References

1. Fire A, Xu S, Montgomery MK, Kostas SA, Driver SE, Mello CC (1998) Potent and specific genetic interference by double-stranded RNA in Caenorhabditis elegans. Nature 391:806–811

2. Ketting RF (2011) The many faces of RNAi. Dev Cell 20:148–161

3. Filipowicz W (2005) RNAi: the nuts and bolts of the RISC machine. Cell 122:17–20

4. Lampson GP, Tytell AA, Field AK, Nemes MM, Hilleman MR (1967) Inducers of interferon and host resistance. I. Double-stranded RNA from extracts of Penicillium funiculosum. Proc Natl Acad Sci USA 58:782–789

5. Gantier MP, Williams BR (2007) The response of mammalian cells to double-stranded RNA. Cytokine Growth Factor Rev 18:363–371

6. Svoboda P, Stein P, Hayashi H, Schultz RM (2000) Selective reduction of dormant maternal mRNAs in mouse oocytes by RNA interference. Development 127:4147–4156

7. Wianny F, Zernicka-Goetz M (2000) Specific interference with gene function by double-stranded RNA in early mouse development. Nat Cell Biol 2:70–75

8. Stein P, Zeng F, Pan H, Schultz RM (2005) Absence of non-specific effects of RNA interference triggered by long double-stranded RNA in mouse oocytes. Dev Biol 286:464–471

9. Tam OH, Aravin AA, Stein P, Girard A, Murchison EP, Cheloufi S, Hodges E, Anger M, Sachidanandam R, Schultz RM, Hannon GJ (2008) Pseudogene-derived small interfering RNAs regulate gene expression in mouse oocytes. Nature 453:534–538

10. Watanabe T, Totoki Y, Toyoda A, Kaneda M, Kuramochi-Miyagawa S, Obata Y, Chiba H,

Kohara Y, Kono T, Nakano T, Surani MA, Sakaki Y, Sasaki H (2008) Endogenous siRNAs from naturally formed dsRNAs regulate transcripts in mouse oocytes. Nature 453:539–543

11. Svoboda P, Stein P, Schultz RM (2001) RNAi in mouse oocytes and preimplantation embryos: effectiveness of hairpin dsRNA. Biochem Biophys Res Commun 287:1099–1104

12. Stein P, Svoboda P, Schultz RM (2003) Transgenic RNAi in mouse oocytes: a simple and fast approach to study gene function. Dev Biol 256:187–193

13. Nagy A, Gertsenstein M, Vintersten K, Behringer R (2003) Manipulating the mouse embryo. A laboratory manual, 3rd edn. Cold Spring Harbor Laboratory Press, Cold Spring Harbor, New York, pp 310–312

14. Kennerdell JR, Carthew RW (1998) Use of dsRNA-mediated genetic interference to demonstrate that frizzled and frizzled 2 act in the wingless pathway. Cell 95:1017–1026

15. Sambrook J, Russell D (2001) Molecular cloning. A laboratory manual, 3rd edn. Cold Spring Harbor Laboratory Press, Cold Spring Harbor, New York

16. Chatot CL, Ziomek CA, Bavister BD, Lewis JL, Torres I (1989) An improved culture medium supports development of random-bred 1-cell mouse embryos in vitro. J Reprod Fertil 86:679–688

Chapter 10

Micro-injection of Morpholino Oligonucleotides for Depleting Securin in Mouse Oocytes

Petros Marangos

Abstract

Gene silencing techniques have brought new insights into mammalian oocyte and embryo development. More specifically, the use of Morpholino oligonucleotides which sterically inhibit translation from target mRNAs thereby compromising gene function, allowed the identification of important oocyte regulators and especially factors involved in meiotic cell cycle control. Here we describe the method of application of Morpholino oligonucleotides in mouse oocyte research.

Key words: Morpholino, Oocyte, RNAi, G2/Prophase arrest, Micro-injection

1. Introduction

The necessity and importance of a protein for a cell system can only be examined by the inhibition or removal of the protein. This approach can also identify the role of a protein. Methods which rely on inhibition often require some knowledge of the protein's function. In cases where the role of a protein regulator is not known or when there is more than one function, approaches which severely reduce or eliminate the protein entirely are indispensable. In the past, the removal of a regulator depended on tedious and time-consuming experimental techniques, such as the immunological removal of the protein from extracts derived from frog eggs and embryos (1) or homologous recombination techniques where the gene of interest is removed from the genome at very early stages of embryonic development (2). In mammalian oocytes, earlier approaches for interrogating gene function involved micro-injection of protein-specific antibodies into live oocytes (3). However, this approach was not very successful or reliable either because the antibodies could not function properly

Hayden A. Homer (ed.), *Mammalian Oocyte Regulation: Methods and Protocols*, Methods in Molecular Biology, vol. 957,
DOI 10.1007/978-1-62703-191-2_10, © Springer Science+Business Media, LLC 2013

within a live oocyte environment or because of their often toxic and off-target effects.

The discovery of mRNA targeting techniques, such as RNA interference (RNAi) and the use of Morpholino oligonucleotides, ushered in a new era in gene silencing in live cell systems. RNAi is a mechanism that exists in cells for physiologically regulating mRNA expression. The role of RNAi is to defend cells against viruses, to direct development, and to regulate gene expression (4). The RNAi pathway is initiated by the enzyme Dicer which identifies double-stranded RNAs (dsRNAs) in the cytoplasm and cleaves them into small RNA molecules called short-interfering RNAs (siRNAs). The siRNAs unwind into single-strand RNAs (ssRNA) and while one ssRNA becomes degraded, the other is incorporated with Dicer into the RNA-induced Silencing Complex (RISC) where it serves as a template for binding to a complementary mRNA causing its cleavage (5).

In mouse oocytes, custom made dsRNAs and siRNAs have been used to target their complementary endogenous mRNA for destruction, allowing the down-regulation of the protein otherwise expressed by the mRNA (6, 7). One disadvantage of dsRNAs is that they are not very stable. Furthermore, the large number and variety of siRNAs being produced after dsRNA cleavage could cause nonspecific mRNA targeting. A more recent mRNA targeting approach, which has proven to be very efficient, is the use of single-stranded Morpholino oligonucleotides (Gene Tools, LLC). Morpholinos (MOs) are short chains of around 25 nucleotides. Each nucleotide is chemically attached to a morpholine ring which confers stability and the formation of oligonucleotides that cannot be degraded by nucleases. The unique structure of MOs blocks its binding to RISC so that unlike dsRNA, it does not induce mRNA degradation. Consequently, an MO bound to an mRNA silences expression through steric hindrance rather than by RNA destruction (http://www.gene-tools.com). An advantage over dsRNA is that MOs are less susceptible to the possible off-target action of dsRNA. MOs are soluble and nontoxic to oocytes or embryos and have been widely used in developmental biology (8). In mouse oocytes, MOs have become a routine tool, especially for the study of cell cycle proteins, such as Mad2 (9), Cdh1 (10), Emi1 (11), Securin (12), and BubR1 (13) (see Note 1). Here we provide a detailed protocol for depleting the key cell-cycle regulator, Securin, using a previously characterized Securin-targeting MO (10, 12, 13).

2. Materials

2.1. Collection and Culture of Oocytes

1. 21–24-day-old female mice. We use the MF-1 strain.
2. Pregnant mare's serum gonadotropin (PMSG) (Intervet).
3. M2 medium, with BSA (Sigma-Aldrich).

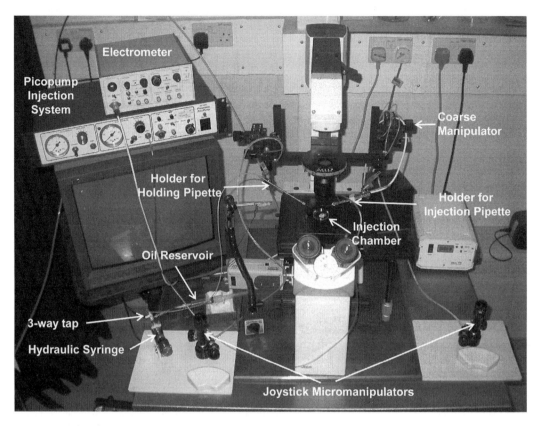

Fig. 1. Micro-injection system composed of Narishige manipulators and a Carl Zeiss Axiovert200 inverted microscope.

4. M16 medium with BSA (Sigma-Aldrich).

5. 200 μM 3-isobutyl-1-methylxanthine (IBMX) (Sigma-Aldrich). Make up 200 mM stock solution by adding 22 mg IBMX to 0.5 ml DMSO. Aliquot and store at –20°C.

6. Plastic Petri dishes, 35 mm diameter (BD Falcon).

7. 0.22 μm filters.

8. Hot-block set to 37°C.

9. Incubator set at 37°C with an ambient atmosphere of 5% CO_2.

10. 27 Gauge needle (BD).

11. Mouth pipette (Sigma-Aldrich).

12. Mineral oil (VWR Dow Corning 200/50cS Fluid).

13. Nuclease-free water (Ambion).

14. Mouse Securin MO 5′-TCA ACA AAG ATA AGA GTA GCC ATT C-3′ (Gene Tools LLC) (see Note 2).

15. Standard Control MO 5′-CCT CTT ACC TCA TTA CAA TTT ATA -3′ (Gene Tools LLC) (see Note 3).

2.2. Micro-injection (Fig. 1)

1. Inverted microscope (Zeiss) equipped with 10× and 20× objectives.

2. Hydraulic micro-manipulators (Narishige).

3. Coarse manipulators (Narishige).

4. Electrometer (Harvard Apparatus).

5. PicoPump pressure injection system (WPI).

6. Holding pipette (Hunter Scientific).

7. Hydraulic syringe for use with holding pipette.

8. Glass capillaries with an internal filament for making injection pipettes (1.5 mm outer diameter and 0.86 mm inner diameter; e.g., GC150F, Harvard Apparatus).

9. Pipette puller (P-30 Vertical Micropipette Puller Sutter Instruments).

10. Holder for the injection pipette containing a side-port for pressure pulse application and a wire for making electrical contact with the solution in the micropipette.

11. Microloader pipette tips for transferring 1–2 μl volumes of solution.

3. Methods

3.1. Mouse Oocyte Collection and Culture

1. Inject, intraperitonealy, 21–24-day MF1 female mice with PMSG (7 IU).

2. After 40–50 h, sacrifice the injected mice and remove the ovaries.

3. Prepare M2 medium containing 200 μM IBMX (M2-IBMX) and prewarm to 37°C on the hot-block. For this, filter 10 ml of M2 medium and add 1 μl of 200 mM IBMX stock solution and warm for about 1 h prior to obtaining the ovaries.

4. Prepare M16 medium containing 200 μM IBMX (M16-IBMX). For this, filter 10 ml of M16 medium and add 1 μl of 200 mM IBMX stock solution. Prewarm to 37°C and equilibrate with CO_2 by placing in the incubator for 2–4 h.

5. Prepare Petri dishes containing 40–80 μl droplets of M2-IBMX or M16-IBMX, cover with mineral oil and maintain warm using the hot-block and incubator, respectively.

6. Place the ovaries from step 2 in a Petri dish containing prewarmed M2-IBMX.

7. Release the cumulus-enclosed germinal vesicle (GV)-stage oocytes from the ovaries by puncturing the surface of the ovary with a 27-guage needle.

8. Collect the oocytes using a mouth pipette and place them in the drops of M2-IBMX (see step 5) (see Note 4).

9. Remove the surrounding cumulus cells from the oocytes by repeated pipetting and transfer them to fresh drops of M2-IBMX (see step 5).

10. Maintain oocytes on the hot-block until micro-injection.

3.2. MO Preparation and Storage

The researcher receives 300 nmol of the MO in lyophilized form.

1. Filter distilled water that will be used for dissolving the MO. Use 0.22 μm filters for complete removal of substances which could impair micro-injection.

2. Pre-heat water to 65°C in order to ensure the complete dissolution of the MO.

3. Make a stock solution of 2 mM MO by adding 150 μl of heated and filtered water to 300 nmol of MO. Ensure that the entire pellet of MO goes into solution by repeated pipetting.

4. Aliquot the MO stock solution and place at –40 or –80°C for long-term storage. It is important to avoid freeze–thaw cycles. Ideally, however, MOs should be freeze-dried for long-term storage.

3.3. Preparation of Micro-injection Rig

1. Insert the holding pipette into the holder to the left side of the inverted microscope. The holder for the holding pipette is connected to a two-way tap by stiff silicone tubing. One output of the tap leads to the hydraulic syringe, whilst the other is connected to a 20 ml syringe containing ~10 ml of mineral oil that acts as an oil reservoir.

2. Using the two-way tap, fill the tubing and holding pipette with mineral oil from the oil reservoir.

3. Prepare an injection chamber by placing a flattened drop of M2-IBMX (~0.2 ml) on to the lid of a 3-cm Falcon culture dish and cover with mineral oil.

4. Place injection chamber onto the stage of the micro-injection microscope.

5. Bring the medium at the base of the chamber into focus using the 10× objective.

6. Lower the holding pipette into the medium in the injection chamber and bring into focus at the base of the chamber using the coarse manipulator.

7. Ensure that the holding system is oil-filled by turning the screw control of the hydraulic syringe so as to expel a tiny volume of oil at the tip of the holding pipette (now immersed in the medium of the injection chamber). In order to confirm that the system is suitably responsive, turn the screw control in the opposite direction when the oil should withdraw into the holding pipette tip along with a small amount of medium.

8. Pull micro-injection pipettes by using a pipette puller and capillaries with an internal filament. The pipette tip diameter should be 0.5–2 μm. The length, from the shoulder to the pipette tip should be about 0.3–0.5 cm (see Note 6).

3.4. Micro-injection of the MO into Mouse Oocytes

1. Heat an aliquot of Securin MO at 65°C for 5 min (see Note 7).

2. Use a micro-loader to transfer approximately 1–2 μl of the Securin MO through the opening at the blunt end of the micro-injection pipette. Advance the micro-loader tip so that it is close to the pipette end before expelling the solution so that the MO solution accumulates within the pipette tip and is free of air bubbles.

3. Attach the MO-loaded injection pipette to its holder so that the wire of the holder slides into the micro-pipette through the back end and tighten the holder. The wire should go most of the way down the inside of the micro-pipette so that the circuit can be completed in preparation for micro-injection (see below). The injection pipette holder should be connected to the output of the PicoPump via stiff silicone tubing and have an electrical connection to the electrometer.

4. Transfer the GV-stage oocytes to the medium within the injection chamber on the microscope stage alongside the tip of the holding pipette.

5. Lower the injection pipette so that the tip dips into the medium and is adjacent to the group of oocytes.

6. Dip the silver wire that is electrically connected with the ground of the electrometer into the medium containing the eggs.

7. Switch the electrometer on when a light will indicate whether an electrical circuit has successfully been made.

8. Move the holding pipette using the fine micromanipulator so that its tip abuts on the first oocyte to be micro-injected.

9. Immobilize the oocyte by applying suction to the holding pipette by turning the screw control of the hydraulic syringe.

10. By using the fine micro-manipulator, advance the tip of the micro-injection pipette through the zona pellucida and against the plasma membrane.

11. Apply a brief overcompensation of negative capacitance by pressing the "buzz" or "zap" button on the electrometer thereby creating a very fine amplitude oscillation that enables the pipette tip to breach the membrane.

12. Inject a small volume of MO solution into the oocyte by applying a brief pressure pulse to the solution in the injection pipette via the PicoPump. Inject a volume equivalent to ~5% of the

oocyte volume estimated from the degree of cytoplasmic displacement (see Note 8).

13. Move the injected oocyte to a clear region of the micro-injection chamber using the holding pipette.

14. Repeat steps 8–13 for each oocyte to be injected making sure to keep the injected oocytes in a separated group from uninjected ones until all injections have been completed.

15. Upon completion of micro-injection, transfer the oocytes to a dish containing fresh drops of M2-IBMX under oil kept warm on the hot-block.

16. Inject a separate group of eggs with a Control MO to act as a negative control.

17. After allowing the oocytes about 30 min to 1 h to recover in M2-IBMX postinjection, transfer them to droplets of M16-IBMX (see step 5 in Subheading 3.1) for longer-term culture (from 24 to 48 h) in the CO_2 incubator to allow time for protein knock-down.

Since *Securin* mRNA is not destroyed following MO treatment, the efficacy of Securin MO cannot be evaluated by measuring mRNA levels but instead requires a quantitative assessment of the levels of the target protein, which in this case is Securin. Evaluation of protein levels can be accomplished through Western blotting and/or Immunofluorescence approaches (see Note 9). For a recent detailed description of immunoblotting of mouse oocytes for Securin, see ref. (14).

4. Notes

1. In somatic cells, following RNAi, complete protein loss is obtained in the generations of cells originating from the initially treated cells. The effectiveness of RNAi approaches involves the culture of treated cells for 3–5 days. Mouse oocytes, however, do not cycle in culture and cannot be cultured at the G2/Prophase-arrested stage, *in vitro*, for more than 24–48 h. Therefore, a protein with known high stability which cannot be degraded within 48 h should not be chosen. In contrast, a target protein of known high turn-over rate is a perfect substrate because the MO will block protein synthesis without affecting rapid protein degradation. APC ligase substrates with significant turn-over, such as Securin (12), Cdh1 (10), or BubR1 (13), have been ideal MO targets, as shown by the complete or near-complete disappearance of the protein within 20 h. Proteins that are at low expression levels at the cell cycle stage of MO introduction into the oocyte (most often

the GV-stage) such as Mad2 (9) are perfect MO targets as well. In this case, the MO will prevent protein synthesis immediately after MO-mRNA binding and the protein will remain in low levels in the following cell cycle stages.

2. An mRNA target should be chosen if it fulfills the guidelines for MO design. MOs can only be purchased by Gene Tools who can also design the MOs at the researcher's request. It is recommended to supply Gene Tools with two or three sequences within the target mRNA for designing the MO and then choose the best designed MO (http://www.gene-tools.com). It is imperative to check that the MO has the following characteristics: (a) 25 nucleotides is the length of maximum efficacy since MOs of less than 20–25 nucleotides of length do not bind efficiently to the target mRNA. (b) MOs are more efficient when the binding site on the target sequence is located within the region extending from the 5'cap to about 25 bases 3' to the AUG translational start site. (c) The selected sequence should have little or no self-complementarity. Self-complementarity can cause intrastrand or interstrand pairing which prevent MOs from binding to the mRNA target. The MO should not form more than 6 continuous base pairs of self-complementarity and 16 continuous intrastrand hydrogen bonds (the number of hydrogen bonds depends on the G and C content). (d) MO GC content should be 40–60% in order to allow strong affinity to the target mRNA. (e) Water solubility is impaired when the MO contains stretches of more than three continuous G or more than 35% of G in total. (f) Either the target mRNA or the nucleotide sequences which are complementary to the MO should be checked for sequence homology with nontarget RNAs through the use of services such as BLAST. It is very important to verify that the MO cannot bind to off-target RNAs causing nonspecific effects and phenotypes.

3. Gene Tools provides standard control MOs. Researchers may also prefer to use, as a negative control MO, the invert of the original target-specific MO. Five mispair oligos are used for specificity controls. In this case, five mispairs should be distributed along the original MO sequence.

4. IBMX is a phosphodiesterase inhibitor which causes the increase of intracellular cyclic AMP levels. This allows the GV-stage oocytes to remain arrested at G2/Prophase (15).

5. Cumulus cells need to be removed in order to distinguish clearly the G2/Prophase arrested oocytes. Furthermore, cumulus cells hamper micro-injection by obscuring the view of the oocyte and by blocking the injection pipettes.

6. The ideal pipette tip size varies depending on the materials being injected and their concentration. A large pipette tip could cause damage to the egg and compromise egg survival

following micro-injection, while a small pipette tip could lead to loading difficulties, or become blocked during micro-injection.

7. Heating allows the annealing of any possible secondary structures or interstrand pairing. It also allows the recovery of MO activity lost during freezing (http://www.gene-tools.com).

8. Different concentrations of Securin MO should be used in order to identify the amount of MO necessary to support mRNA silencing without causing nonspecific effects.

9. MO efficacy detection involves the use of antibodies for Immunofluorescence and Western blotting experiments. Ideally, well-characterized primary antibodies recognizing the target protein should, already, be in existence. Otherwise, the researcher should take into account, before ordering an MO, the fact that custom antibodies would have to be produced. If MO micro-injection leads to complete or almost complete disappearance of the protein target, the MO is efficient. However, this is not necessarily proof that any observed phenotypes are related to target protein loss rather than nonspecific or toxic effects of the MO. For determining that a phenotype is a direct result of the target protein being destroyed, an approach often used in oocytes is to express the target protein from an exogenous mRNA in MO-injected oocytes. If the physiological phenotype is rescued by the exogenous mRNA this is strong proof of MO specificity. For the rescue experiment, the MO should be co-injected with the in vitro synthesized mRNA corresponding to the target protein. It is preferred that a non-mouse mRNA is used to avoid MO binding to the exogenous mRNA. Another approach is to use a different MO sequence which targets the same mRNA but at a different site from the original one. If both sequences give identical phenotypes this is further proof of MO specificity.

If there is no change in the oocyte's function or the protein levels, the MO is not effective or the protein is too stable to be degraded. In this case, unfortunately, there is no evidence that the MO blocks the target mRNA efficiently. A second MO or alternative approaches of gene silencing such as dsRNA would need to be used in order to determine whether the experimental project can be pursued further.

References

1. Maller JL (1985) Regulation of amphibian oocyte maturation. Cell Differ 16:211–221

2. Recillas-Targa F (2006) Multiple strategies for gene transfer, expression, knockdown, and chromatin influence in mammalian cell lines and transgenic animals. Mol Biotechnol 34:337–354

3. Simerly C, Balczon R, Brinkley BR, Schatten G (1990) Microinjected centromere [corrected] kinetochore antibodies interfere with chromosome movement in meiotic and mitotic mouse oocytes. J Cell Biol 111:1491–1504

4. Dykxhoorn DM, Lieberman J (2005) The silent revolution: RNA interference as basic

biology, research tool, and therapeutic. Annu Rev Med 56:401–423

5. Tijsterman M, Plasterk RH (2004) Dicers at RISC; the mechanism of RNAi. Cell 117:1–3

6. Svoboda P, Stein P, Hayashi H, Schultz RM (2000) Selective reduction of dormant maternal mRNAs in mouse oocytes by RNA interference. Development 127:4147–4156

7. Wianny F, Zernicka-Goetz M (2000) Specific interference with gene function by double-stranded RNA in early mouse development. Nat Cell Biol 2:70–75

8. Corey DR, Abrams JM (2001) Morpholino antisense oligonucleotides: tools for investigating vertebrate development. Genome Biol 2: REVIEWS1015

9. Homer HA, McDougall A, Levasseur M, Murdoch AP, Herbert M (2005) Mad2 is required for inhibiting securin and cyclin B degradation following spindle depolymerisation in meiosis I mouse oocytes. Reproduction 130:829–843

10. Reis A, Chang HY, Levasseur M, Jones KT (2006) APCcdh1 activity in mouse oocytes prevents entry into the first meiotic division. Nat Cell Biol 8:539–540

11. Marangos P, Verschuren EW, Chen R, Jackson PK, Carroll J (2007) Prophase I arrest and progression to metaphase I in mouse oocytes are controlled by Emi1-dependent regulation of APC(Cdh1). J Cell Biol 176:65–75

12. Marangos P, Carroll J (2008) Securin regulates entry into M-phase by modulating the stability of cyclin B. Nat Cell Biol 10:445–451

13. Homer H, Gui L, Carroll J (2009) A spindle assembly checkpoint protein functions in prophase I arrest and prometaphase progression. Science 326:991–994

14. Homer H (2011) Evaluating spindle assembly checkpoint competence in mouse oocytes using immunoblotting. Methods Mol Biol 782:33–45

15. Tsafriri A, Chun SY, Zhang R, Hsueh AJ, Conti M (1996) Oocyte maturation involves compartmentalization and opposing changes of cAMP levels in follicular somatic and germ cells: studies using selective phosphodiesterase inhibitors. Dev Biol 178(2):393–402

Chapter 11

Measuring Transport and Accumulation of Radiolabeled Substrates in Oocytes and Embryos

Jay M. Baltz, Hannah E. Corbett, and Samantha Richard

Abstract

Radiolabeled compounds that are substrates for transmembrane transporters can be used to study transport and metabolism in mammalian oocytes and preimplantation embryos. Because even very small amounts of radioisotopes can be detected, these techniques are feasible to use with only a few oocytes or embryos, even down to the level of single oocytes or embryos. Here, we describe the methods for determining the transport and accumulation of radiolabeled compounds into oocytes and preimplantation embryos and the determination of the rate of saturable transport via specific transporters in the plasma membrane.

Key words: Embryo, Kinetics, Oocyte, Preimplantation, Radioisotope, Radiolabeled, Transport

1. Introduction

Methods for investigating physiological processes in mammalian oocytes and preimplantation embryos can be employed routinely only if they can feasibly be used with small groups of oocytes or embryos, due to the low numbers of oocytes or embryos that can be practically obtained. Detection methods for radiolabeled compounds are extremely sensitive, and thus have been used successfully in studies of mammalian oocytes and preimplantation embryos, principally for elucidating their mechanisms of transmembrane transport of a wide array of substrates (1–15). Thus, they can be used for determining the presence of specific transporters, elucidating the identity of transporters that carry a given substrate in oocytes or preimplantation embryos, and measuring their kinetic parameters (16). Depending on the specific activity of the radiolabeled compound and the rate at which it is taken up into the cells, measurements of transport kinetics and characteristics can be carried out using very small groups of oocytes or embryos or even

Hayden A. Homer (ed.), *Mammalian Oocyte Regulation: Methods and Protocols*, Methods in Molecular Biology, vol. 957, DOI 10.1007/978-1-62703-191-2_11, © Springer Science+Business Media, LLC 2013

single embryos (17). The main transport mechanisms investigated have been those for amino acids and various metabolic substrates (18), as well as a few other types of transport, e.g., by inorganic ion transporters (1). The basic paradigm consists of incubating oocytes or embryos with the radiolabeled compound for a precise period, washing them completely free of any significant external labeled compound, and then measuring the intracellular amount of the compound using scintillation counting. This protocol assumes that the techniques for obtaining and handling mammalian oocytes and embryos are already known and concentrates only on those aspects specific to using radiolabeled substances to measure their transport into mammalian oocytes or embryos.

2. Materials

2.1. Media Stock Solutions

For mouse oocytes and embryos, modified KSOM (mKSOM) and modified Hepes-KSOM (Hepes-mKSOM) media are used (see Note 1), based on KSOM mouse embryo culture medium (19). Stock solutions should be made with sterile embryo-grade water that is either produced and validated for embryo culture in-house or purchased as embryo-tested. Embryo-tested chemicals are used (or, if not available, cell culture grade). Except for stocks 1–4, which must be made fresh (19), the remainder can be stored for up to 1 month under the conditions specified. Larger or smaller volumes may be made of each stored stock, depending on usage (the 100 ml mKSOM preparation described here in item 1 in Subheading 2.2 consumes ~1–3.5% of each stored stock).

1. 1.0 M Na lactate: add 1.87 g Na lactate (60% syrup) to a 15 ml culture tube and add water to 10 ml total. Mix well. Filter-sterilize using a sterile 10 ml syringe and syringe filter into a sterile 15 ml culture tube. Make fresh each time media are made.

2. 1.0 M $NaHCO_3$: add 0.84 g $NaHCO_3$ to a 15 ml culture tube and add water to 10 ml total. Mix until dissolved. Filter-sterilize using a sterile 10 ml syringe and syringe filter into a sterile 15 ml culture tube. Make fresh each time media are made.

3. 100 mM Na pyruvate: add 0.11 g Na pyruvate to a 15 ml culture tube and add water to 10 ml total. Mix until dissolved. Filter-sterilize using a sterile 10 ml syringe and syringe filter into a sterile 15 ml culture tube. Make fresh each time media are made.

4. 100 mM glucose: add 0.18 g glucose to a 15 ml culture tube and add water to 10 ml total. Mix until dissolved. Filter-sterilize using a sterile 10 ml syringe and syringe filter into a sterile 15 ml culture tube. Make fresh each time media are made.

5. 5.0 M NaCl: add 58.4 g NaCl to 150 ml water in a 250 ml sterile plastic container. Heat in microwave until warmed. Let some of the salt dissolve, leaving the solution until cooled. Remove 100 ml of the liquid (measured with volumetric flask). The container will now have ~50 ml of liquid plus the remaining undissolved salt. Add water up to ~90 ml, and completely dissolve the remaining salt. Transfer to volumetric flask and add water up to 100 ml. Combine both flasks (200 ml total) into a 250 ml plastic filter unit and filter-sterilize. The stock may be stored for up to 1 month at 4°C.

6. 1.0 M KCl: add 1.49 g KCl to a graduated cylinder and add water to 20 ml total. Mix until dissolved. Filter-sterilize using a sterile 30 ml syringe and syringe filter into a sterile 50 ml culture tube. The stock may be stored for up to 1 month at 4°C.

7. 100 mM KH_2PO_4: add 0.272 g to a graduated cylinder and add water to 20 ml total. Mix until dissolved. Filter-sterilize using a sterile 30 ml syringe and syringe filter into a sterile 50 ml culture tube. The stock may be stored for up to 1 month at 4°C.

8. 100 mM $MgSO_4$: add 0.247 g $MgSO_4 \cdot 7H_2O$ to graduated cylinder and add water to 10 ml total. Mix until dissolved. Filter using a sterile 20 ml syringe and syringe filter into a sterile 15 ml culture tube. The stock may be stored for up to 1 month at 4°C.

9. 10 mM EDTA: add 0.038 g tetrasodium EDTA to graduated cylinder and add water to 10 ml total. Mix until dissolved. Filter-sterilize using a sterile 20 ml syringe and syringe filter into a sterile 15 ml culture tube. The stock may be stored for up to 1 month at 4°C.

10. 30 mM Penicillin G: add 0.224 g penicillin G (potassium salt) to a graduated cylinder and add water to 20 ml total. Mix until dissolved. Filter-sterilize using a sterile 30 ml syringe and syringe filter into a sterile 50 ml culture tube. The stock may be stored for up to 1 month at –20°C, if aliquoted into smaller sterile tubes (completely thaw and mix before use and do not refreeze).

11. 10 mM Streptomycin: Add 0.291 g Streptomycin SO_4 to a graduated cylinder and add water to 20 ml total. Mix until dissolved. Filter-sterilize using a sterile 30 ml syringe and syringe filter into sterile 50 ml culture tube. The stock may be stored for up to 1 month at –20°C, if aliquoted into smaller sterile tubes (do not refreeze).

12. 100 mM $CaCl_2$: add 1.47 g $CaCl_2 \cdot 2H_2O$ to a graduated cylinder and add water to 100 ml total (or use sterile plastic container to dissolve, then measure with volumetric flask). Mix until dissolved. Filter-sterilize using a 150 ml plastic filter unit. The stock may be stored for up to 1 month at 4°C.

13. 1.0 M Hepes: add 4.77 g Hepes to a graduated cylinder and add water to 20 ml total. Mix until dissolved. Filter-sterilize using a sterile 30 ml syringe and syringe filter into a sterile 50 ml culture tube. The stock may be stored for up to 1 month at 4°C. A greater volume of this stock (e.g., 100 ml) should be prepared if Hepes-KSOM will be used frequently or also used for oocyte and embryo isolation.

2.2. Embryo and Oocyte Media

1. mKSOM: Media should be made using appropriate sterile techniques. Place about 50 ml of embryo-grade water in a 150 ml sterile plastic container (filter receptacle). Add the indicated volumes of the specified stocks: 1.9 ml NaCl, 0.25 ml KCl, 0.35 ml KH_2PO_4, 0.2 ml $MgSO_4$, 1.0 ml Na lactate, 0.2 ml glucose, 0.2 ml Na pyruvate, 2.5 ml $NaHCO_3$, 0.1 ml EDTA, 0.53 ml penicillin, and 0.30 ml streptomycin. Mix well. Add 1.7 ml $CaCl_2$ stock and mix (see Note 2). Transfer to a graduated cylinder and add embryo-grade water to bring to 100 ml total volume. Mix and transfer back to the plastic container. Add 100 mg polyvinyl alcohol (PVA, 30–70 kDa average molecular weight). Mix well using a magnetic stir bar for approximately 1 h until the PVA is completely dissolved and then filter-sterilize (see Note 3) using a 150 ml filter unit. The medium can be stored at 4°C for up to 2 weeks.

2. Concentrated mKSOM (70mKSOM; see Note 4): Prepare 70mKSOM as for mKSOM, but after transferring to the graduated cylinder, add embryo-grade water to only 70 ml total volume. Add 100 mg PVA. Follow the protocol in item 1 to dissolve PVA and filter-sterilize. Store as specified for mKSOM.

3. Hepes-mKSOM: Add the water and stocks to a container exactly as for mKSOM in item 1, except using only 0.4 ml $NaHCO_3$ stock and adding 2.1 ml Hepes stock. Follow the protocol for mKSOM through the addition of PVA. After the PVA is dissolved, continue to stir moderately and adjust pH to about 7.4 at room temperature (see Note 5) by adding concentrated NaOH dropwise while stirring and monitoring pH with a pH electrode. Continue stirring until pH is stabilized at about 7.5. Filter-sterilize and store as for mKSOM.

4. 100 mM unlabeled compound stock solution (see Note 6): Make approximately 10 ml of a 100 mM solution of the same compound whose transport is to be measured by dissolving it in embryo-grade water. Filter-sterilize using a syringe filter into a sterile 15 ml plastic culture tube. The solution can be aliquoted into sterile ultracentrifuge tubes and frozen at –20°C until needed. It should be made fresh, however, if the compound should not be frozen.

5. Oil for overlay: Filter-sterilize embryo-tested mineral oil in a 500 ml filter unit. Because of the high viscosity of oil, this will

take a considerable amount of time, and preparing the oil should be done well ahead of time and the oil stored. Add a small amount of phenol red to about 10 ml of mKSOM (the amount of powder added should be just enough to turn the mKSOM a deep red) and filter-sterilize using a sterile 20 ml syringe with a syringe filter into a sterile 15 ml culture tube. Filter the 10 ml of KSOM with phenol red a second time using a new syringe, filter, and culture tube directly into the filtered oil under sterile conditions (see Note 7). Cap and mix by gently inverting three to five times and let it sit overnight for the mKSOM to settle to the bottom before using. Filtered oil is stored at room temperature and should be used within a month of filtration.

6. Filter-sterilized embryo-grade water: Just before each experiment, filter-sterilize about 5–10 ml of embryo-grade water using a syringe filter into a sterile 15 ml sterile plastic culture tube. Keep capped at room temperature until used.

2.3. Dishes, Tubes, and Oocyte or Embryo Handling

1. Embryo culture dishes: 35×10 mm diameter plastic sterile culture dishes validated for embryo culture.

2. Medium plastic culture dishes: 60×15 mm diameter sterile culture dishes validated for embryo culture.

3. Large plastic Petri dish lids: lids from 100×15 mm sterile polystyrene Petri dishes (the dishes themselves can also be used if the increased height of the rim is not a problem for the user).

4. 1.5 ml microcentrifuge tubes.

5. Flame-pulled Pasteur pipets for embryo handling: Over a small alcohol, natural gas, or butane flame, melt and pull glass Pasteur pipets to form a uniformly thinned portion of the terminal section. Break off to provide a tip that is flat and perpendicular to the long axis and has an inner diameter slightly larger than the diameter of an oocyte or embryo. The thinned section should be approximately 4 cm in length and have little taper (i.e., it should ideally be a uniform cylinder with inner diameter slightly larger than the oocyte or embryo), so that embryos can be transferred with as small an amount of liquid as possible. When used, a pulled pipet should be attached to a flexible synthetic rubber or similar tube through which suction can be controlled either by mouth or by a hand-held controller (if mouth-operated pipetting is used, a trap must be inserted in-line so that droplets of the solution with radioisotopes can never be communicated to the operator). A number of pipets should be pulled at once and stored, since multiple pulled pipets will be needed for each experiment. Below, Pasteur pipet always refers to a pulled pipet.

2.4. Scintillation Counting

1. Vials: 7 ml polyethylene scintillation vials with screw caps.
2. Scintillation fluid: 4 ml per vial of scintillation cocktail (e.g., Fisher Scintiverse).
3. Liquid scintillation counter.

---

3. Methods

3.1. Standard Curve

1. A standard curve for conversion to molar amount of radiolabeled compound (see Note 8) should be constructed within the same week that scintillation counting is performed on the oocyte or embryo samples. The same reagents, particularly the batch of radiolabeled compound stock, that are used in the measurements with oocytes or embryos must be used for the standard curve. If a radioisotope with short half-life is used, the calibration curve should be constructed on the same day that the oocyte or embryo samples are processed for measurements.

2. Prepare two 15 ml culture tubes, one with 5.0 ml and the other with 7.5 ml water, using the filtered water that is used to make the media.

3. Add 1 μl of the radiolabeled compound stock supplied by the manufacturer to each tube to produce 1:5,000 and 1:7,500 dilution stocks. Cap and mix by vortexing.

4. Perform five serial twofold dilutions of each stock. The use of two independent offset dilutions reduces the chance that an error will go undetected. For each stock, prepare five 1.5 ml microcentrifuge tubes each containing 0.5 ml water. Using a micropipettor with appropriate tips, transfer 0.5 ml of the stock to the first tube, cap, and vortex. Transfer 0.5 ml from that tube into the next tube, and repeat, until serial dilutions are completed for all five tubes, for both stocks. Pipet up and down several times in the tube for each transfer to ensure that the pipet tip is emptied. Use a new pipet tip for each transfer. This will produce dilutions (relative to the initial manufacturer's stock) of 1:5,000, 1:7,500, 1:10,000, 1:15,000, 1:20,000, 1:30,000, 1:40,000, 1:60,000, 1:80,000, 1:120,000, 1:160,000, and 1:240,00 (see Note 9).

5. Place 5 μl of each of the 12 dilutions into a separate scintillation vial and add 4 ml of scintillation fluid. In a 13th tube, place 5 μl of the water used for dilution (for background). Cap and mix each vial by vortexing. Use a new pipet tip for each.

6. Place tubes in a scintillation counter and process for measurements according to the manufacturer's directions using settings appropriate for the radioisotope used. The output is usually expressed in counts per minute (CPM).

7. Subtract the value measured for the background (in the vial containing only water) from each of the other measurements.

8. Plot the calibration curve and fit a line by linear least squares regression. The points should not deviate from linear, all points should be close to the line, and the regression should yield an r^2 value of at least 0.98 (>0.99 is the usual value) and pass very close to the origin. The parameters (slope, intercept) of the regression are used in a linear equation to convert the measured radioactivity to molar concentration of radiolabeled compound (see Note 10).

3.2. Preparation of Wash and Incubation Drops

1. Calculate the volume (V_R) of the radiolabeled compound stock solution (supplied by the manufacturer) that would need to be added to 1 ml medium to produce the desired final concentration of radiolabeled compound (see Note 11). The concentration of the stock radiolabeled compound will be specified by the manufacturer and differs for each lot (see Note 12). For most compounds, the final working concentration will be on the order of 1 μM (see Note 13).

2. Prepare medium to which the radiolabeled compound will be added, starting with a microcentrifuge tube containing 700 μl of 70mKSOM (concentrated mKSOM). Add filtered embryo-grade water to a total volume that equals 1 ml minus V_R. This will be used to make the incubation medium used to determine the amount of radiolabeled compound taken up by the oocytes or embryos.

3. Prepare a second tube containing 700 μl of 70mKSOM. Add 50 μl of the 100 mM unlabeled compound stock solution. Add filtered embryo-grade water to a total volume that equals 1 ml minus V_R. This will be used to make the incubation medium to determine the nonspecific (i.e., non-saturable) uptake of radiolabeled compound in the presence of a large excess (5 mM) of unlabeled compound (see Note 14).

4. Prepare wash dishes by placing three 50 μl drops of mKSOM separately onto a small plastic culture dish. Cover the drops with oil using a plastic disposable pipet, so that the oil surface is approximately 3 mm above the tops of the drops of medium. Cover dishes and place them in an incubator at 37°C with an appropriate gas mixture validated for embryo culture (e.g., 5%CO_2, since mKSOM is bicarbonate-buffered). Equilibrate for at least 2 h (up to overnight). One wash dish will be needed for each separate drop in which the oocytes or embryos will be incubated with radiolabeled compound.

5. Prepare each dish in which embryos are to be incubated with the radiolabeled compound by placing one drop of the medium prepared in step 2 onto a small plastic culture dish.

The drop should have a volume of 50 µl minus the volume of radiolabeled substance to be added ($=V_R/20$) (see Notes 15 and 16). Only one drop should be placed in each dish, and as many separate dishes prepared as will be needed. Cover the drop with oil using a plastic disposable pipet, so that the oil surface is approximately 3 mm above the top of the drop of medium. Cover the dishes, mark them for identification, and place them in an incubator at 37°C with an appropriate gas mixture validated for embryo culture. Equilibrate for at least 2 h (up to overnight).

6. If the nonspecific transport is to be assessed (see Note 14), prepare drops using the medium made in step 3 using the same procedure as in step 5.

3.3. Incubation of Oocytes or Embryos with the Radiolabeled Compound

1. Place a 15 or 50 ml plastic culture tube with an appropriate amount of Hepes-mKSOM into ice at least 15 min before starting and set it aside. Approximately 2.5 ml ice-cold medium will be needed per drop of radiolabeled compound into which embryos will be placed.

2. Place three large drops of room temperature Hepes-mKSOM onto a medium culture dish, just before starting with each group of oocytes or embryos. These will be used as preliminary wash drops.

3. Using a flame-pulled Pasteur pipet, transfer a group of oocytes or embryos (see Note 17) through the three Hepes-mKSOM preliminary wash drops sequentially, moving them around in each drop to wash them free of any substances remaining from their isolation from the female tract. The oocytes or embryos should not be drawn into the Pasteur pipet beyond the tip region, so that as little liquid as possible is carried from one drop to the next. After placing the oocytes or embryos in a drop, expel any liquid remaining in the pipet from the previous drop before continuing to the next drop.

4. Remove an equilibrated mKSOM wash dish from the incubator immediately before use. Using the same technique as in the previous step, wash the oocytes or embryos sequentially through the three drops of equilibrated mKSOM in the wash dish under oil.

5. Expel any remaining medium from the Pasteur pipet.

6. Remove one of the equilibrated dishes with radiolabeled compound (prepared in step 5 or 6 in Subheading 3.2) from the incubator. Use the Pasteur pipet to transfer the washed group of oocytes or embryos into the 50 µl drop containing radiolabeled compound. As little medium as possible should be transferred with the embryos to avoid diluting the radiolabeled compound. Ensure that the oocytes or embryos are mixed into

the drop (using the pipet to move them around) so that they are immediately exposed to the radiolabeled compound. As soon as the oocytes or embryos are in the drop with the radiolabeled compound, make a note of the time or set a timer. Replace the dish into the incubator.

7. Using a new pulled Pasteur pipet each time, repeat through step 6 for each equilibrated 50 μl drop with radiolabeled compound (see Note 18).

8. Incubate the oocytes for precisely the fixed time chosen for the incubation period (see Note 19).

9. Within approximately the last minute of the incubation period, prepare a small plastic culture dish with approximately 2 ml of ice-cold Hepes-mKSOM (from the tube kept on ice) and also 5 wash drops of ice-cold Hepes-mKSOM on the lid of a large plastic Petri dish (see Note 20).

10. At the end of the culture period, immediately transfer the oocytes or embryos into the dish with 2 ml ice-cold Hepes-mKSOM. Using the Pasteur pipet with which they were transferred, mix the oocytes or embryos around in the large volume of medium to dilute the radiolabeled compound as much as possible.

11. Transfer the embryos sequentially through the five drops of cold medium on the Petri dish lid, changing to a new Pasteur pipet between each drop in order to dilute out as much of the external radiolabeled compound as possible by the final drop.

12. Using a new Pasteur pipet, transfer the embryos from the last wash drop into the bottom of an embryo scintillation tube. Add 4 ml of scintillation fluid.

13. Using the same pipet, transfer a similar amount of medium from the final wash drop that had contained the embryos into another scintillation tube, and add 4 ml of scintillation fluid. This will be used to determine the background for the oocytes or embryos that came from that wash drop.

3.4. Determining the Amount of Radiolabeled Compound Taken up by Oocytes or Embryos

1. Transfer the scintillation tubes to a liquid scintillation counter and process for measurements according to the manufacturer's directions using settings appropriate for the radioisotope used. The output is usually expressed in CPM (see Note 21).

2. For each measurement of radiolabeled compound in a sample with oocytes or embryos, subtract the background determined by the measurement of its paired wash drop sample. This yields the net amount of radioactivity that is associated specifically with the oocytes or embryos (see Note 22).

3. Convert the measurement of radiolabeled compound after background subtraction (e.g., in CPM) to molar amount of radiolabeled compound using the standard curve (see Note 23).

Units of femtomoles or picomoles are usually appropriate. Divide this quantity by the number of oocytes or embryos in the tube to yield the molar amount per oocyte or embryo.

4. If the rate of transport is to be calculated, divide the molar amount per oocyte or embryo by the incubation period (yielding, for example, fmoles or picomoles per embryo or oocyte per min; see Note 24).

5. Dispose of the samples in scintillation fluid according to acceptable procedures taking into account that they are not only radioactive waste but likely also classified as hazardous waste due to the scintillation fluid.

4. Notes

1. This protocol is written assuming that mouse oocytes and embryos are being used, and thus details are given for mouse embryo culture media. For other species, appropriate media can be substituted. In mKSOM and Hepes-mKSOM, glutamine has been omitted and bovine serum albumin (BSA) was replaced by PVA. Glutamine is omitted since it is transported by embryos and can therefore interfere with results if amino acid transport is being assessed. PVA is used as the macromolecular component of the media to prevent embryos sticking to culture dishes and during manipulation. It is used in place of BSA, since BSA may contain impurities that could affect results. An important feature of media used during the incubation of oocytes or embryos with radiolabeled substrates is that any component of the medium that could interfere with the transporter or other processes being studied is omitted. Only completely defined media without the addition of undefined components such as serum should be used. Specialized media (e.g., omitting a single inorganic ion) needed for studying transport physiology can be produced by substituting appropriate stock solutions (e.g., choline chloride instead of NaCl, etc.).

2. In order to prevent the precipitation of calcium phosphate or other insoluble calcium compounds, it is standard practice in making media that the $CaCl_2$ stock is added with stirring after most of the water has been added.

3. Prior to filter-sterilizing mKSOM and Hepes-mKSOM, the osmolarity should be checked if desired. mKSOM and Hepes-mKSOM should have an osmolarity of about 250 mOsM and 240 mOsM, respectively.

4. By using concentrated 70mKSOM, which is so designated because it is concentrated mKSOM made by adding water only

to 70% of its normal final volume, it is possible to make additions to culture media without diluting mKSOM. 70mKSOM is used to allow the radiolabeled compound and any inhibitors or excess amino acids to be added from stock solutions to the medium. After these additions, the volume will be adjusted with water to produce normal-strength medium, as specified later in the protocol.

5. The pH is adjusted to about 7.4 at room temperature if the Hepes-KSOM is to be used at room temperature. If used at 37°C, pH at room temperature is adjusted to 7.5 to compensate for the temperature-dependence of Hepes buffering and thus produce a final pH of about 7.3–7.4 at 37°C.

6. This solution is needed if the amount of saturable transport vs. total transport is to be assessed. A large excess of the unlabeled compound is then present in some of the incubation drops to act as a competitive inhibitor and determine the nonspecific, non-saturable residual amount of uptake of the radiolabeled compound. This solution can be omitted if not needed. If the unlabeled compound is not soluble at 100 mM, the protocol should be adjusted to accommodate adding a larger volume of a more dilute stock.

7. mKSOM containing phenol red is added to mineral oil in order to equilibrate the oil with medium (so it will not draw water out of the microdrops in culture) and to provide a warning of oil contamination. If the oil is contaminated by microorganisms, the phenol red will turn yellow and should be discarded. Oil should always be removed from the stock with a sterile pipet using sterile precautions, avoiding touching the mKSOM at the bottom. Each new batch of mineral oil should be tested by culturing 1-cell embryos to the blastocyst stage. The double filtering of the mKSOM is carried out to reduce the chance of contamination during the long-term room temperature storage.

8. Appropriately radiolabeled compounds can be obtained from any of several sources, such as Perkin-Elmer New England Nuclear, American Radiolabeled Chemicals, Movarek Biochemicals and Radiochemicals, or MP Biomedicals. The concentration, specific activity, and solvent in which the compound is supplied will differ for different chemical compounds and manufacturers, and specific activity will also be different for each batch. This will have to be taken into account when designing experiments. Commonly, organic compounds are radiolabeled with ^{3}H or [14]C. For the same compound, ^{3}H-labeled compounds have approximately 1,000-fold higher specific activity than a ^{14}C-labeled analog, and the former would thus be highly preferred. If only low specific activity compounds are available, they may still be usable if the

concentration used can be increased sufficiently. Detection can also be improved by increasing the number of oocytes or embryos used per measurement and the time of incubation with the radiolabeled compound. Among other commonly used radioisotopes are ^{35}S and ^{32}P which both have very high specific activities and are easily detected.

9. The starting dilutions may need to be adjusted depending on the specific activity of the radiolabeled compound and the rate at which it is taken up into the oocytes or embryos. The range of dilutions should completely span the amounts of radiolabeled compounds that will be measured in oocytes or embryos.

10. Typically, the stock of radiolabeled compound obtained from the manufacturer will specify the specific activity (usually in Curies or milliCuries per millimole: Ci/mmole or mCi/mmole), and the activity concentration in Ci/ml or mCi/ml. The molar concentration of radiolabeled compound in the stock is thus obtained by dividing the activity concentration by the specific activity (e.g., Ci/ml divided by Ci/mmole, yielding mmoles/ml or M).

11. The amount of radiolabeled compound stock that is added to the 700 μl of 70mKSOM is that which will produce the desired final concentration of radiolabeled compound in 1 ml of final volume. The amount of radiolabeled stock added (V_R) should be much less than 300 μl, to accommodate adding additional substances such as inhibitors from stocks where necessary. After all additions are completed, the final volume is brought to 1 ml in each case using the appropriate amount of water.

12. If the molar concentration in the supplied stock is found to be too low (i.e., the volume that would need to be added is not much less than 300 μl), it can often be concentrated by partial evaporation or by evaporating to dryness and then resuspending in a smaller volume. If the latter, a test should be done to determine that close to 100% of the compound is actually resuspended after being dried (some types of tubes exhibit more irreversible binding for some compounds than others, which has to be determined empirically). Also, if the radiolabeled compound stock is supplied in ethanol or an ethanol solution rather than water, it must be ensured that the final concentration of ethanol will not exceed about 0.2% in the final medium. If this is not the case, the stock solution can possibly be concentrated sufficiently by evaporation of the solvent. If that is not possible or desirable, employ the alternate procedure described in Note 16, below. Similar considerations would apply if the compound is supplied in other solvents that could affect oocyte or embryo viability.

13. For many studies in which the transport of a substance is to be investigated, the final concentration of radiolabeled compound

will be on the order of 1 μM (3, 15). However, in some cases, it will need to be varied. For example, if experiments are being done with very few oocytes or embryos, using small, growing oocyte, or if the available compound has low specific activity, the concentration will need to be increased (usually the upper practical limit due to cost will be on the order of 10–100 μM). Conversely, if the accumulation of the compound over a long time is to be assessed, the concentration may need to be decreased to avoid toxic effects of radioactive decay. In this case, the total concentration of compound can be kept constant by mixing the radiolabeled compound with a suitable amount of the same, unlabeled compound to decrease the effective specific activity. Finally, if the concentration of the compound is to be varied (e.g., to determine transport kinetics), then higher concentrations can be similarly produced by mixing radiolabeled compound with unlabeled. To ensure that higher concentrations of radiolabeled compound are not toxic, particularly for longer incubations, toxic effects should be assessed by suitable culture experiments.

14. This is only needed if the experiment includes determining the non-saturable component of transport. If not, it can be omitted.

15. Because the medium prepared in steps 1–3 was made so that when V_R is added, it would produce 1 ml of mKSOM, each drop will need to have $V_R/20$ of radiolabeled compound stock added to bring its total volume to 50 μl. To validate that the correct concentration of radiolabeled compound is present, a specific volume of the drop (e.g., 1–10 μl) can be removed (after the oocytes or embryos have been removed) and processed for scintillation counting.

16. If the stock of radiolabeled compound as supplied by the manufacturer is in an ethanol:water mix or ethanol (or another non-water solvent) such that too much ethanol (or other solvent) would be added to the medium in order to obtain the desired concentration of radiolabeled compound, an alternative procedure can be used. For each 50 μl drop needed, add the volume of radiolabeled compound stock needed for a 50 μl drop to a microcentrifuge tube, and evaporate to dryness. Add 50 μl of the final medium (e.g., regular strength mKSOM or mKSOM plus inhibitors, etc.) to each tube, and resuspend the compound completely. It should be tested that the compound can be completely resuspended and the expected concentration of radiolabeled compound is present in the incubation drop (see Notes 12 and 15). Place the contents of each tube, with the resuspended radiolabeled compound, onto a dish and cover with oil and equilibrate them as in step 5.

17. For most transport experiments, 15 embryos per drop are sufficient. It is possible, however, to measure transport into

single oocytes or embryos if the rate of transport is sufficiently robust and the specific activity of the compound high enough. Conversely, for compounds that are transported only slowly or where the radiolabeled compounds have low specific activity, larger groups of oocytes or embryos may be needed.

18. To ensure that each group of embryos or oocytes is exposed for the same length of time to the radiolabeled isotope and that the washing steps take an equal length of time, the groups of embryos should be begun at staggered times, so that processing of one group can be completed before the next group's incubation has been completed. The dishes should be premarked for identification, so that time does not need to be taken to mark dishes during the experiment.

19. The length of time that the oocytes or embryos are incubated with radiolabeled compound depends on the experimental protocol and other parameters. A detailed discussion of experimental design is beyond the scope of this protocol, but the excellent and extensive discussion by Van Winkle should be consulted (16).

20. Ice-cold Hepes-mKSOM is used here to ensure that uptake of the radiolabeled compound is stopped immediately at the end of the incubation period. The dishes can be placed on a bed of ice if there is concern that it may not remain cold during the wash steps.

21. The average number of CPM should be determined over a sufficient period to ensure that a good average is obtained for the CPM. For most experiments, a 5 min counting period for each vial should be sufficient. It should be kept in mind that the error in counting radioactive decays varies with the square root of the total number of counts (i.e., $C = CPM \times t$, where t is the total time of counting) with the error being approximately $\pm 1.96\sqrt{C}$. The experimental conditions ($t =$ counting period in scintillation counter, amount of radiolabeled compound, length of incubation, etc.) should be set such that at least several hundred total counts are detected in the embryos in groups where uninhibited transport is occurring. For example, 200 CPM with a counting period of 5 min equals 1,000 total counts, for an intrinsic counting error of ~6%.

22. The signal in the oocyte or embryo samples should be at least ten times greater than the background that is subtracted. For the best measurements, the background should ideally not exceed a few percent of the measured values for oocytes or embryos (except where transport has been inhibited or at the lower end of the range when measuring transport as a function of substrate concentration).

23. All measurements for radiolabeled compounds should fall well within the range of points used to construct the standard curve

for calibration. A new standard curve should be constructed with a wider range if measurements fall outside.

24. In experiments designed to determine rates of transport, it is important to have done preliminary experiments to ensure that the length of the incubation time used in experiments falls within the range of time where the amount of radiolabeled compound taken up by the oocytes or embryos is increasing linearly with time, ensuring that measurements will yield a good approximation of the actual instantaneous rate of transport.

Acknowledgements

These techniques as carried out in JMB's laboratory were developed and used in projects that were funded by the Canadian Institutes of Health Research (CIHR) including operating grants and the Program on Oocyte Health. H.E.C. and S.R. were supported by scholarships from the CIHR Institute of Human Development, Child and Youth Health Training Program in Reproduction, Early Development and the Impact on Health (REDIH). We thank Ms. Megan Kooistra for helping validate and edit these protocols.

References

1. Van Winkle LJ, Campione AL (1991) Ouabain-sensitive Rb+uptake in mouse eggs and preimplantation conceptuses. Dev Biol 146:158–166

2. Van Winkle LJ, Campione AL, Gorman JM, Weimer BD (1990) Changes in the activities of amino acid transport systems b0,+ and L during development of preimplantation mouse conceptuses. Biochim Biophys Acta 1021: 77–84

3. Van Winkle LJ, Campione AL, Farrington BH (1990) Development of system B0,+ and a broad-scope Na(+)-dependent transporter of zwitterionic amino acids in preimplantation mouse conceptuses. Biochim Biophys Acta 1025:225–233

4. Van Winkle LJ, Haghighat N, Campione AL, Gorman JM (1988) Glycine transport in mouse eggs and preimplantation conceptuses. Biochim Biophys Acta 941:241–256

5. Van Winkle LJ, Campione AL, Gorman JM (1988) Na+-independent transport of basic and zwitterionic amino acids in mouse blastocysts by a shared system and by processes which distinguish between these substrates. J Biol Chem 263:3150–3163

6. Van Winkle LJ, Patel M, Wasserlauf HG, Dickinson HR, Campione AL (1994) Osmotic regulation of taurine transport via system beta and novel processes in mouse preimplantation conceptuses. Biochim Biophys Acta 1191: 244–255

7. Colonna R, Mangia F (1983) Mechanisms of amino acid uptake in cumulus-enclosed mouse oocytes. Biol Reprod 28:797–803

8. Haghighat N, Van Winkle LJ (1990) Developmental change in follicular cell-enhanced amino acid uptake into mouse oocytes that depends on intact gap junctions and transport system Gly. J Exp Zool 253:71–82

9. Higgins BD, Kane MT (2003) Inositol transport in mouse oocytes and preimplantation embryos: effects of mouse strain, embryo stage, sodium and the hexose transport inhibitor, phloridzin. J Reprod Fertil 125:111–118

10. Pelland AM, Corbett HE, Baltz JM (2009) Amino acid transport mechanisms in mouse oocytes during growth and meiotic maturation. Biol Reprod 81:1041–1054

11. Tartia AP, Rudraraju N, Richards T, Hammer MA, Talbot P, Baltz JM (2009) Cell volume regulation is initiated in mouse oocytes after ovulation. Development 136:2247–2254

12. Steeves CL, Hammer MA, Walker GB, Rae D, Stewart NA, Baltz JM (2003) The glycine neurotransmitter transporter GLYT1 is an organic

osmolyte transporter regulating cell volume in cleavage-stage embryos. Proc Natl Acad Sci USA 100:13982–13987

13. Dawson KM, Collins JL, Baltz JM (1998) Osmolarity-dependent glycine accumulation indicates a role for glycine as an organic osmolyte in early preimplantation mouse embryos. Biol Reprod 59:225–232

14. Anas MK, Lee MB, Zhou C, Hammer MA, Slow S, Karmouch J, Liu XJ, Broer S, Lever M, Baltz JM (2008) SIT1 is a betaine/proline transporter that is activated in mouse eggs after fertilization and functions until the 2-cell stage. Development 135:4123–4130

15. Anas MKI, Hammer MA, Lever M, Stanton JA, Baltz JM (2007) The organic osmolytes betaine and proline are transported by a shared system in early preimplantation mouse embryos. J Cell Physiol 210:266–277

16. Van Winkle LJ (1999) Transport kinetics. In: Van Winkle LJ (ed) Biomembrane transport. Academic, San Diego, pp 65–131

17. Hammer MA, Kolajova M, Leveille M, Claman P, Baltz JM (2000) Glycine transport by single human and mouse embryos. Hum Reprod 15:419–426

18. Van Winkle LJ (2001) Amino acid transport regulation and early embryo development. Biol Reprod 64:1–12

19. Lawitts JA, Biggers JD (1993) Culture of preimplantation embryos. Methods Enzymol 225:153–164

Chapter 12

Immunofluorescence Staining of Spindles, Chromosomes, and Kinetochores in Human Oocytes

Solon Riris, Suzanne Cawood, Liming Gui, Paul Serhal, and Hayden A. Homer

Abstract

Understanding how human oocytes execute chromosome segregation is of paramount importance as errors in this process account for the overwhelming majority of human aneuploidies and increase exponentially with advancing female age. The spindle is the cellular apparatus responsible for separating chromosomes at anaphase. For accurate chromosome segregation, spindle microtubules must establish appropriately configured attachments to chromosomes via kinetochores. With regard to understanding the mechanistic basis for human aneuploidies therefore, it will be important to explore the molecular underpinnings of spindle structure and the interaction of its microtubules with chromosomes in human oocytes. Here we describe a technique for simultaneously immunolabelling chromosomes, spindle microtubules and kinetochores in human oocytes.

Key words: Mammalian oocyte, Meiosis, Spindle, Kinetochore, Immunostaining, Confocal microscopy, Chromosome alignment

1. Introduction

The vast majority of human aneuploidy is the consequence of errors arising specifically during the first meiotic division (meiosis I) in oocytes (1). Significantly, these female meiosis I errors exhibit an exquisite vulnerability to aging culminating in high rates of embryonic aneuploidy for older women with knock-on effects in terms of reduced fertility and IVF success rates and a heightened predisposition to miscarriage (2). Not surprisingly therefore, understanding the mechanisms regulating chromosome segregation in mammalian oocytes has been the focus of intense investigation (3). Due to the extreme paucity of human oocytes available for research however, the bulk of work in this field has concentrated on the

Hayden A. Homer (ed.), *Mammalian Oocyte Regulation: Methods and Protocols*, Methods in Molecular Biology, vol. 957,
DOI 10.1007/978-1-62703-191-2_12, © Springer Science+Business Media, LLC 2013

mouse model. Although rigorous genetic interrogation in mouse oocytes have yielded invaluable insight into female meiosis I regulation, it will be important to extend this work, at least in part, to human oocytes.

A fundamental prerequisite for accurate chromosome segregation in any cell-type is that chromosomes acquire properly configured attachments with spindle microtubules (4). Failure to establish proper connections between chromosomes and microtubules predisposes to aberrant chromosome sharing when chromosomes get pulled apart by the spindle at anaphase. The interface between microtubules and chromosomes is a multi-protein super-complex that assembles on centromeric DNA known as the kinetochore (5), which can readily be localized in immunostained oocytes using anti-centromere antibodies (ACA) (6). By studying human oocytes immunostained for chromosomes and spindles, earlier results showed a high proportion of severely misaligned chromosomes in oocytes obtained from older women (7). This suggests that mechanisms responsible for establishing kinetochore-microtubule attachments might falter with advancing age and contribute to the increased rates of chromosomal mis-segregation observed in older women.

Quite apart from the difficulties presented by limited availability, other properties of human oocytes make them a challenging system in which to undertake high-resolution imaging especially for relatively small subcellular structures such as kinetochores. One of these is nonspecific background "noise" due to the oocyte's very large volume, which for a mouse oocyte is about 40–50 times that of a somatic cell. This challenge is even greater for a human oocyte as its diameter is up to 50% greater than that of a mouse oocyte added to which, its spindle is considerably smaller than a mouse spindle (see Fig. 1). Here we describe a simple method for simultaneously

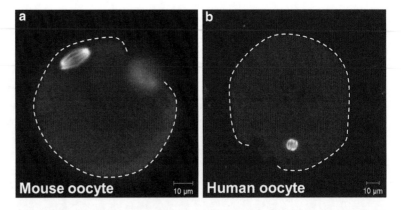

Fig. 1. Relative sizes of human and mouse oocytes and spindles. Shown is a mouse oocyte (**a**) and a human oocyte (**b**) immunostained for spindle microtubules. Note that human oocytes are larger than mouse oocytes, whereas human spindles are smaller than mouse spindles. *Dashed white line* outlines the oocyte. Scale bars, 10 μm.

immunolabelling chromosomes, microtubules, and kinetochores in human oocytes that takes account of these specific obstacles.

2. Materials

2.1. Human Oocyte Handling

1. Human oocytes (see Note 1).
2. Proprietary human oocyte culture medium (see Note 2).
3. Stereo microscope (e.g., Leica MDG 30).
4. 35 × 10 mm polystyrene Petri dishes.
5. Embryo-tested light mineral oil.
6. Bunsen burner.
7. Flame-pulled glass Pasteur pipettes (230 mm) for mouth pipette assembly (see Note 3).
8. Mouth pipette assembly consisting of a flexible silicone rubber nosepiece and hard plastic mouthpiece connected by latex tubing. The assembly is completed by inserting a flame-pulled glass pipette (see step 7 and Note 3) into the wider end of a 1,000 µl plastic pipette tip, the tip of which is inserted into the flange of the rubber nosepiece.

2.2. Permeabilization, Fixation, and Blocking

1. Magnetic stirrer.
2. pH meter.
3. 50 ml plastic conical tubes with screw-top cover.
4. Rocking platform.
5. 0.22 µm syringe-driven filter unit and 10 ml syringe.
6. PHEM buffer: 60 mM PIPES, 25 mM HEPES, 10 mM EGTA, 4 mM $MgSO_4 \cdot 7H_2O$, pH 6.9. To make 100 ml of buffer, add 1.81 g of PIPES, 0.6 g of HEPES, 0.38 g of EGTA, and 0.098 g of $MgSO_4 \cdot 7H_2O$ to 80 ml of dd H_2O and stir. Adjust pH to 6.9 using several drops of 4 N NaOH during which time constituents will solubilize. Top up volume to 100 ml total with dd H_2O. Store for 1–2 weeks at 4°C.
7. 0.25% Triton X-100 in PHEM. For 10 ml, add 25 µl of Triton X-100 to 10 ml of PHEM in a 50 ml conical tube and dissolve on a rocking platform (see Note 4).
8. 3.7% paraformaldehyde (PAF) in PHEM. For 10 ml, weigh out 0.37 g of PAF in a 50 ml conical tube and add 10 ml of PHEM. After dissolving, filter using 10 ml syringe and 0.22 µm filter.
9. PBB: 0.5% bovine serum albumin (BSA) in phosphate-buffered saline (PBS). For 100 ml, add 0.5 g of BSA to 100 ml of PBS and dissolve on the rocking platform.

10. 0.25% Triton X-100 in phosphate-buffered saline (PBS). For 10 ml, add 25 μl of Triton X-100 to 10 ml of PBS in a 50 ml conical tube and dissolve on a rocking platform.

11. Blocking solution: 3% BSA in PBS with 0.05% Tween-20. For 10 ml, weigh out 0.3 g of BSA in a plastic container, add 10 ml of PBS and 5 μl of Tween-20 and dissolve on rocking platform.

2.3. Immunostaining and Imaging

1. Hot block set to 37°C.

2. Primary antibody solutions (see Note 5):

    (a) Spindle microtubules: 1:800 Mouse anti-β tubulin (Sigma) in blocking solution.

    (b) Kinetochores: 1:40 human ACA (ImmunoVision) in blocking solution.

3. Secondary antibody solutions (see Note 6):

    (a) 1:200 Alexa Fluor 633 goat anti-mouse (Invitrogen).

    (b) 1:200 Alexa Fluor 546 goat anti-human (Invitrogen).

4. DNA staining solution: 1 μg/ml Hoechst 33342 in PBB.

5. Wash solution: PBB containing 0.05% Tween-20. For 100 ml, add 0.5 g of BSA and 5 μl of Tween-20 to 100 ml of PBS and dissolve on the rocking platform.

6. 35 mm Glass bottom culture dish with 14 mm microwell containing No. 0 coverglass (e.g., Mak Tek P35G-0-14-C).

7. Confocal microscope equipped with appropriate laser lines and filter-sets for three-color imaging. Here we describe the use of a Zeiss LSM 510 META equipped with a C-Apochromat 63×/1.2 NA water immersion objective and four lasers.

3. Methods

3.1. Pre-permeabilization, Fixation, Permeabilization, and Blocking

1. Add 2–3 ml of PHEM to a 35 mm Petri dish and pre-warm to 37°C on hot-block.

2. Prepare separate Petri dishes containing 2–3 ml of 0.25% Triton X-100 in PHEM, 3.7% PAF in PHEM, PBB, 0.25% Triton X-100 in PBS and blocking solution.

3. Wash oocytes out of culture medium by transferring them by mouth-pipetting into pre-warmed PHEM for ~5 s.

4. Pre-permeabilize oocytes by transferring to 0.25% Triton-X in PHEM for ~10–15 s (see Note 7).

5. Immediately transfer oocytes to 3.7% PAF and incubate for 30 min at room temperature (RT) to fix (see Note 8).

6. Transfer oocytes to PBB for 5 min.

7. Permeabilize oocytes by transferring to 0.25% Triton X-100 in PBS for 15 min.

8. Block nonspecific binding sites by transferring oocytes to blocking solution and maintain overnight at 4°C (see Note 7).

3.2. Immunostaining and Confocal Imaging

1. After overnight blocking at 4°C, remove oocytes from the refrigerator and allow to equilibrate with RT.

2. Prepare three staining chambers by adding 10–15 μl microdrops of either primary antibody solution (see Note 9) or secondary antibody solution or DNA staining solution to the lid of a Petri dish and cover with mineral oil (see Note 10). Prewarm the primary and secondary antibody solutions to 37°C by placing dish lids on hot-block ~30 min prior to use.

3. Transfer one to five oocytes to each micro-drop of primary antibody solution and incubate on hot-block for 1 h (see Note 11).

4. Wash oocytes in 2–3 ml of wash solution for 5 min (×3).

5. Transfer one to five oocytes to each micro-drop of secondary antibody solution and incubate on hot-block for 1 h (see Note 11). Cover to keep oocytes in the dark.

6. Wash oocytes in 2–3 ml of wash solution for 5 min (×3) ensuring that oocytes are maintained in the dark.

7. Transfer one to five oocytes to each micro-drop of DNA staining solution and incubate at RT for 30–60 s. Cover to keep oocytes in the dark.

8. Wash oocytes in 2–3 ml of PBB for 5 min.

9. Prepare an imaging chamber by making up 1–2 μl micro-drops within the micro-well of the glass bottom dish and covering with mineral oil (see Note 12).

10. Transfer one to three oocytes to each micro-drop in the imaging chamber.

11. Select the 63×/1.2 NA water immersion objective on the Zeiss LSM 510 META and locate individual oocytes within micro-drops.

12. Activate the 633 nm Helium/Neon2, the 543 nm Helium/Neon1, and the 364 nm UV laser lines.

13. For imaging spindle microtubules, kinetochores and DNA (see Fig. 2), make the following selections using the Zeiss LSM software:

    (a) For Alexa Fluor 633-labeled spindle microtubules, select the HFT UV/488/543/633 dichroic and the 650 nm long-pass emission filter.

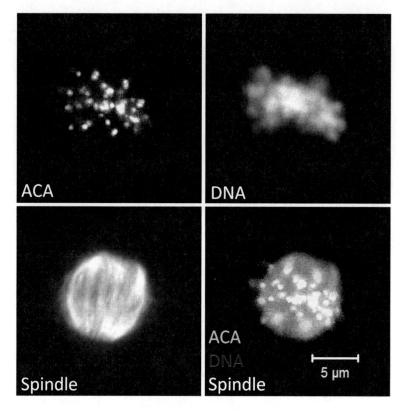

Fig. 2. A metaphase II-arrested human oocyte immunostained for kinetochores (ACA), chromosomes (DNA), and spindle microtubules. Note that this is an enlarged image of the spindle-containing region shown in Fig. 1b. Scale bar, 5 μm.

(b) For Alexa Fluor 546-labeled kinetochores, select the HFT 488/543 dichroic and the 560–615 nm band-pass emission filter.

(c) For Hoechst 33342-labeled DNA, select the HFT UV/488 dichroic and the 385–470 nm band-pass emission filter.

14. Use the zoom function to select the sub-volume of the oocyte containing the spindle and chromosomes and set the upper and lower limits in the Z-axis.

15. Acquire image stacks for each channel at 1–2 μm intervals from which projections can be derived.

16. Process confocal images using Zeiss LSM Image Browser software.

17. Following imaging, oocytes can be stored in the dark at 4°C in droplets of PBB under mineral oil.

4. Notes

1. The vast majority of human oocytes used for research are "spare" oocytes arising from patients having IVF treatment. The two commonest categories of spare oocytes are failed-to-fertilize metaphase II-arrested oocytes and oocytes that are found to be immature (that is, prophase I-arrested with an intact germinal vesicle or GV) at the time of oocyte retrieval and therefore unsuitable for fertilization. Such GV-stage oocytes can then be matured in vitro and used for immunostaining either during meiosis I or at metaphase II if they succeed in completing maturation in vitro. An important prerequisite for research involving human oocytes is that the project has undergone institutional ethical board review and that oocytes are only used from couples who have given explicit consent.

2. Human oocytes obtained following transvaginal ultrasound-guided oocyte retrieval are routinely placed into any one of a number of commercially available IVF culture media (e.g., Vitrolife and SAGE®) that are specifically designed for human oocytes.

3. In order to narrow the bore of the glass pipette for mouth pipetting, hold the two extremes of the narrow aspect of the glass pipette between thumb and forefinger of each hand and heat in the flame of the Bunsen burner. When the glass begins to melt, remove the glass from the flame and rapidly draw the glass out horizontally by pulling hands apart. Break off the distal portion of the pipette so as to leave a suitable working length. The aim is to create a narrow-bore pipette that is of slightly larger caliber than the oocyte's diameter (the diameter of a human oocyte is ~120 μm).

4. Triton X-100 is very viscous so we cut off the narrow end of the pipette tip to create a larger aperture that facilitates drawing up the necessary volumes.

5. Here we specify dilutions and incubation conditions for a set combination of commercially available antibodies. Alternative antibodies can of course be used but will require optimization to determine concentrations, incubation times, temperatures, etc.

6. Here we suggest using a combination of Alexa Fluor 633, Alexa Fluor 546, and Hoechst 33342 (UV). However, any other combination of fluorophores can be used provided each of their excitation/emission parameters is sufficiently separate from one another.

7. Human oocytes are very large cells. Therefore, a key objective when labeling small subcellular structures such as kinetochores

is to eliminate as much nonspecific background "noise" as possible. For this reason, we employ a relatively long pre-permeabilization step as well as overnight blocking.

8. Take care when transferring oocytes from the pre-permeabilization solution into 3.7% PAF fixative as the mixture of the two solutions creates considerable surface tension that can lead to rapid and unpredictable oocyte displacement and loss. To minimize this, transfer the oocyte in as little pre-permeabilization solution as possible.

9. We suggest small volumes since with Tween-20, drops of larger volume have a tendency to disperse over the surface of the lid thereby leading either to oocyte loss or to the mixing of samples.

10. Two options are available for staining. Firstly, both primary antibodies can be combined in a single incubation step followed by a single incubation with combined secondary antibodies. Secondly, oocytes can be incubated with one primary antibody followed by the corresponding secondary antibody after which the process is repeated for the second antibody pairing. In our hands, either approach works well with the first option having the clear advantage of simplicity and speed. The user will need to determine which works best in their hands especially if antibodies from sources other than those described herein are used.

11. It is very important to transfer the oocytes in as little volume of blocking solution as possible. Remember that with micro-drops of only 10–15 μl volume, any significant additional volume of blocking solution that accompanies the oocyte during transfer will significantly alter the final antibody concentration.

12. Using micro-drops of very small volume aids the ability to localise of oocytes on the microscope especially with the relatively limited field of view provided by the 63× objective. Other protocols describe mounting oocytes on a glass slide under a sealed glass cover-slip. The benefit of the technique described here is that oocytes can be "rolled" to change their position thereby allowing the plane of their spindles to be adjusted so that the spindle's long axis is parallel to the surface of the dish. Another potential benefit is that following imaging, oocytes can be subjected to a further round of staining for detecting another antigen if so desired.

Acknowledgments

This work was funded by a Wellcome Trust Clinical Fellowship to H.H. (082587/Z/07/Z). We are extremely grateful to Yoshanta Wade for her tireless work looking after the IVF patients.

References

1. Hassold T, Hunt P (2009) Maternal age and chromosomally abnormal pregnancies: what we know and what we wish we knew. Curr Opin Pediatr 21:703–708

2. Homer H (2007) Ageing, aneuploidy and meiosis: eggs in a race against time. In: Hillard T (ed) Yearbook of obstetrics and gynaecology, vol 12. RCOG Press, London, pp 139–158

3. Homer H (2011) New insights into the genetic regulation of homologue disjunction in Mammalian oocytes. Cytogenet Genome Res 133:209–222

4. Tanaka TU (2010) Kinetochore-microtubule interactions: steps towards bi-orientation. EMBO J 29:4070–4082

5. Cheeseman IM, Desai A (2008) Molecular architecture of the kinetochore-microtubule interface. Nat Rev Mol Cell Biol 9: 33–46

6. Homer H, Gui L, Carroll J (2009) A spindle assembly checkpoint protein functions in prophase I arrest and prometaphase progression. Science 326:991–994

7. Volarcik K, Sheean L, Goldfarb J, Woods L, Abdul-Karim F, Hunt P (1998) The meiotic competence of in-vitro matured human oocytes is influenced by donor age: evidence that folliculogenesis is compromised in the reproductively aged ovary. Hum Reprod 13:154–160

Chapter 13

Studying the Roles of Aurora-C Kinase During Meiosis in Mouse Oocytes

Kuo-Tai Yang, Yi-Nan Lin, Shu-Kuei Li, and Tang K. Tang

Abstract

We previously isolated Aurora-C (*Aurkc/Aie1*) in a screen for kinases expressed in mouse sperm and eggs. Aurora-C kinase was reported to be a chromosomal passenger protein that plays critical roles in chromosome alignment, segregation, kinetochore-microtubule attachment, and cytokinesis in female mouse meiosis. This chapter describes experimental approaches for examining the subcellular localization and function of Aurora-C kinase during female mouse meiosis, presenting detailed methods for introducing exogenous Aurora-C wild-type and kinase-dead mutant mRNAs into mouse oocytes by cytosolic microinjection, and preparing whole-mount meiotic oocytes and chromosome spreads for confocal immunofluorescence microscopy.

Key words: Aurora, Oocytes, Meiosis, Whole-mount, Chromosome spreads, Microinjection

1. Introduction

Aurora kinases comprise a family of serine/threonine kinases that are pivotal to the successful execution of cell division. In mammals, there are three highly related Aurora kinases (Aurora-A, -B, and -C) that share sequence homology in their central catalytic kinase domains (1). Aurora-A and -B, which exhibit different subcellular localization, are known to play distinct roles in mitosis (2–4). Aurora-A, located at centrosomes and mitotic spindle poles, is required for centrosome maturation, bipolar spindle assembly, and chromosome segregation (2). Aurora-B, a member of chromosome passenger complex (CPC), localizes to the centromere and spindle midzone and midbody during mitosis. Aurora-B forms a complex with INCENP, survivin, and borealin that is essential for proper chromosome segregation, kinetochore–microtubule interaction, and cytokinesis (3, 4). Aurora-C (mouse Aie1/human

Hayden A. Homer (ed.), *Mammalian Oocyte Regulation: Methods and Protocols*, Methods in Molecular Biology, vol. 957,
DOI 10.1007/978-1-62703-191-2_13, © Springer Science+Business Media, LLC 2013

AIE2) was first identified in our laboratory in a screen for kinases expressed in sperm and eggs (5). Human Aurora-C had also been independently cloned, and its gene (*AURKC/STK13*) has been mapped to chromosome 19q13.3-ter (6). Unlike Aurora-A and -B, which are ubiquitously expressed in somatic tissues, particularly in mitotically dividing cells, Aurora-C is predominantly expressed in the germ cells, mainly meiotic spermatocytes (7) and oocytes (8). The slow progress in studying the localization and function of Aurora-C during the past few years was due to lack of specific antibodies, very limited meiotic germ cells, and reliable methods for the preparation and characterization of meiotic germ cells for such a study.

To address these limitations, we have generated specific antibodies that recognize mouse Aurora-C (7). Immunofluorescence analyses revealed that endogenous Aurora-C is present at the centromeres in MI and MII, and relocates to the spindle midzone and midbody during anaphase I–telophase I and anaphase II–telophase II transitions in spermatocytes (7) and oocytes (8). The role of Aurora-C in meiotic divisions was functionally analyzed by microinjecting kinase-dead Aurora-C mRNA into mouse oocytes. These experiments showed that a deficiency of Aurora-C kinase caused not only chromosome misalignment, premature chromosome separation, and abnormal kinetochore–microtubule interaction, but it also led to cytokinesis failure and the production of large polyploid oocytes (8). Interestingly, a recent report described a homozygous mutation in the *Aurora-C* gene that led to the production of large-headed, multiflagellar, polyploid spermatozoa (9). Our findings provided a possible explanation for how Aurora-C deficiency causes polyploid spermatozoa in humans. Here, we provide detailed methods for analyzing Aurora-C function in meiosis.

2. Materials

2.1. Analysis and Production of In Vitro-Transcribed mRNA

1. 3 M sodium acetate, pH 5.2.
2. mMESSAGE mMACHINE T7 Ultra Kit (Ambion, Applied Biosystems) (see Note 1).
3. 70% Ethanol prepared from nuclease-free water.
4. 10× MOPS buffer: 200 mM MOPS (Sigma-Aldrich), 80 mM sodium acetate, and 10 mM EDTA (pH 8.0) in distilled water. Store at room temperature (RT), protected from light. Prepare 1× MOPS fresh in distilled water as a running buffer before electrophoresis.
5. 1% Formaldehyde/agarose gel: Dissolve 0.5 g agarose (Invitrogen) in 36 ml distilled water and cool to 60°C, then

add 5 ml of 10× MOPS buffer and 9 ml of 37% formaldehyde (Merck).

6. Ethidium bromide (50 μg/ml).

7. UV illumination (BioRad).

8. NanoDrop 1000 spectrophotometer (Thermo Scientific).

2.2. Collection and Microinjection of Mouse Germinal Vesicle Stage Oocytes

2.2.1. Follicle Maturation

1. Three-week-old female mice of the C57BL/6 strain.

2. Sterile normal saline (0.9% NaCl): Sterilize by filtering through 0.45-μm nylon filter (Millipore).

3. PMSG (pregnant mare's serum gonadotrophin): 5 IU in 150 μl normal saline for each superovulated female.

4. 1-ml syringes and 27-gauge needles for intraperitoneal injection.

2.2.2. Collection and Culture of GV Oocytes

1. Stereomicroscope (e.g., Zeiss, Stemi SV6).

2. Surgery package, including fine scissors, fine and blunt forceps, a pair of sharp pointed forceps, and 27-gauge needles for dissecting ovaries.

3. 35-mm Petri dish (Corning).

4. M2 medium (Sigma-Aldrich): for oocyte collection.

5. 100 mM IBMX: Dissolve 22.2 mg IBMX (3-isobutyl-1-methylxanthine, Sigma-Aldrich), a phosphodiesterase inhibitor that arrests oocytes at the prophase stage, in 10 ml DMSO and sterilize by filtering through a 0.22-μm nylon filter (Millipore) to yield a 100 mM (22.2 mg/ml), 1,000× sterile stock solution. Aliquot and store at −20°C.

6. M16 medium (Sigma-Aldrich): for oocyte culture.

7. Paraffin or mineral oil (Merck).

8. CO_2 incubator (37°C, 5% CO_2).

9. Mouth transfer pipette: composed of an aspirator mouthpiece (general use 200-μl tip with filter), plastic tube, and a transfer pipette holder.

10. TW100-6 borosilicate glass capillaries (World Precision Instruments, Inc.): for preparation of transfer and holding pipettes only.

2.2.3. GV Oocytes Microinjection

1. Differential interference contrast (DIC) microscope with 10×, 20×, and 40× magnifications (Nikon, ECLIPSE TE3000) and a heating device containing a warming platform fitted to the microscope stage (CO 102, Linkam Scientific).

2. Mechanical and electronic micromanipulators (Nikon, Narishige).

3. Microinjection device (FemtoJet Express, Eppendorf) and accompanying air compressor.

4. Air and oil pipette holders.

5. Micropipette puller (Model P-97, Sutter Instrument Co.).

6. Microforge MF-900 (Narishige).

7. 1B100F-6 borosilicate glass capillaries with filament (World Precision Instruments, Inc.): for preparation of injection needles only.

8. Cover of 35-mm Petri dish (Corning): for use as a microinjection chamber.

9. Paraffin or mineral oil (Merck).

2.3. Fixation and Immunostaining of Oocytes

1. 1× PBS: Dilute from 10× PBS (27 mM KCl, 14.7 mM KH_2PO_4, 1.37 M NaCl, 77 mM Na_2HPO_4) stock in distilled water and autoclave before use.

2.3.1. Fixation of Whole-Mount Oocytes

2. PHEM buffer: 60 mM PIPES (piperazine-N,N'-bis[2-ethanesulfonic acid]), 25 mM HEPES, 10 mM EGTA, and 4 mM $MgCl_2$ (pH 6.9) in distilled water. Store at RT.

3. Fixation buffer: PHEM buffer containing 2% formaldehyde (Merck) and 0.2% Triton X-100 (Merck).

4. Permeabilization buffer: 0.1% Triton X-100 in PBS containing 0.3% bovine serum albumin (BSA).

5. Wash buffer I: 0.01% Triton X-100 in 1× PBS.

2.3.2. Preparation of Chromosome Spreads

1. Tyrode's solution (Sigma-Aldrich).

2. M2 medium (Sigma-Aldrich).

3. Hypotonic solution: 50% BSA in distilled water. Aliquot and store at 4°C.

4. 70 mM sodium tetraborate buffer: Dissolve 0.267 g sodium tetraborate (Merck) in 10 ml distilled water.

5. 1% Paraformaldehyde solution for chromosome spreading: 1% paraformaldehyde (Sigma-Aldrich), 5 mM sodium tetraborate, and 0.15% Triton X-100 (Merck) in distilled water. Adjust to pH 9.2 with 5 N NaOH.

6. 12-Well culture plate (Coring): for washing slides.

7. Microscope slides and 12-mm coverslips (Marienfeld).

2.3.3. Immunofluorescence Staining

1. Antibody dilution buffer I (ADB-I): 3% BSA in 1X PBS for whole-mount oocytes. Store at 4°C.

2. Antibody dilution buffer II (ADB-II): 10% normal goat serum (NGS) in 1× PBS (for chromosome spreads). Store at 4°C.

3. Wash buffer I (WB-I): 0.01% Triton X-100 in 1X PBS (for whole-mount oocytes). Store at RT.

4. Wash buffer II (WB-II): 0.1% Tween 20 in 1× PBS (for chromosome spreads). Store at RT.

5. Primary and secondary antibodies. Aliquot and store at 4°C.

6. DAPI (4, 6-diamidino-2-phenylindole). Aliquot and store at 4°C.

7. Mounting medium for fluorescence (Vectashield H-1000, Vector Laboratories, Inc.). Aliquot and store at 4°C.

8. Microscope slides and 12-mm coverslips (Marienfeld).

3. Methods

3.1. Preparation of In Vitro-Transcribed RNA for Microinjection

We cloned our genes into the pcDNA3.1(+) vector containing the T7 promoter and used the mMESSAGE mMACHINE T7 Ultra kit for in vitro mRNA synthesis, following the manufacturer's protocol. This kit is useful for obtaining high quality and longer products of transcribed RNAs.

3.1.1. Linearization of Template DNA

1. Digest 10 μg of plasmid DNA in a 1.5-ml microcentrifuge tube with an appropriate restriction enzyme that cuts downstream of the insert for transcription termination.

2. After overnight digestion, terminate the reaction by adding 1/10 volume of 3 M sodium acetate and 2 volumes of ethanol (for DNA precipitation).

3. Pellet the precipitated, linearized DNA in a microcentrifuge at $16,060 \times g$ for 15 min and remove the supernatant. Briefly spin again and remove the residual fluid completely.

4. Resuspend the DNA pellet in nuclease-free water at a concentration of approximately 1 μg/μl (see Note 2).

3.1.2. In Vitro Transcription

1. Assemble the transcription reaction in a 1.5 ml microcentrifuge tube on ice by sequentially adding 5 μl of nuclease-free water, 10 μl of T7 2× NTP/ARCA, 2 μl of 10× T7 Reaction Buffer, ~1 μg of linearized template DNA, and 2 μl of T7 Enzyme Mix in a final volume of 20 μl (see Note 1).

2. Mix thoroughly (see Note 3); spin down the reaction mixture briefly and then incubate at 37°C for 2 h (see Note 4).

3. Place the reaction tube on ice; add 1 μl of TURBO DNase (see Note 1), mix well, and then incubate at 37°C for 15 min.

4. Perform tailing reaction by sequentially adding 36 μl of nuclease-free water, 20 μl of 5× E-PAP Buffer, 10 μl of 25 mM $MnCl_2$, and 10 μl of ATP Solution to the reaction tube and mixing well. Add 4 μl of E-PAP Enzyme to the reaction mixture after removing and transferring 2.5 μl of the mixture to a new 1.5-ml microcentrifuge tube containing 7.5 μl of Formaldehyde Load Dye (minus-enzyme control RNA sample).

5. Gently, but completely, mix the 100 µl final reaction mixture and incubate at 37°C for 45 min.

6. Stop the reaction and precipitate the transcribed RNA by adding 50 µl of LiCl Precipitation Solution (see Note 1). Mix thoroughly and chill at −20°C for at least 30 min.

7. Pellet the transcribed RNAs in a microcentrifuge at $16,060 \times g$, 4°C for 15 min. Remove the supernatant, wash the pellet with 1 ml of 70% ethanol, and recentrifuge at $16,060 \times g$, 4°C for 10 min.

8. Remove the 70% ethanol carefully, spin briefly, and remove the residual fluid completely.

9. Resuspend the transcribed RNA in nuclease-free water completely, and transfer 1 µl of RNA sample into a new 1.5 ml microcentrifuge tube containing 3 µl of Formaldehyde Load Dye (tailing reaction RNA).

10. Add 0.5 µl of ethidium bromide (50 µg/ml) to both minus-enzyme control and tailing reaction RNA samples, heat at 75°C for 10 min, and then chill on ice.

11. Separate RNA samples on 1% formaldehyde/agarose gels in 1× MOPS running buffer and visualize by UV illumination to validate the quality and integrity of transcribed mRNA. The tailed RNA should be more than 50 bp longer than the corresponding minus-enzyme control in which RNA was not tailed.

12. Determine RNA concentration using a NanoDrop 1000 spectrophotometer, adjust the final concentration to 1 µg/µl, and store the RNA samples at −80°C.

3.2. Collection of GV Oocytes and RNA Microinjection

3.2.1. Harvesting GV Oocytes from Ovaries

1. Hormonally prime 3-week-old female mice by intraperitoneal injection of 5 IU of PMSG.

2. Prepare a culture medium dish containing 5–8 µl of 16 drops of M16 medium and cover with mineral oil to prevent evaporation (Fig. 1a). Place the dish in a CO_2 incubator and allow to equilibrate for at least 2 h.

3. Prepare a dish containing approximately 3 ml of M2 medium containing 100 µM IBMX for collection of GV oocytes.

4. Sacrifice a donor female mouse by cervical dislocation 48 h after intraperitoneal PMSG injection.

5. Open the abdomen with a fine forceps and scissors, and move the viscera up to expose the reproductive tract.

6. Hold the uterine horn with forceps, transect the mesovarium, and then dissect the ovary away from the surrounding fat and oviducts.

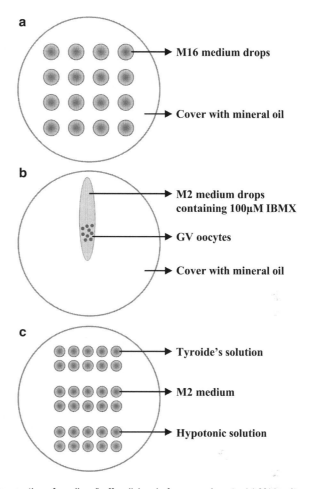

Fig. 1. Preparation of medium/buffer dishes before experiments. (**a**) M16 culture medium dish: used for culturing oocytes during meiotic maturation. (**b**) Microinjection chamber/ dish: used for in vitro micromanipulation of oocytes. (**c**) Fixative dish: for preparation of oocyte chromosome spreads.

7. Transfer ovaries into the previously prepared dish containing IBMX-treated M2 medium. Under a dissecting microscope, segment ovaries using a fine forceps and a 27-gauge needle (Fig. 2a).

8. After dissection, immobilize the ovarian fragments with a fine forceps and with a 27-gauge needle, gently tear open the expanded follicles to release the enclosed cumulus–oocyte complexes (Fig. 2b).

9. Mechanically denude the surrounding cumulus cells from immature germinal vesicle (GV) oocytes using a mouth transfer pipette (see Note 5 and Fig. 2c–e). In our hands, we routinely obtain 30–50 fully-grown GV-stage oocytes from a single hormonally primed mouse.

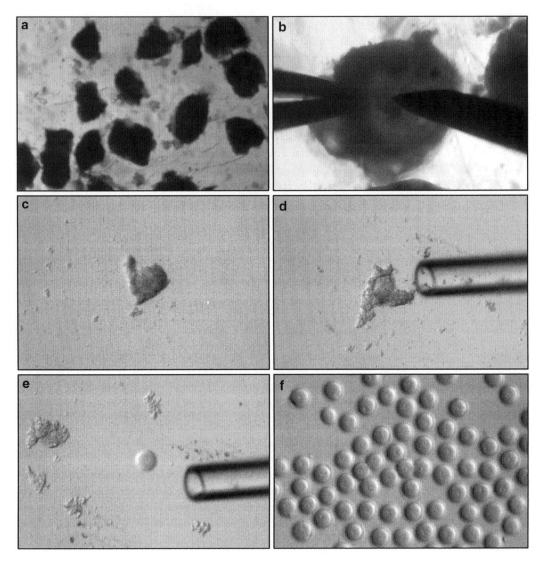

Fig. 2. Isolation of GV-stage mouse oocytes from ovaries. (**a**) Fragments of dissected ovaries. (**b**) Tearing open the antral follicle to release contents. (**c**) Cumulus–oocyte complex. (**d**) Mechanical removal of surrounding cumulus cells by repeatedly aspirating and expelling the oocyte through a finely drawn glass pipette controlled by mouth suction. (**e**) Denuded GV oocyte. (**f**) Denuded GV oocytes selected for cytosolic microinjection.

3.2.2. Cytosolic Microinjection of Transcribed mRNA

1. Thaw the in vitro-transcribed RNA on ice and dilute with nuclease-free water before cytosolic microinjection. The appropriate concentration for each injected RNA needs to be determined empirically. A final RNA concentration of ~0.5 µg/µl is suggested for an initial trial.

2. Prepare an injection chamber by placing a 20 µl drop of M2 medium containing 100 µM IBMX onto the surface of an upturned Petri dish lid and cover medium with mineral oil to prevent evaporation (Fig. 1b).

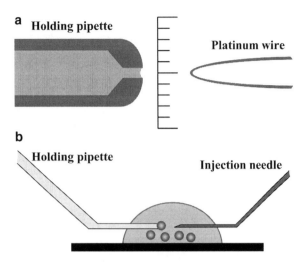

Fig. 3. (a) Schematic showing holding pipette preparation and specifications. (b) Schematic depicting a side view of the microinjection chamber/dish.

3. Transfer ~30 of the healthiest looking GV oocytes (see Note 6 and Fig. 2f) obtained as described in Subheading 3.2.1 to the medium in the microinjection chamber.

4. Place the chamber containing GV oocytes onto the microscope warming platform and focus using the 40× objective.

5. Load the holding pipette (see Note 7 and Fig. 3a) into its holder on the microinjection rig. Lower the holding pipette into the medium within the injection chamber so that it is brought into the same focal plane as that of the oocytes.

6. Fill the tip of an injection needle (see Note 8) by dipping the blunt end into the diluted RNA solution and load the needle into its holder on the microinjection rig. Lower the needle tip into the medium within the injection chamber so that it is brought into the same focal plane as that of the holding pipette.

7. Carefully break off the tip of the injection needle by gently advancing it against the holding pipette.

8. Stabilize an oocyte using the holding pipette and adjust the focus to clearly visualize the oocyte cytoplasm. The GV oocyte must be placed close to the injection needle to minimize injection damage (Fig. 3b).

9. Use the micromanipulator to slowly advance the injection needle through the zona pellucida and oolemma and into the oocyte cytoplasm. Press the button of the FemtoJet Express injector to inject the mRNA, which will lead to dispersal of the surrounding ooplasm. Remove the injection needle tip quickly from the oocyte cytoplasm; a visible dispersal of ooplasm will be evident (see Note 9).

10. Move the injected oocyte to another area away from the uninjected oocytes in the chamber, and repeat steps 8 and 9.

11. When the injection of oocytes is complete, raise the holding pipette and the injection needle, and place the microinjection chamber into a 37°C, humidity-controlled, 5% CO_2 incubator for 2 h (at least 1 h) before washing the injected oocytes into IBMX-free medium.

3.3. Fixation and Immunofluorescence Staining of Oocytes

Before fixing oocytes, wash them into IBMX-free M16 culture medium and place in a 37°C, humidified, 5% CO_2 incubator. Oocytes will then spontaneously resume meiosis I marked by breakdown of the GV.

3.3.1. Fixation and Immunofluorescence Staining of Whole-Mount Oocytes

1. Transfer meiotic oocytes (see Note 10) into a 35-mm Petri dish by mouth pipetting and fix them by incubating in PHEM buffer containing 2% formaldehyde and 0.2% Triton X-100 at 37°C for 1 h (see Note 11).

2. After fixation, permeabilize oocytes by incubating with 0.1% Triton X-100 in PBS containing 0.3% BSA at 37°C for 40 min.

3. Incubate the permeabilized oocytes in ADB-I at 4°C for 1 h to block nonspecific binding sites before performing immunofluorescence staining.

4. Prepare solutions of primary antibodies diluted in ADB-I.

5. Incubate oocytes with the appropriate primary antibodies at 4°C overnight.

6. Wash the oocytes three times for 10 min each in WB-I at RT to remove unbound primary antibodies.

7. Prepare a solution containing DAPI (1:500) and the appropriate secondary antibodies diluted in ADB-I. Secondary antibodies conjugated with Alexa 488, Alexa 568, or Alexa 647 are recommended for their enhanced photostability.

8. Incubate oocytes with the secondary antibody/DAPI solution at 4°C overnight.

9. Wash oocytes three times for 10 min each with WB-I buffer at RT to remove unbound secondary antibodies.

10. Prepare clean slides and coverslips, drop 2 µl of antifade mounting solution onto the slide, transfer the washed oocytes into the center of the drop, and cover with a coverslip. Avoid the production air bubbles.

11. Seal the edges of coverslips with nail polish (see Note 12) before acquiring immunofluorescence images.

1. Prepare a dish containing a number of drops of Tyrode's solution, M2 medium, and hypotonic solution, and cover with mineral oil (Fig. 1c).

2. Incubate oocytes in the Tyrode's solution at RT until the disappearance of the zona pellucida (ZP) is observed under a stereomicroscope. If the ZP still remains after several minutes, transfer the oocytes to another drop of Tyrode's solution and repeat step 2.

3. Transfer the ZP-free oocytes into M2 medium drops and wash several times to remove Tyrode's solution. (The Tyrode's solution is harmful to ZP-free oocytes if left in contact for too long.)

4. Transfer oocytes into the hypotonic solution and incubate at 37°C for 10 min until the ZP-free oocytes begin to swell.

5. Drop 15 µl of a 1% paraformaldehyde fixative solution onto the center of a coverslip, and then transfer a swollen oocyte into the fixative drop until the swollen oocyte ruptures and chromosomes are dispersed (see Note 13).

6. Air-dry the fixative on coverslips for about 2 h to allow chromosomes to adhere (see Note 14), and then briefly wash coverslips three times in PBS. Incubate coverslips in PBS containing 10% NGS and place them in a humidified chamber at 4°C for 1 h before performing immunofluorescence staining.

7. Prepare solutions of primary antibodies diluted in ADB-II.

8. Drop 10 µl of primary antibody solution onto a Parafilm membrane. Place the coverslips with chromosome spreads face down on top of the primary antibody drops and incubate in a humidified chamber at 4°C overnight.

9. Place coverslips in the wells of a 12-well culture plate and wash three times (5 min each) in WB-II at RT to remove unbound primary antibodies.

10. Prepare a solution containing DAPI (1:500) and the appropriate secondary antibodies (e.g., Alexa 488-, Alexa 568-, or Alexa 647-conjugated) diluted in ADB-II.

11. Drop 10 µl of secondary antibody/DAPI solution onto a Parafilm membrane. Place the coverslips with chromosome spreads face down on top of the secondary antibody drops and incubate in a humidified chamber at 4°C overnight.

12. Wash the coverslips as described in step 9 to remove unbound secondary antibodies.

13. Drop 2 µl of antifade mounting solution onto a clean slide, and place the coverslips with chromosome spreads face down on the drop. Avoid the production of air bubbles.

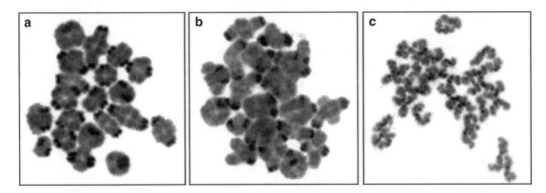

Fig. 4. Chromosome spread patterns in MI-stage mouse oocytes. Shown are chromosome spreads that were prepared from uninjected MI oocytes (control, **a**) or MI oocytes that were injected with mRNA encoding either Aurora-C-WT (**b**) or Aurora-C-kinase dead (**c**). Bivalent chromosome with paired homologous were commonly observed in spread chromosomes prepared from uninjected (**a**) or Aurora-C-WT-injected (**b**) oocytes. In contrast, most chromosomes in Aurora-C-kinase dead-injected oocytes were univalent, implying premature chromosome segregation in meiosis I (**c**).

14. Remove excess antifade mounting solution by suction and seal the edges of coverslips with nail polish before acquiring immunofluorescence images (Fig. 4) with a fluorescence microscope (e.g., Zeiss LSM 700 laser-scanning confocal microscope).

4. Notes

1. We generally use the T7 promoter-based pcDNA3.1 vector from Invitrogen for subcloning and producing the transcripts. Other in vitro transcription systems may be for T3 or SP6 promoter-based vectors. The in vitro transcription system should include the reagents for poly (A) tailing to produce functional transcripts for injection. All of regents listed in the in vitro transcription section are supplied in the mMESSAGE mMACHINE T7 Ultra Kit.

2. It is not necessary to quantify the resuspended DNA. In general, the DNA (10 µg) is dissolved in 10 µl of nuclease-free water and the concentration is assumed to be 1 µg/µl.

3. Gently flick the microcentrifuge tube or slowly pipette the reaction mixture up and down several times. Do not mix the reactions vigorously by vortexing as this will significantly reduce the transcription efficiency, quality, and yield of transcribed RNA.

4. Typically, an 80% yield is achieved after a 1-h incubation. For maximum yield, we recommend an incubation time of

2 h or longer. We usually obtain 40–50 μg RNA, which should ideally be used within 1 month (if stored at –80°C) for more consistent results.

5. Transfer pipette preparation: Soften the glass capillary in a fine flame until the glass becomes soft. Withdraw the glass capillary from the heat and quickly pull both sides to produce a tube with an internal diameter of 150–200 μm. Snap the glass with an oilstone to yield a neat break.

6. The best (i.e., healthiest) GV oocytes are those that show an intact zona pellucida, clear and plump cytoplasm, a visible nucleolus, and well-defined nuclear membrane. Exclude lysed GV oocytes and those with irregular cytoplasm and no clear nucleus.

7. Holding pipettes with an internal diameter of 80–120 μm are prepared as described in the transfer pipette preparation method (see Note 5 above). Clamp the holding pipette on a microforge close to the heated platinum wire until the tip begins to melt. Allow the pipette tip to shrink to a diameter of 10–15 μm. Using a microforge with the platinum wire heated to 60°C, make a bend 2–3 mm away from the tip at an angle of 15–20°.

8. The parameters for injection-needle preparation are as follows: heat, 300; pull, 200; vel, 150; ant time, 150. Make a bend 1–2 mm away from the tip at an angle of 15–20° using a microforge with the platinum wire heated to 60°C as described in Note 7 for holding pipettes.

9. GV oocytes can be harmed if left in the chamber for a long time. Thus, the number of oocytes that can be handled simultaneously in the chamber for microinjection depends on the speed and skill of the technician performing the injections.

10. The arrested GV oocytes are washed out of IBMX and then cultured in M16 medium to resume meiotic progression. Approximately at 6, 8, and 12 h after release from IBMX induced arrest, the meiotic oocytes progress to prometaphase-I, metaphase-I, and Telophase-I stages, respectively. The exact timing for fixation of oocytes at specific stages to study other genes of interest needs to be determined empirically.

11. An alternative method for fixing whole-mount oocytes: Fix and permeabilize oocytes simultaneously in PHEM buffer containing 4% formaldehyde and 0.5% Triton X-100 at RT for 1 h, followed by immunofluorescence staining with primary antibodies. This protocol is useful for reducing fixation time, and the results are as good as those obtained with the original fixation method.

12. Do not suction off the overflow antifade mounting medium before sealing; otherwise, the oocytes are easily lost.

13. The fixative solution should cause the swollen oocytes to rupture within 5 s; metaphase chromosomes will be dispersed and adhere to slides.

14. Do not over air-dry; otherwise, the post-immunostaining morphology of chromosome spreads will be poor.

References

1. Nigg EA (2001) Mitotic kinases as regulators of cell division and its checkpoints. Nat Rev Mol Cell Biol 2(1):21–32

2. Andrews PD, Knatko E, Moore WJ, Swedlow JR (2003) Mitotic mechanics: the auroras come into view. Curr Opin Cell Biol 15(6):672–683

3. Carmena M, Earnshaw WC (2003) The cellular geography of aurora kinases. Nat Rev Mol Cell Biol 4(11):842–854

4. Ruchaud S, Carmena M, Earnshaw WC (2007) Chromosomal passengers: conducting cell division. Nat Rev Mol Cell Biol 8(10):798–812

5. Tseng TC, Chen SH, Hsu YPP, Tang TK (1998) Protein kinase profile of sperm and eggs: cloning and characterization of two novel testis-specific protein kinases (AIE1, AIE2) related to yeast and fly chromosome segregation regulators. DNA Cell Biol 17(10): 823–833

6. Bernard M, Sanseau P, Henry C, Couturier A, Prigent C (1998) Cloning of STK13, a third human protein kinase related to Drosophila aurora and budding yeast Ipl1 that maps on chromosome 19q13.3-ter. Genomics 53(3):406–409. doi:S0888-7543(98)95522-7

7. Tang CJ, Lin CY, Tang TK (2006) Dynamic localization and functional implications of aurora-C kinase during male mouse meiosis. Dev Biol 290(2):398–410

8. Yang KT, Li SK, Chang CC, Tang CJ, Lin YN, Lee SC, Tang TK (2010) Aurora-C kinase deficiency causes cytokinesis failure in meiosis I and production of large polyploid oocytes in mice. Mol Biol Cell 21(14):2371–2383. doi:E10-02-0170

9. Dieterich K, Soto Rifo R, Karen Faure A, Hennebicq S, Amar BB, Zahi M, Perrin J, Martinez D, Sele B, Jouk P-S, Ohlmann T, Rousseaux S, Lunardi J, Ray PF (2007) Homozygous mutation of AURKC yields large-headed polyploid spermatozoa and causes male infertility. Nat Genet 39(5):661–665

Chapter 14

Chromosome Spreads with Centromere Staining in Mouse Oocytes

Jean-Philippe Chambon, Khaled Hached, and Katja Wassmann

Abstract

This chapter describes a technique for performing chromosome spreads from mouse oocytes. It is based on a previously described protocol (Hodges and Hunt, Chromosoma 111: 165–169, 2002), which we have modified. Chromosomes are stained with either Propidium Iodide or Hoechst. This spreading technique allows for simultaneous immunostaining of proteins associated with chromosomes. It is very useful to stain spreads with CREST serum which labels kinetochores, to be able to distinguish bivalents (chromosome pairs), dyads or univalents (paired sister chromatids), and single sister chromatids without ambiguity.

Key words: Chromosome Spreads, CREST, Meiosis, Bivalent, Univalent, Sister chromatid, Cohesion, Mouse oocytes

1. Introduction

In metaphase of meiosis I, pairs of homologous chromosomes (each chromosome consisting of two sister chromatids) are aligned at the metaphase plate. The sister chromatids forming a chromosome are held together by a protein complex known as the Cohesin complex (2). Kinetochores (sites where attachment of the bipolar spindle occurs before the metaphase to anaphase transition) of sister chromatids are mono-oriented, which means that they are attached to the same pole. Chromosome pairs are held together through chiasmata, which allow the establishment of tension generating attachments. Chromosomes are segregated from each other in anaphase of meiosis I, whereas in meiosis II, sister chromatids are oriented in a bipolar manner to opposite poles and segregated such as in mitosis (3). This is made possible through the stepwise removal of Cohesin in meiosis: In meiosis I, only Cohesin on arms is removed, and centromeric Cohesin is protected to maintain sister chromatids together until meiosis II. It is only in meiosis II that

Hayden A. Homer (ed.), *Mammalian Oocyte Regulation: Methods and Protocols*, Methods in Molecular Biology, vol. 957, DOI 10.1007/978-1-62703-191-2_14, © Springer Science+Business Media, LLC 2013

centromeric Cohesin is deprotected and removed, allowing for the separation of sisters to take place (3). How chromosomes are correctly oriented and attached in meiosis I and II in mammalian oocytes is still an open question.

In mammalian oocytes, missegregations occur at high rates for reasons that are not entirely clear, and are therefore subject of research in different laboratories. In humans, it has been estimated that 20% of oocytes are aneuploid (4). As an example, missegregations in female meiosis I are responsible for 90% of trisomies 21 in humans (4). Furthermore, an age-dependent increase in chromosome missegregations in female meiosis I is observed (4). Age-related increase of aneuploidies in oocytes is at least partially due to a weakening of Cohesion between sister chromatids, and a lower chiasmata count (5–8). Studies addressing how chromosome segregation is regulated in meiosis rely in part on chromosome spreading techniques such as the one described here, that allow to distinguish single sister chromatids from paired sisters, and paired chromosomes without any doubt. Simultaneous staining of kinetochores with a CREST serum improves our ability to analyze chromosome spreads crucially. The ploidy of the spread can be assessed with high confidence. Furthermore, indications on chromosome attachment in meiosis I (bipolar or monopolar) can be obtained, when used carefully and in a statistically meaningful manner. Cohesion of chromosome arms or the centromere region, and chiasmata count can be determined (Fig. 1) (e.g., see (9–11)). The technique described here is also useful to determine localization of proteins involved in protection of centromeric cohesin during meiosis I (Sgo2 and PP2A), Cohesin complex proteins such as Rec8, or Spindle Assembly Checkpoint proteins such as Mad2 (11) and Bub1 (1).

---

2. Materials

2.1. Culture and Collection of Oocytes and Zona Pellucida Removal

1. 9–14 week-old mice of the CD-1 (Swiss) strain (Janvier, France).

2. M2 medium (Sigma or Millipore) containing 100 μg/ml dbcAMP (dibutyryl cyclic AMP) where indicated.

3. Embryo culture certified mineral oil (Sigma).

4. 60 mm easy grip polystyrene tissue culture dishes (Falcon).

5. Binocular microscope with heating stage set to 38°C.

6. Tissue culture incubator, set to 38°C.

7. Mouth pipette apparatus: Pasteur pipet with a drawn out tip (use a flame, and manually break off the very end of the tip to obtain a tip diameter small enough for one oocyte to enter

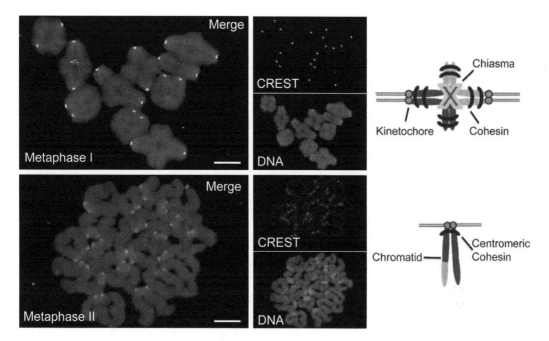

Fig. 1. Chromosome spreads in metaphase I (*upper panel*) and metaphase II (*lower panel*). Centromere staining allows to distinguish individual sister kinetochores in metaphase II but not metaphase I. A schematic presentation of the chromosome figures observed in meiosis I and II is shown next to the spreads. Important note: all mouse chromosomes are telocentric.

pasteur pipet without force). Connect pasteur pipet to a plastic tubing and mouth piece. This is used to collect and transfer oocytes after harvesting.

8. 21G needle mounted on a 5 ml syringe.

9. Tyrode's acid solution (12): 137 mM NaCl, 2.7 mM KCl, 1.8 mM $CaCl_2 \cdot 2H_2O$, 0.5 mM $MgCl_2 \cdot 6H_2O$, 5.5 mM D-glucose, 0.09 mM PVP (Polyvinylpyrrolidone).

Bring up to final volume with Brown's water (B. Brown Medical France).

Prepare 50 ml of the Tyrode's acid solution and store in 1 ml aliquots at −20°C.

2.2. Chromosome Spreading and Immunofluorescence

1. 8 well 8 mm Microscope slides (Thermo Scientific) (see Note 1).

2. Binocular microscope with a stage at room temperature.

3. Precise pH meter (± 0.01 pH).

4. 24 × 50 mm cover slips.

5. Citifluor.

6. Nail polish.

7. PBS (Phosphate-Buffered Saline): 140 mM NaCl, 1.9 mM NaH$_2$PO$_4$, 8.9 mM Na$_2$HPO$_4$, pH 7.4.

8. 16% Paraformaldehyde solution in PBS, pH 9.2
Prepare 20 ml of 16% paraformaldehyde stock solution by adding 3.2 g of paraformaldehyde to filtered sterilized water (leave enough volume to add 10× PBS and for adjusting pH!) and heat carefully to 55–60°C (see Note 2), ensuring that the temperature of the solution does not rise above 65°C. Add 10 M NaOH until the solution becomes clear. Filter solution through filter paper and add 10× PBS for a final concentration of 1× PBS. Adjust pH to 9.2 and add H$_2$O for a final volume of 20 ml. Aliquot in 1 ml volumes and store at –20°C.

9. Chromosome spreading solution (1): 1% Paraformaldehyde, 0.15% Triton X100, 3 mM DTT
Prepare 10 ml of chromosome spreading solution in filter-sterilized, deionized H$_2$O. Paraformaldehyde is added last. Adjust pH of spreading solution to pH 9.2–9.3 with 1 M NaOH or 1 M HCl, at room temperature. The solution should be used immediately, and remaining solution should be thrown away.

10. Blocking solution: 3% BSA in PBS.

11. Hoechst solution: 5 μg/ml Hoechst 33342 (Sigma) in PBS (see Note 3).

12. Propidium Iodide solution: 2 μg/ml Propidium Iodide (Molecular Probes) in PBS (see Note 3).

13. Antibody solution: Use antibodies at the appropriate dilution in Blocking solution. For labeling CREST, use the following dilutions:

Human anti-CREST antibody (HCT-100, Immunovision) 1:150.

Cy2 or Cy3 secondary anti-human antibody (Interchim) 1:200.

3. Methods

3.1. Oocyte Culture

1. Prewarm 500 μl of PBS and 2 ml dbcAMP-containing M2 medium in a 38°C water bath.

2. Make up tissue culture dishes containing small droplets (around 50–100 μl) of dbcAMP-treated M2 medium and untreated M2 medium under mineral oil and maintain at 38°C.

3. Sacrifice non-hormonally primed female mice by cervical dislocation, open up the peritoneal cavity, and dissect out the ovaries.

4. Place ovaries in prewarmed PBS in tissue culture dish.

5. Transfer ovaries into a drop of 500 μl pre-warmed M2 medium with dbcAMP.

6. Obtain germinal vesicle (GV)-stage oocytes by puncturing follicles of ovaries with the syringe-mounted needle on the heated stage of the binocular microscope (see Note 4). GV is the term used to refer to the large nucleus of the oocyte. An intact GV is a morphological indicator that the oocyte is arrested at the dictyate stage of prophase I.

7. Collect oocytes that have been released from ovaries by mouth pipetting. Oocytes should not be aspirated with undue force, but should smoothly enter the pipet with no more force than used for normal breathing. Be careful to select only oocytes that are fully grown (fully grown oocytes exhibit a smooth and full appearance with a centrally located GV).

8. Transfer all GV-stage oocytes into droplets in M2 medium with dbcAMP covered with mineral oil.

9. Transfer oocytes into droplets of M2 medium without dbcAMP. In order to ensure that dbcAMP is removed as completely as possible, transfer oocytes sequentially through four to five droplets of M2 medium with as little medium carry over as possible. Maintain oocytes in the incubator at 38°C.

10. After 90 min, select the oocytes that have undergone GV breakdown (GVBD) and discard the rest. GVBD marks the resumption of meiosis I following which extrusion of the first Polar Body should occur within about 7.5–9 h in this strain background (see Note 5).

3.2. Zona Pellucida Removal

1. Make up a dish containing droplets (around 50–100 µl) of Tyrode's solution. In the same dish place a droplet of M2 medium. Cover with mineral oil.

2. Transfer oocytes into droplets of Tyrode's solution with as little medium carry-over as possible (see Note 6). Ensure that oocytes are moved continuously from one drop to the next by mouth pipetting in order to avoid them sticking to the surface of the Petri dish once the zona pellucida starts to disappear (see Note 7) (Fig. 2a).

3. Following zona dissolution, transfer oocytes immediately to a fresh drop of M2 medium (in the same Petri dish) in order to remove remnants of tyrode's acid solution.

4. Transfer oocytes to a fresh droplet of M2 medium under oil for further culture and recovery in the incubator. No tyrode's acid solution should be carried over. At this point, no zona pellucida should be visible (Fig. 2a).

3.3. Spreading (See Note 8)

1. Place a drop (~100 µl) of chromosome spread solution in one 8 mm well of an 8-well printed slide.

2. Transfer no more than three zona-free oocytes with as little M2 medium as possible into the drop of chromosome spread

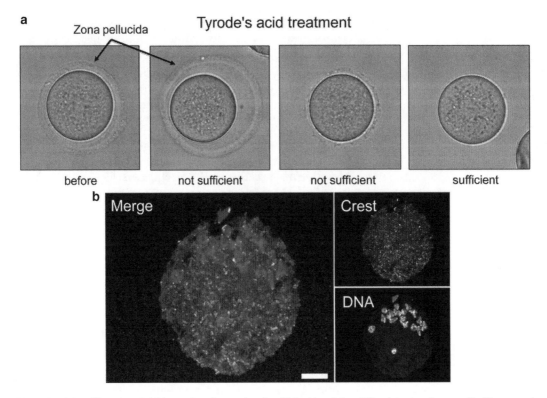

Fig. 2. Tyrode's acid treatment. (**a**) Examples of correct, or insufficient tyrode's acid treatment and zona pellucida removal. (**b**) Consequences of insufficient removal of the zone pellucida on subsequent chromosome spread and CREST staining.

solution. The transfer should be done very quickly, using a microscope without a heating stage, or with the heating stage turned off.

3. Observe oocyte dissolution under the microscope, which should happen in less than 5 s.

4. Leave the slide to air dry at room temperature (without moving slide around unnecessarily).

5. When slides are dry either proceed immediately with immunostaining or transfer slides to storage at –20°C, where they can be kept for weeks.

3.4. Immunolabeling and Image Acquisitions

1. Wash slides three times for 3 min and two times for 10 min with 1× PBS at room temperature to remove dirt and residual cellular material (see Note 9).

2. Pre-incubate slides with 100 µl blocking solution to block nonspecific antibody binding sites.

3. Add 50 µl blocking solution containing anti-CREST antibody to the well and incubate for 2–3 h to stain the centromeres (see Note 10).

4. Wash spreads three times for 10–15 min in 1× PBS.

5. Incubate spreads with 50 μl secondary antibody solution for 60–90 min.

6. Wash spreads three times for 10–15 min in 1× PBS.

7. Incubate spreads for 10 min with 50 μl Hoechst or Propidium iodide solution to stain chromosomes.

8. Remove chromosome staining solution, by washing spreads once for not more than 3 min with 1× PBS.

9. Add 2 μl of mounting medium into each well containing a chromosome spread (see Note 11).

10. Slowly lower coverslip onto wells to equally disperse mounting medium without generating air-bubbles.

11. Seal first the corners and then the edges of the cover slip with nail polish.

12. Analyze oocyte spreads by fluorescence microscopy (see Notes 12 and 13)

4. Notes

1. These printed slides define individual wells in which oocytes are spread. It is important that slides are chosen with well diameters of at least 8 mm. Using printed slides greatly enhances the number of oocytes that can be analyzed on one slide, significantly reduces the amount of antibody solution that has to be used, and facilitates the search for spreads when doing acquisitions. Slides are cleaned with ethanol and air-dried prior to use.

2. Paraformaldehyde is toxic. It can cause severe skin irritation upon contact and is a suspected cancer hazard. Protective equipment should be worn, and the powder should be handled under a chemical fume hood.

3. Hoechst and Propidium Iodide are potentially mutagenic and carcinogenic and should be handled with proper protective equipment.

4. Oocytes are kept at 38°C during all microscopic manipulations.

5. We always verify how chromosome segregations take place in controls in meiosis I and after activation in meiosis II in order to confirm that our manipulations have not interfered with correct chromosome segregation. Less than optimal medium, or temperature, or even too harsh handling of oocytes by the user translates into aberrant chromosome segregation, even though Polar Body extrusion may take place normally.

6. This is a critical step, as the presence of the zona pellucida will delay oocyte dissolution in the spreading solution and thereby affect the quality of the chromosome spread. Also, inadequate oocyte dissolution will make subsequent antibody labeling almost impossible. Zona pellucida removal requires some practice. Don't be discouraged!

7. Usually, the zona pellucida becomes detached from the oocyte before disappearing completely. It is very important to leave oocytes sufficiently long in tyrode's acid solution to completely detach the zona pellucida, but not too long as this risks compromising the integrity of the oocyte. In the beginning it is preferable to treat no more than four oocytes at the time.

8. We use prometaphase I (around 4 h after GVBD), metaphase I, anaphase I, metaphase II, and anaphase II oocytes from different mouse strains for chromosome spreads. Usually, the zona pellucida is removed 30–45 min before spreading, to give oocytes enough time to recover. For spreads to analyze meiosis II sister chromatid separation, the zona pellucida is removed from oocytes in CSF arrest, and oocytes are left to recover for 45 min. Then, in order to overcome CSF arrest, oocytes are activated by transfer into M2 medium containing strontium (Strontium is added just before use to a final concentration of 10 mM, from a 1 M stock solution stored at –20°C). After 40 min incubation, sister chromatids would have segregated, but not yet become decondensed, and can be used for spreading. The activation protocol and timing of anaphase II onset after activation has to be determined empirically for different strains of mouse.

9. To wash spreads and incubate them with antibodies, solutions are carefully pipetted into the wells. We try to dissolve oocytes in the center of the wells and to add washing or antibody solutions from the periphery.

10. To study localization of another protein at centromeres simultaneously with CREST, sequential labeling is recommended to avoid steric hindrance. When staining for proteins that are also present in the cytoplasm, special care has to be taken to distinguish the real signal of the protein associated with chromosomes from background signal due to cytoplasm that may have collapsed onto chromosomes (In this case, often the whole chromosome is stained). It is absolutely essential to include negative controls and to compare the signal obtained by this spreading technique to the signal obtained in whole oocyte fixation immunofluorescence studies.

11. Alternatively, Hoechst or Propidium iodide can be added to the mounting medium.

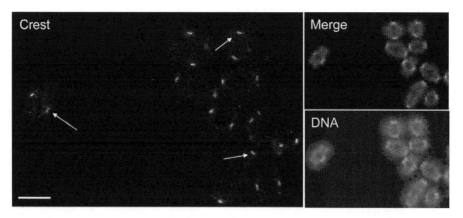

Fig. 3. Mono-oriented sister kinetochores. Depending on the spread, sister kinetochores may sometimes be discernible as two closely approximated *dots* (as in the cases highlighted by *arrows*), even though they are mono-oriented in meiosis I.

12. Oocyte spreads are analyzed by fluorescence microscopy, using a 63× or 100× objective for acquisitions. It is recommended to first search for the location of the spread chromosomes under low magnification (e.g., 20× objective), as it is often difficult to find them. For best acquisitions, we use a Confocal Leica SP5/ AOBS microscope and take acquisitions using the smallest useful z-step size, and then either select the individual z-section that provides the best image or obtain a z-projection of a series of z-sections.

13. High quality chromosome preparations are required for best immunolabelling results essential for proper analysis of cohesion. In particular, remains of cytoplasm and other cellular material will reduce the labeling signal and generate high levels of background (Fig. 2b). Ideally, individual chromosomes should be spread apart sufficiently so that their number and morphology can be assessed easily. The CREST labeling is a direct means of detecting whether sister kinetochores are attached to the same pole (monopolar) or to the opposite pole (bipolar), when used carefully. Usually, sister kinetochores in meiosis I appear as a single CREST signal, but as can be seen in Fig. 3, two signals very closely approximated to one another can occasionally be distinguished. Notably however, in such instances, the two sister kinetochore dots are very close together, contrasting with metaphase II spreads in which sister kinetochores are clearly separated. Spreads in metaphase I or II reflect the tension that has been applied to the chromosomes, because treatment with nocodazole (which depolymerizes spindles) in metaphase II prior to spreading reduces the distance between sister kinetochores to what is observed in metaphase I (J.P. Chambon and S. Touati, unpublished observation).

Acknowledgments

We thank S. Touati for comments on the manuscript. K.H. was supported through a PhD fellowship by the Ministère de la Recherche and ARC. Research in the group of K.W. is supported by a grant from La Ligue Nationale Contre Le Cancer (Comité d'Ile de France), the UPMC and CNRS.

References

1. Hodges CA, Hunt PA (2002) Simultaneous analysis of chromosomes and chromosome-associated proteins in mammalian oocytes and embryos. Chromosoma 111:165–169

2. Nasmyth K, Haering CH (2009) Cohesin: its roles and mechanisms. Annu Rev Genet 43:525–558

3. Petronczki M, Siomos MF, Nasmyth K (2003) Un menage a quatre: the molecular biology of chromosome segregation in meiosis. Cell 112:423–440

4. Hassold T, Hunt P (2001) To err (meiotically) is human: the genesis of human aneuploidy. Nat Rev Genet 2:280–291

5. Revenkova E, Herrmann K, Adelfalk C, Jessberger R (2010) Oocyte cohesin expression restricted to predictyate stages provides full fertility and prevents aneuploidy. Curr Biol 20:1529–1533

6. Chiang T, Duncan FE, Schindler K, Schultz RM, Lampson MA (2010) Evidence that weakened centromere cohesion is a leading cause of age-related aneuploidy in oocytes. Curr Biol 20:1522–1528

7. Lister LM, Kouznetsova A, Hyslop LA, Kalleas D, Pace SL, Barel JC, Nathan A, Floros V, Adelfalk C, Watanabe Y, Jessberger R, Kirkwood TB, Hoog C, Herbert M (2010) Age-related meiotic segregation errors in mammalian oocytes are preceded by depletion of cohesin and Sgo2. Curr Biol 20:1511–1521

8. Hodges CA, Revenkova E, Jessberger R, Hassold TJ, Hunt PA (2005) SMC1beta-deficient female mice provide evidence that cohesins are a missing link in age-related nondisjunction. Nat Genet 37:1351–1355

9. Daniel K, Lange J, Hached K, Fu J, Anastassiadis K, Roig I, Cooke HJ, Stewart AF, Wassmann K, Jasin M, Keeney S, Toth A (2011) Meiotic homologue alignment and its quality surveillance are controlled by mouse HORMAD1. Nat Cell Biol 13:599–610

10. Niault T, Hached K, Sotillo R, Sorger PK, Maro B, Benezra R, Wassmann K (2007) Changing Mad2 levels affects chromosome segregation and spindle assembly checkpoint control in female mouse meiosis I. PLoS One 2:e1165

11. Hached K, Xie SZ, Buffin E, Cladiere D, Rachez C, Sacras M, Sorger PK, Wassmann K (2011) Mps1 at kinetochores is essential for female mouse meiosis I. Development 138:2261–2271

12. Nicolson GL, Yanagimachi R, Yanagimachi H (1975) Ultrastructural localization of lectin-binding sites on the zonae pellucidae and plasma membranes of mammalian eggs. J Cell Biol 66:263–274

Chapter 15

Preparation of Mammalian Oocytes for Transmission Electron Microscopy

Lynne Anguish and Scott Coonrod

Abstract

The visualization of subcellular organelles and structures is a valuable tool for understanding cellular changes that occur in oocytes and early embryos as a result of genetic alterations, incubation conditions, drug treatments, and many other manipulations. Preparing oocytes for transmission electron microscopic analysis can be challenging as these cells cannot be visualized without a microscope and they are more susceptible to mechanical disruption during manipulation. Here we describe methods for immobilizing oocytes on either a solid surface or within a matrix and then document our embedding techniques which work well for preserving the ultrastructure of the mouse oocyte.

Key words: Transmission electron microscopy, Epoxy embedding, Oocytes, Primary aldehyde fixation, Secondary osmium tetroxide fixation, Uranyl acetate en bloc staining, Early embryos, Oocyte ultrastructure

1. Introduction

Preparing mammalian oocytes and embryos for transmission electron microscopy (TEM) presents many challenges due to their small size and susceptibility to degradation at room temperature (1). Due to the toxicity of most of the chemicals required for proper fixation and embedding, transferring single or groups of oocytes from solution to solution under a dissecting microscope on the bench is impractical and dangerous. All manipulations (beginning with primary fixation) should be done under a proper functioning fume hood. The key to easily handling oocytes or embryos for TEM processing is to immobilize them on a solid support or embed them in a solid matrix which is still permeable to the processing chemicals.

The cationic polymer polylysine is frequently used to coat glass or plastic surfaces in tissue culture systems to attach non- or slightly

Hayden A. Homer (ed.), *Mammalian Oocyte Regulation: Methods and Protocols*, Methods in Molecular Biology, vol. 957, DOI 10.1007/978-1-62703-191-2_15, © Springer Science+Business Media, LLC 2013

adherent cells for growth of monolayers (2). In aqueous solution, it contains a positively charged hydrophilic amino group which electrostatically binds to the solid surface and negatively charged molecules in the cell membrane. As with tissue culture cells, it can also be used to immobilize oocytes or embryos on plastic or glass coverslips for fixation and embedding for TEM (3). The advantage of using this method is that the oocytes all end up on the same plane on the block face which allows for simultaneous sectioning of up to five or six oocytes.

As an alternative, a number of researchers have used agar to embed oocytes after primary (4, 5) or secondary (6, 7) fixation, often by concentrating the oocytes at the tip of a tube by centrifugation. Others have used a protein matrix produced by aldehyde treatment of bovine serum albumin and centrifugation (8). In addition to the coverslip adherence technique, here, we also present a method which utilizes low melting point agarose to capture the oocytes (without centrifugation) after a brief fix in paraformaldehyde. This modified technique likely reduces any artifactual effects that higher heat and centrifugal force may have on oocyte ultrastructure.

The actual processing of oocytes after immobilization is fairly standard with an aldehyde primary fixation, osmium tetroxide secondary fix, optional en bloc staining with uranyl acetate, graded ethanol dehydration series, and embedding in epoxy resin. There are many variations of this basic protocol in the literature and they could also be used. We have found that the steps we present here have given us excellent preservation of oocyte and early embryo ultrastructure with minimal artifactual aberrations. For further reading on TEM techniques, principles, and theory, we would recommend referring to the texts by Borzzola and Russell (9) or Hayat (10).

2. Materials

Prepare all solutions using ultrapure water (deionized MilliQ water with a resistance of 18 MΩ cm at 25°C) and EM grade chemical reagents. EM grade reagents can be purchased from Electron Microscopy Sciences, Ted Pella, Inc., and others. Many of the chemicals are quite toxic, heavy metals, or radioactive (uranyl acetate) and should be worked with in a fume hood only with proper personal protective equipment (nitrile gloves and lab coat). Waste should be collected, held, and disposed of according to waste disposal regulations in your area or institution. Obtain and read all MSDS documents for the chemicals you will be using.

2.1. Primary Fixative: 0.1 M Sodium Cacodylate (pH 7.4), 0.01 M Magnesium Chloride, 2.5% Glutaraldehyde, 4% Paraformaldehyde, 0.1% Tannic Acid

1. 0.4 M Sodium Cacodylate Buffer, pH 7.4 (see Note 1): Dissolve 21.4 g sodium cacodylate in 250 ml of MilliQ water. The pH should be close to 7.4 without adjustment.

2. 2 M Magnesium chloride-hexahydrate (see Note 2): Dissolve 10.165 g $MgCl_2$ $(6H_2O)$ in approximately 15–20 ml of MilliQ water in a 25 ml volumetric or graduated cylinder. When nearly dissolved, add water to a volume of 25 ml and mix until completely dissolved. *Note: These first two reagents can be stored at 4°C for up to 1 year.*

3. 10% glutaraldehyde (EM grade).

4. 16% paraformaldehyde (EM grade).

5. Tannic acid (EM grade).

6. To prepare 50 ml of primary fix, mix together 12.5 ml 0.4 M sodium cacodylate buffer, 12.5 ml 10% glutaraldehyde solution, 0.05 g tannic acid, 250 µl 2 M magnesium chloride, and 12.5 ml 16% paraformaldehyde solution. When thoroughly mixed, add MilliQ water to bring the volume to 50 ml. Mix and filter through a 0.2–0.45 µm filter. It is best to prepare the primary fix fresh for each fixation (within 24 h of use). However, it can be stored at 4°C and used for up to 1 week if several fixations are to be carried out during that time period (see Note 3).

2.2. Secondary Fix: 1–1.5% Osmium Tetroxide, 0.1 M Sodium Cacodylate (pH 7.4)

1. 4% aqueous osmium tetroxide: Purchase sealed vials of 4% aqueous EM grade osmium tetroxide, from a commercial supplier, preferably in small volumes (such as 5 ml per vial). *Never open a vial or work with osmium outside of a fume hood!!*

2. 0.4 M Sodium Cacodylate Buffer (pH 7.4): See item 1, Subheading 2.1.

3. To prepare 4 ml of the secondary fix: *Prepare in a capped tube in a fume hood!* To make 4 ml of 1.5% osmium, mix 1.5 ml of 4% osmium, 1 ml of 0.4 M cacodylate buffer, and 1.5 ml MilliQ water. To make 1% osmium, mix 1 ml of 4% osmium with 1 ml of 0.4 M cacodylate buffer and 2 ml of MilliQ water. Place cap on tube, invert to mix thoroughly. Secondary fix should be prepared immediately before use and any remaining solution discarded as it will quickly oxidize (see Note 4).

4. Potassium ferrocyanide (optional): Potassium ferrocyanide can be added to the secondary fix at a concentration of 1.5% if visualization of glycogen is desired. Dissolve 60 mg of potassium ferrocyanide in 4 ml of secondary fixative (see Note 5).

2.3. Uranyl Acetate En Bloc Stain (Optional)

1. 2% uranyl acetate: Dissolve 1.0 g in 50 ml MilliQ water and store in the dark at 4°C for up to a month. Immediately before use, filter a small aliquot of the uranyl acetate solution to remove undissolved salts using a 0.2 µm filter.

Although uranyl acetate contains radioactive isotopes and must be disposed of as radioactive waste, no special permits are required to use it. Uranyl salts are photolabile and should not be used (or stored) in bright light. All gloves, pipettes, tubes, etc. that come in contact with uranyl acetate should be disposed of as *radioactive waste*, not chemical waste, unless your institution has other specific guidelines for disposal (see Note 6).

2.4. Graded Ethanol Dehydration Solutions

1. 50, 70, 85, and 95% ethanol: Make 10 ml of each of the listed ethanol concentrations as follows; 5 ml ethanol plus 5 ml MilliQ water, 7 ml ethanol plus 3 ml MilliQ water, 8.5 ml ethanol plus 1.5 ml water, 9.5 ml ethanol plus 0.5 ml water, respectively. The ethanol used to make these solutions should be EM grade 100% (200 proof) absolute ethanol. Pharmtec and Koptec make an absolute ethanol product that works well for this purpose. Solutions should be made no more than 1 week in advance and kept at 4°C (in a refrigerator rated for flammable compounds). A volume of 10 ml is adequate for most oocyte preps (unless you have more than ten samples). The solutions should be prepared in screw top tubes and the cap sealed with parafilm. An opened bottle of 100% ethanol should also be sealed with parafilm and stored in a dessicator at room temperature as 100% ethanol is hygroscopic.

2.5. Epoxy Resins for Embedding

Work with resins and their components only in a fume hood. Resin should be made fresh for each step of the infiltration or embedding procedure. The resin components are best added by weight, but the volume equivalents are given as well. Resin components should be stored at room temperature. However, if necessary, they can be warmed individually up to 60°C to decrease viscosity so that the volume can be measured (see Note 7 before starting).

1. LX112 Resin (Ladd Research, Williston, VT, Cat# 21210).

2. Dodecenyl succinic anhydride (DDSA): Resin component (EM grade).

3. Methyl-5-norbornene-2,3-dicarboxylic anhydride (NMA): Epoxy hardener (EM grade).

4. 2, 4, 6-Tri(dimethylaminomethyl) phenol (DMP-30): Resin accelerator (EM grade).

5. Complete Epoxy Resin: To make approximately 18 ml of resin, place the following in a screw-capped disposable 30–50 ml *polypropylene or glass* tube or vial (*warning: polystyrene will dissolve, do not use!*): 9.7 g (9.2 ml) LX112, 3.2 g (3.5 ml) DDSA, and 5.9 g (5.5 ml) NMA. (Alternatively, you can also follow the LX-112 manufacturer's instructions using an approximate ratio of Solution A:Solution B of 4:6.) Seal capped tube with parafilm, place on an end-over-end or nutating mixer, and

allow to mix thoroughly for about 10 min or shake vigorously by hand for at least 1 min. Add 310 μl of the accelerator, DMP-30, return to mixer (or shake by hand) and mix thoroughly. Note: Use a wide orifice pipette tip or cut the end off of a standard tip as the accelerator is viscous. When preparing resin for the final embedding step place the tubes of thoroughly mixed resin in a centrifuge and spin at approximately $300 \times g$ for 5 min to remove bubbles. This step is not needed for infiltration steps.

2.6. Other Materials

1. Poly-L or poly-D-Lysine coated Thermanox (Nunc) 22 mm coverslips or MatTek dishes (MatTek Corporation, Ashland, MA): These products can either be purchased pre-coated or you can coat them in bulk using a solution of 1 mg/ml aqueous poly-L-lysine (PLL). Place coverslips in a dish containing ~10 ml PLL and rock for at least 30 min at room temperature. Wash coverslips at least ten times in MilliQ water and either air-dry or store at 4°C in 100% ethanol (allow to air-dry before use).

2. 10× Dulbecco's phosphate buffered saline (PBS) pH 7.2–7.4: (commercially available). 1× is made by adding 1 ml of 10× to 9 ml of MilliQ water. Store at 4°C.

3. 4% paraformaldehyde in PBS: Make by adding 5 ml of 16% paraformaldehyde and 2 ml of 10× PBS to 13 ml of MilliQ water. Store at 4°C.

4. 2% SeaPlaque low melting point agarose (Lonza Cat #50101) in MilliQ water: Add 0.08 g agarose to 4 ml water. Dissolve in a microwave oven or water bath >65°C. Aliquot 1 ml amounts into 1.5 ml microcentrifuge tubes and store at 4°C up to 6 months.

5. Dissecting microscope with epi-illumination.

6. Plastic disposable pipettes (5, 10, and 25 ml) and pipetting device.

7. Transfer pipettes (3 ml) (VWR Cat #16001-176 or other supplier).

8. An electric curing oven that is small enough to be placed inside a fume hood and can maintain a temperature of 60°C.

9. 200 and 1,000 μl pipetters and tips.

10. 35 mm petri dishes with 100 or 150 mm petri dishes, or 6- or 12-well plates.

11. Beem capsules (Electron Microscopy Sciences, Hatfield, PA, Cat #7000).

12. Flat molds (Electron Microscopy Sciences, Hatfield, PA, Cat # 70900).

13. PTFE spray dry release agent for coating molds or Beem capsules.

14. 6″ wood applicator sticks.

15. Parafilm M.

16. Glass screw top 20 ml scintillation vials.

17. Nutating or end-over-end mixer.

18. Fume hood.

19. Nitrile gloves and solvent-resistant lab coat.

20. Solvent-resistant permanent marker pens or permanent markers and clear tape.

21. Razor blade.

22. Slide duplicating mold (Electron Microscopy Sciences, Hatfield, PA, Cat # 70172).

23. Ziplock bags, quart to gallon size.

24. Ice bucket with crushed ice.

25. PLL-coated microscope slides.

26. Brightfield microscope with a 20–40× objectives.

27. Gilder thinbar 200 mesh grids (Electron Microscopy Sciences, Cat # GT200).

28. Reynold's lead citrate or Sato's calcined lead citrate stains.

3. Methods

All procedures are carried out in a fume hood at room temperature or on ice as specified. Minimum protection of nitrile gloves pulled over the cuffs of a lab coat should be worn for all procedures. It is best to label vials or dishes with permanent marker and place a piece of clear tape over the label as many of the solutions used in specimen preparation will dissolve the marker. Solvent-resistant markers are also commercially available.

3.1. Immobilization of Oocytes

Note: For a detailed description of the oocyte/embryo isolation procedure please *see Manipulating the Mouse Embryo* (1). Prior to preparation for EM, isolated oocytes (maintained in serum-containing collection medium) may be freed of cumulus cells using hyaluronidase if they are not relevant to the study as it is easier to see the oocytes in the plastic resin without them. Oocytes should then be washed from the serum-containing collection media through at least two washes (of 1–2 ml volume) in serum-free PBS, as the serum prevents oocytes from adhering tightly to the PLL-coated slides. Early embryos up to the blastocyst stage can be used in these protocols as well, but for brevity, here we will use the term "oocytes" to encompass all stages. The above procedures are carried out at room temperature. Immobilization can be performed using one of the following methods.

3.1.1. Immobilization on PLL-Coated Thermanox (Nunc) Coverslips (See Note 8)

1. Place coated 22 mm coverslips into a 35 mm petri dish (or well of a 6- or 12-well plate, depending on the sample number).

2. Add a microdrop (about 10 µl) of PBS to the coverslip.

3. Place oocytes in groups of three to ten in the microdrop of PBS on the coverslip. Up to four to five drops can be placed on each coverslip and we would recommend several drops with at least 10 oocytes per coverslip as cells can be physically dislodged and lost during fixation and embedding procedures (*don't place your eggs all in one basket!*).

4. Allow the oocytes to settle for 2–5 min with the dish covered to prevent evaporation. Most of the oocytes will adhere to the coverslip within this time, however if a few are not, they can be discarded.

5. Add an approximately equal volume of primary fix to the PBS drop and allow to equilibrate for 15 min at room temperature. You may then continue with primary fixation below (Subheading 3.2).

The above steps can also be used for immobilizing oocytes on poly-D-lysine coated MatTek dishes which have glass coverslip bottoms and can be purchased in a number of configurations already coated with lysine. The major drawback with the MatTek dishes is that the glass must be completely removed from the embedding resin before sectioning (see Note 9).

3.1.2. Immobilization by Embedding Oocytes in Agarose (See Note 10)

1. Remelt an aliquot of 2% SeaPlaque agarose by placing the tube in a microwave, water bath, or dry heat block at >65°C.

2. When agarose is melted, place tube in a dry heat block, set to 40°C, and allow agarose to cool to 40°C.

3. Place several wide bore 200 µl pipette tips in the heat block to warm.

4. Prepare a petri dish containing 50 µl droplets of PBS.

5. Transfer a group of oocytes to the PBS droplet.

6. Add 50 µl of 4% paraformaldehyde in PBS to the PBS droplet containing the oocytes. Allow to fix for 15 min. This quick fixation step before adding agarose helps to reduce heat-induced artifacts.

7. Remove approximately ten oocytes from the fix in a volume of about 10 µl and place in the bottom of a thin-walled PCR tube.

8. Using the heated pipette tips from step 3, pipette 10–20 µl of agarose into the bottom of the PCR tube with the oocytes.

9. Allow the agarose to solidify.

10. Cut off the tip of the tube with the agarose plug and oocytes using a razor blade and flip the plug of agarose out of the tip with a small pointed object such as a syringe needle.

11. With the plug on the stage of a dissecting microscope, trim off the excess agarose so that the plug is no more than 1 mm in at least one dimension (see Note 11). Several plugs from the same sample can be placed in a glass scintillation vial in 2–5 ml of primary fix (See Subheading 3.2). All of the following processing steps should be carried out with the vials on a nutating rotator to allow for even and rapid penetration of chemicals into the samples.

3.2. Primary Fixation

3.2.1. Oocytes on Coverslips or Dishes

1. After the 15 min equilibration in 50% primary fix, gently flood dishes with 2 ml of fix by placing the pipette tip against the side of the dish and slowly allowing the fix to enter the plate. Never pipette fluids into the dish quickly or into the center of the dish as this can dislodge oocytes that are adhered to the coverslip or dish.

2. Allow oocytes to fix at room temperature for 2–3 h. Do not rock or shake dishes.

3. After primary fixation you may elect to go on to secondary fixation or store samples in primary fix overnight as follows. Dishes (35 mm) can be placed in a 100 or 150 mm petri dish as secondary containment and the larger dish placed in a ziplock bag. If you have several samples and are using 12- or 6-well plates, place the plate in a ziplock bag. The bag can then be placed at 4°C overnight being careful not to tip plate contents or disrupt oocytes. Do not store longer than 18 h in the primary fix as leaching of the cytoplasmic components may occur and membrane lipids and lipid droplets are not immobilized during primary fixation (see Note 12).

4. Remove primary fix into a chemical waste container with a transfer pipette.

5. Wash by gently adding 2–3 ml 0.1 M sodium cacodylate buffer to dishes or vials and allow to stand for 15 min. Vials may be returned to the nutating mixer for washes. Washing may be performed directly after the 3 h primary fixation or after overnight storage at 4°C. Remove wash into chemical waste container with a transfer pipette.

6. Repeat washes three more times and prepare secondary fix during this time period. Go to Subheading 3.3 for secondary fixation.

3.2.2. Oocytes in Agarose Plugs

1. Add 2–3 ml of fix to the scintillation vial. Agarose plugs may be placed directly in fix in scintillation vials without the 15 min equilibration in 50% fix, as the oocytes were stabilized by the 2% paraformaldehyde treatment before embedding in agarose.

2. Place the vial in a rack on a nutating mixer in the hood for 3 h at room temperature. Vials can then be placed in a secondary container or ziplock bag and stored at 4°C overnight as above.

Do not store longer than 18 h in the primary fix as leaching of the cytoplasmic components may occur and membrane lipids and lipid droplets are not immobilized during primary fixation (see Note 12).

3. Perform washes as described in steps 4–6 in Subheading 3.2.1 above.

3.3. Secondary Fixation

1. Add approximately 1–2 ml of secondary fix to dishes or vials and allow to fix for 1 h at room temperature or 4°C overnight. Vials do not need to be on the mixer if left overnight in the cold. If placed in the cold overnight, dishes, plates, and vials should be placed in secondary containment as for primary fix. The concentration of osmium tetroxide in the fix should be 1% for oocytes in dishes and 1.5% for oocytes embedded in agarose.

2. Remove secondary fix into a chemical waste container with a transfer pipette.

3. Wash by adding 2–3 ml ice-cold MilliQ water (see Note 13) to dishes or vials and allow to stand for 15 min ensuring that vials and dishes are kept on ice. Vials should be placed in a small container with ice on the nutating mixer for washes if possible.

4. Remove wash into chemical waste container with a transfer pipette.

5. Repeat washes three more times.

After washing, en bloc staining (Subheading 3.4), dehydration, and plastic infiltration should all be performed on the same day to avoid artifacts.

3.4. En Bloc Staining with Uranyl Acetate

Although the presence of phosphate ions will precipitate uranyl ions in this step, the PBS used in oocyte immobilization should be well leached out of the sample by this time (see Note 14).

1. Immediately after the final MilliQ water wash, add 1–2 ml of freshly filtered (0.2 µm) cold 2% uranyl acetate (UA) to each dish or vial and place at 4°C in the dark for 2 h (in secondary containment as above).

2. During UA staining, make up the dilutions of ethanol and place on ice along with a tube each of MilliQ water and of 100% ethanol.

3. Remove UA to a separate waste container with a transfer pipette to discard as radioactive waste.

4. To wash, add 3 ml ice-cold MilliQ water to each dish or vial and allow to stand for 15 min on ice. (Vials may be shaken on ice as in step 3 in Subheading 3.3.)

5. Discard washes into the UA waste container.

6. Repeat one further wash.

3.5. Dehydration and Resin Infiltration

1. Immediately after removing the final wash from en bloc uranyl acetate staining, flood dishes and vials sequentially with 2–3 ml of ice-cold 50%, 70%, 85%, 95% and 100% ethanol for 15 min each.

2. Remove the used ethanol and discard as chemical waste.

3. During the ethanol treatments, prepare epoxy resin (without accelerator) and mix thoroughly to a volume sufficient for 1 ml per dish or vial.

4. Remove dishes or vials from ice after the cold 100% ethanol treatment and briefly allow them to warm to room temperature.

5. Treat with 100% ethanol at room temperature (RT) for 15 min.

6. During this time, add an equal volume of 100% ethanol to the epoxy resin prepared in **step 3** and vigorously shake to mix thoroughly.

7. Remove the RT 100% ethanol from dishes or vials, add approximately 2 ml of 50% resin/50% ethanol, and allow to infiltrate the sample for 2 h at room temperature. Shake vials on nutating mixer for all resin infiltrations.

8. During this time, prepare another batch of fresh epoxy resin (without accelerator) to a volume sufficient for 2 ml per dish or vial.

9. Remove the 50% resin to chemical waste and replace with 100% resin (without accelerator). Allow to infiltrate overnight at room temperature.

10. The next morning prepare another batch of resin *PLUS* accelerator in the same volume as used in step 3.

11. Remove the resin without accelerator from the vials or dishes and replace with about 2 ml of the resin plus accelerator per sample. Allow to infiltrate at room temperature for 2 h.

3.6. Epoxy Resin Embedding

Lightly spray molds with a Teflon-based dry release agent. This will allow the cured resin to be easily removed from the mold. Fill molds with freshly prepared epoxy resin plus accelerator which has been centrifuged to remove bubbles (see item 5, Subheading 2.5). A tiny piece of paper with the sample number or other distinguishing marks written in #2 pencil can be embedded in the plastic in the mold (in an area away from the oocytes) to identify the samples.

3.6.1. Thermanox Coverslips

1. Gently remove coverslips from the 35 mm dish (or 6- or 12-well plate) and invert over a coverslip well (oocytes toward the mold) in a resin-filled slide duplicating mold (see Note 15). This should be done gently and slowly to avoid dislodging oocytes and the creation of air bubbles.

2. Place molds in a curing oven at 60°C for 24 h.

3. Remove the molds from the oven and allow to cool.

4. After cooling, pop the embedded coverslips off of the mold.

5. Examine each coverslip on a dissecting microscope for groups of oocytes and place a dot over the groups (on the cover slip side of the plastic) using a fine-tipped marker.

6. Trim off the plastic around the groups of oocytes to make tiny squares using a single-edged razor blade (or other fine cutting tool such as a jeweler's saw or Dremel-type tool equipped with a fine diamond cutting blade). Often, the small coverslip piece attached to the square will pop off of the resin. The oocytes will be at the resin face, not on the coverslip so any pieces that fall off can be discarded.

7. Place the small squares with the oocytes at the tip of a flat mold well (with the oocytes facing the tip) and fill the well with fresh accelerator-containing resin.

8. Using a wood applicator stick or similar object, reorient the squares to the tip (if they move) and remove bubbles.

9. Place molds in a curing oven at 60°C for 24–48 h.

3.6.2. MatTek Dishes

1. Remove infiltrated epoxy and replace with fresh epoxy plus accelerator to a depth of 1–2 mm.

2. Cure in the oven and cool as in steps 2 and 3 in Subheading 3.6.1 above.

3. Snap the plastic part of the dish off of the square glass coverslip and dissolve the glass using hydrofluoric acid (HF) (see Note 9).

4. Trim the remaining epoxy disk, re-embed the squares, and cure as in steps 5–9 in Subheading 3.6.1 above.

3.6.3. Agarose-Immobilized Oocytes

1. Remove each small piece of agarose plug from the infiltrating resin with a large bore pipette tip and place at the tip of a flat mold well or Beem capsule.

2. Fill the flat mold wells with resin and treat as in steps 7–9 in Subheading 3.6.1 above.

3. Place a small drop of accelerator-containing resin in the tip of the Beem capsules and allow to cure as in steps 8 and 9 in Subheading 3.6.1. This prevents the small plug from floating up from the tip.

4. After cooling, fill the rest of the capsule with fresh resin and cure as before.

3.7. Sectioning, Staining, and Microscopy

Below we briefly describe our preferences for procedures we use beyond embedding; however, a detailed discussion of these methods is outside of the scope of this paper. Our sectioning techniques, using an ultramicrotome and diamond or glass knives, postsectioning staining of grids and use of TEM are fairly standard and closely follow those used for most other tissues prepared for TEM. For detailed discussion on post-embedding techniques, please refer to the following EM literature: Borzzola and Russell (9) or Hayat (10).

1. Cut semithin sections at ~100–150 nm.

2. Place cut sections on a PLL-coated microscope slide on a slide warmer and stain with conventional section stains (10, pp. 363–366) for about 2 min. It is important to stain approximately every tenth section when first cutting into the sample in order to not miss the oocyte.

3. Examine these sections on a brightfield microscope with a 20–40× objective to determine if the oocyte has been sectioned.

4. Use Gilder thinbar 200 mesh grids to pick up groups of sections. The holes in these grids are 113 μm but the bars are only 12 μm as opposed to 35 μm in conventional grids. This allows for adequate section support, with less grid obscuring the samples. We do not use formvar coated grids.

5. Stain the sections on the grids with 2% aqueous UA for 20 min.

6. Wash grids in three changes of MilliQ water.

7. Stain with lead citrate for 7 min followed by washing in five changes of MilliQ water. We prefer to use a modified Sato's calcined lead citrate (11) as it is far more stable than other lead stains. (The method for making calcined lead citrate is outlined in the Electron Microscopy Sciences catalog "Technical Tips" with their listing for lead citrate trihydrate.)

8. For electron microscopic imaging, magnifications of 11–26,000× are a good starting point to image oocyte internal structures. Figures 1–5 show typical organelles and their organization within oocytes and two-cell embryos using the methods described in this paper.

4. Notes

1. Sodium cacodylate buffer is preferred over phosphate buffers as phosphate ions cause precipitation of the uranyl ions when en bloc staining with uranyl acetate is performed. Sodium cacodylate should be weighed out in a fume hood or a dust mask worn as it contains arsenic. Do not breathe salts or let buffer come into contact with skin.

2. Magnesium or calcium ions in the primary fix help to preserve membranes and DNA. We prefer to use magnesium as its salts are more readily dissolved in water than calcium salts.

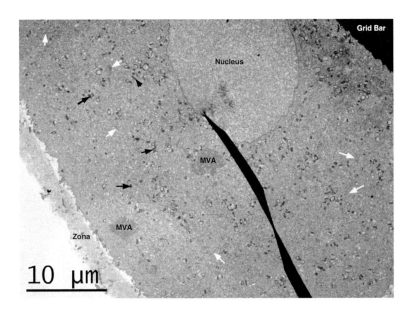

Fig. 1. Cross section of a GV oocyte at low magnification. Note the nucleus (Nucleus) with evenly distributed heterochromatin surrounded by the nuclear membrane. In great abundance in the cytoplasm are mitochondria (*black arrows*) and cytoplasmic lattices (*white arrows*), a cytoplasmic structure found only in oocytes and early embryos. Multivesicular aggregates (MVA) can be seen in the cytoplasm to the left of the nucleus. The oocyte is surrounded by the matrix-like zona pellucida (Zona), which can be seen at the bottom left of the figure. The *large black line* in the cytoplasm is a fold in the section. *Note*: It is difficult to image an entire oocyte without seeing the grid bar, as oocytes are approximately 100 μm and the spaces between the grid bars on a 200 mesh thin bar grid are 113 μm.

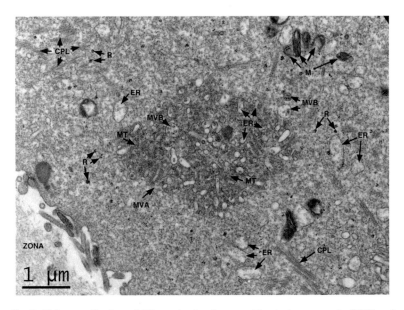

Fig. 2. High magnification of GV oocyte showing a multivesicular aggregate (MVA) and surrounding organelles. The aggregate contains multivesicular bodies (MVB), an abundance of endoplasmic reticulum (ER) and many clusters of microtubules (MT). In the surrounding cytoplasm are groups of ribosomes (polysomes) (R), mitochondria (M), and cytoplasmic lattices (CPL). At this magnification, the typical "railroad track" appearance of the CPLs can be seen. The endoplasmic reticulum (ER) appears swollen and filled with accumulated product.

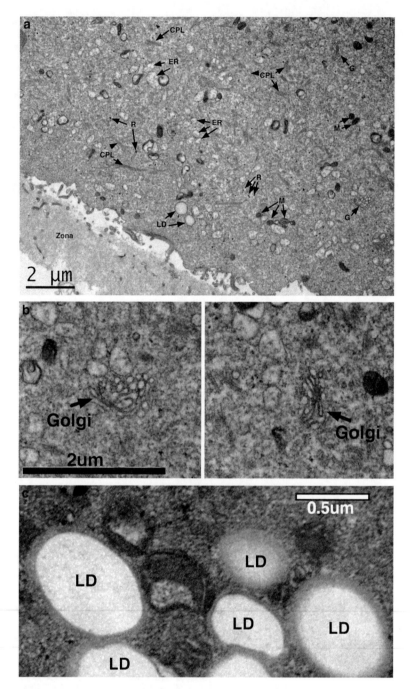

Fig. 3. Cortical region of GV oocyte. (**a**) Organelles are well distributed throughout the cytoplasm, including in the cortical region. Swollen ER (ER), ribosomes (R), cytoplasmic lattices (CPL), mitochondria (M), and lipid droplets (LD) are abundant. A golgi apparatus (G) is also visible. (**b**) The golgi apparatus appears smaller than seen in many somatic cells. (**c**) A higher magnification of lipid droplets (LD) from a 2-cell embryo reveals the lack of a distinct membrane, however there is a fuzzy electron dense layer on the hydrophilic/ hydrophobic interface of the droplet where bipolar molecules such as adipophilin accumulate. In oocytes (**a**) and two-cell embryos, the lipids within the droplets have been extracted (and, thus are seen as *white*) in the dehydration process. However, lipid droplets contain a variety of lipids and in cases where the lipids are oxidized by osmium (primarily unsaturated fatty acids), the droplets appear *black*.

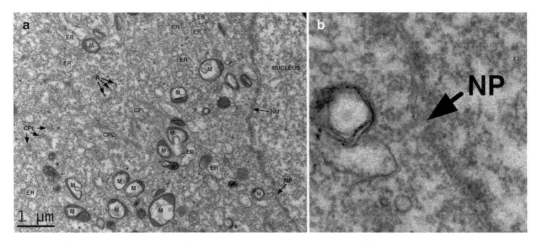

Fig. 4. Nuclear region of GV oocyte. (**a**) Abundant organelles appear to be well distributed as in the cortical region of the oocyte (Fig. 3). Note the double-layered nuclear membrane (NM) and the presence of a nuclear pore (NP). (**b**) Higher magnification of nuclear pore.

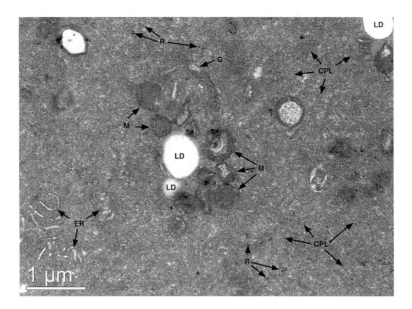

Fig. 5. Two-cell embryo. As shown here, the cytoplasm of mature metaphase II-arrested oocytes and early cleavage division embryos appears much denser than in GV oocytes. In addition, the cytoplasm is filled with cytoplasmic lattices (CPL) which appear as "brushstrokes on a canvas." The ER is not as swollen with product as in GV oocytes and often is seen in clusters instead of being more evenly distributed.

3. 10% Glutaraldehyde and 16% paraformaldehyde should be purchased as EM grade from a supplier to ensure consistency and purity. Both aldehydes will form polymers and degradation products with age and open vials or bottles should be stored at 4°C or lower (–20 or –80°C preferred). Any aldehyde solution showing a dark, as opposed to light yellow, color should be disposed of. Tannic acid promotes stabilization and

visualization of microtubules, cytoplasmic lattices, membranes, and microfilaments in the oocyte. It can be omitted if desired.

4. The glass vial with leftover 4% osmium can be sealed with parafilm and the whole vial placed in a screw top 50 ml disposable polypropylene centrifuge tube which can then be capped and sealed with parafilm. Osmium should never be stored directly in plastic but can be stored at 4°C in the dark and used within a few months. Any osmium tetroxide solution that has darkened beyond its usual light yellow color has oxidized and should be disposed of. Osmium preferentially binds to, and immobilizes, cell membrane lipids and lipid droplets. It will produce a brown color in the oocytes, making them easier to visualize in the plastic before sectioning.

5. When added to the secondary fix, potassium ferrocyanide will reduce osmium tetroxide producing a dark brown color. It preferentially binds to, and helps both stabilize and visualize, glycogen and membrane-containing structures. A major drawback with this reagent is that it extracts ribosomes and, thus these organelles are difficult to visualize when this fix is used.

6. If possible, uranyl acetate should be weighed in a fume hood. Additionally, nitrile gloves, a lab coat, and dust mask should be worn when working with this reagent. The scale and immediate area around it should be wiped with a damp paper towel which should then be disposed of as radioactive waste, as should used weigh boats, spatulas (disposable), tubes, syringe filters, syringes, etc. Non-disposable tubes or instruments in contact with uranyl acetate should be rinsed a minimum of three times (and the rinse collected as radioactive waste) before being placed through the regular laboratory wash cycle.

7. Epon-like epoxy resins provide a high contrast for visualization of oocyte ultrastructure when used for embedding in TEM. One of their drawbacks, however, is that these epoxys require an intermediary solvent after ethanol dehydration that is miscible with the resin to interface between the hydrophilic sample and the hydrophobic resin. Propylene oxide and acetone have been the most common solvents used for this purpose. However, these solvents also dissolve plastic coverslips and tissue culture dishes. Therefore, if you want to use plasticware to immobilize oocytes for processing, an alternative method must be used. We have found that the epon-alternative resin LX-112 (Ladd Research, Williston, VT, Cat# 21210) does not require the intermediary solvent and gives very consistent results. We use this resin at ratio that gives a medium to hard block (see manufacturer's instructions). The other components (DDSA, NMA, and DMP-30) can be purchased from Ladd Research or any other supplier of EM grade chemicals.

8. Given that Thermanox coverslips do not need to be separated from the resin after embedding, this is our preferred method

for preparation. These coverslips can be easily cut through with a glass or diamond knife or razor blade to trim the block before thin sections are cut. Often the coverslip piece will just pop off of the resin face as it is being trimmed, leaving the oocyte at the surface of the block ready for thick or thin sectioning.

9. Any glass chips or dust left on the resin block can ruin a glass or diamond knife edge. Previous reports indicated that removal of glass from the resin block can be obtained by placing the glass on solid dry ice or liquid nitrogen and then popping it off of the resin. However, we have not had consistent results using this procedure. Dissolving the glass with hydrofluoric acid (HF) works extremely well but in many institutions special training or a permit is required to use HF as it is extremely dangerous. Our institution has a special lab that will dissolve glass with HF for us for a small fee. If this is available for you, or you are set up to work with HF, glass coverslips give a clean, smooth surface to the resin block for sectioning.

10. The advantage of this method is that it prevents loss of oocytes by physical disruption during processing. If you have a limited number of oocytes to work with, this method may be the best for you. We use SeaPlaque agarose with a low melting point (Lonza, Rockland, ME, cat# 50101) as it can be maintained at a cooler temperature (35–40°C) after melting without re-solidifying. This limits artifacts within the oocyte which may develop from exposure to higher temperatures.

11. We prefer to use double-edged razor blades over single-edged blades or scalpel blades since the blade is thinner. Thicker blades tend to compress the sample before cutting and this could split the agarose plug rather than trimming it.

12. It is important to perform a complete primary fix at room temperature before placing oocytes at 4°C as microtubules are known to depolymerize at 4°C (12) and we have found that cellular structures and proteins migrate toward the oocyte cortex without a room temperature fixation first (13).

13. Since the oocytes are completely fixed at this point, they are no longer affected by the osmolarity of the fluid that they are in and, therefore, MilliQ water may be used as a wash.

14. Uranyl acetate en bloc staining greatly enhances the preservation and visualization of membranes, DNA, mitochondria, nucleoproteins, and phospholipids. It is used routinely prior to embedding in many tissues due to its superior ultrastructural preservation after aldehyde/osmium fixation.

15. Do not overfill the coverslip well. You should try to fill only the well part to a depth of 1–2 mm, not the whole slide mold. This way the resin layer will not be too thick to trim for re-embedding.

Acknowledgement

This work was supported by NICHD grant HD38353-07A2 to S.A.C.

References

1. Nagy A et al (2003) Manipulating the mouse embryo, a laboratory manual, 3rd edn. Cold Spring Harbor Press, Cold Spring Harbor, NY

2. Freshney RI (2005) Culture of animal cells, a manual of basic technique, 5th edn. Wiley, Hoboken, p 111

3. Plancha CE, Carmo-Fonesca M, David-Ferreira JF (1989) Cytokeratin filaments are present in golden hamster oocytes and early embryos. Differentiation 42:1–9

4. Hyttel P, Madsen I (1987) Rapid method to prepare mammalian oocytes and embryos for transmission electron microscopy. Acta Anat 129:12–14

5. Wu C et al (2006) Effects of cryopreservation on the developmental competence, ultrastructure and cytoskeletal structure of porcine oocytes. Mol Reprod Dev 73:1454–1462

6. Talbot P, DiCarlantonio G (1984) The oocyte-cumulus complex: ultrastructure of the extracellular components in hamsters and mice. Gamete Res 10:17–142

7. Nogues C et al (1993) A simple method for processing individual oocytes and embryos for electron microscopy. J Microsc 174:51–54

8. Britton AP, Moon YS, Yuen BH (1991) A simple handling technique for mammaliam oocytes and embryos during preparation for transmission electron microscopy. J Microsc 161:497–499

9. Borzzola JJ, Russell LD (1992) Electron microscopy, principles and techniques for biologists, 2nd edn. Jones and Bartlett, Sudbury, MA

10. Hayat MA (2000) Principles and techniques of electron microscopy, biological applications, 4th edn. Cambridge University Press, New York

11. Hanaichi T et al (1986) A stable lead by modification of Sato's method. J Electron Microsc 35:304–306

12. Melki R et al (1989) Cold polymerization of microtubules to double rings: geometric stabilization of assemblies. Biochemistry 28:9143–9152

13. Morency E, Anguish L, Coonrod S (2011) Subcellular localization of cytoplasmic lattice-associated proteins is dependent upon fixation and processing procedures. PLoS One 6(e17226):1–10

Chapter 16

Measuring Ca²⁺ Oscillations in Mammalian Eggs

Karl Swann

Abstract

At fertilization mammalian eggs are activated by a prolonged series of oscillations in the intracellular free Ca²⁺ concentration. These oscillations can be monitored with any number of Ca²⁺-sensitive fluorescent dyes. The oscillations last for several hours at fertilization and so there are some considerations with mammalian eggs that make them distinct from somatic cells that are commonly used in Ca²⁺ imaging experiments. I describe the use of two particular dyes that can be loaded into mouse eggs and that give the most valuable results. The first one is PE3 which can be loaded by incubation with the AM form of the dye which is membrane permeable. The other is rhod dextran which requires microinjection. Either one of these dyes offers advantages over the more commonly used fura2. I describe the way that the fluorescence from dye-loaded eggs is measured with a conventional epifluorescence microscope and a CCD camera.

Key words: Ca²⁺, Egg, Fluorescence, Oscillations, Mouse

1. Introduction

Mammalian eggs (or mature oocytes arrested at metaphase of meiosis II) are activated to begin development at fertilization by a prolonged series of oscillations in the intracellular free Ca²⁺ concentration (1, 2). Many parthenogenetic stimuli also activate development by causing a Ca²⁺ increase (1). This can be in the form of a single large Ca²⁺ rise, as with Ca²⁺ ionophores, or a series of oscillations, as with Sr²⁺ containing media (see Note 1). These Ca²⁺ oscillations are both necessary and sufficient for egg activation at fertilization. There is considerable interest in the mechanisms generating these oscillations, how the oscillations affect cell cycle proteins during activation, as well as in the effect that different patterns of oscillations have upon embryo development (2). Hence it is important to monitor Ca²⁺ in activating mammalian eggs.

Ca²⁺ has been measured in a variety of mammalian eggs using the photoprotein aequorin (3), Ca²⁺-sensitive electrodes (4), and

Hayden A. Homer (ed.), *Mammalian Oocyte Regulation: Methods and Protocols*, Methods in Molecular Biology, vol. 957, DOI 10.1007/978-1-62703-191-2_16, © Springer Science+Business Media, LLC 2013

using a range of fluorescent dyes that include fura2 (5), indo-1 (6), fluo3 (7), fura red (8), rhod 2 (6), Ca^{2+} green (9), Oregon Green BAPTA dextran (10), fura dextran (11), and PE3 (12). The methods described in this chapter will use fluorescent dyes to measure Ca^{2+}. In choosing a dye and method for measuring Ca^{2+} in eggs one has to take into account a number of factors that are distinct from the more common laboratory protocols involved in measuring Ca^{2+} in somatic cells. The most significant issues in mammalian eggs are related to the fact that Ca^{2+} oscillations are rather slow and last much longer than signals in somatic cells, such that one generally has to monitor Ca^{2+} for many hours in eggs if the full response is to be measured. The issues that arise from making prolonged recordings are the compartmentalization of the fluorescent dye over time, and the photo-damage that can occur in the egg with prolonged exposure to excitation light. Compartmentalization of fluorescent dyes is evident from confocal images of some of the so-called AM loading methods. Fluo3 AM is a particularly problematic dye in that it loads into many organelles and vesicles in the cytoplasm of mouse eggs (13). However, nearly all the AM-loaded dyes have this problem. Compartmentalization can probably account for the trend in some traces for the resting Ca^{2+} to appear to increase with time, and the amplitude of Ca^{2+} transients to appear to decrease with time. Such effects are not seen with the protein aequorin or with dextran-linked Ca^{2+} dyes which stay in the cytoplasm. Photo-damage to eggs can only be avoided by sampling with less light at less regular intervals or by using longer wavelength excitation dyes. This is one reason to use a cooled CCD camera to measure the fluorescence. The protocols I present will provide some tips to mitigate these issues. A final issue to consider when imaging eggs using fluorescent dyes is cellular autofluorescence. The autofluorescence from eggs arises mostly from NADH and FAD (6). NADH is fluorescent throughout the cytoplasm and emits in the blue region of the spectrum, whereas FAD is localized in the mitochondria and gives rise to green fluorescence. The autofluorescence from NADH and FAD changes in response to Ca^{2+} oscillations (6), so one has to be wary of potential artifacts caused by autofluorescence rather than Ca^{2+} itself.

In this chapter, I will discuss some of the considerations for the effective monitoring of these oscillations. The methods described are based upon an imaging system that uses a cooled CCD camera (see Fig. 1), which is becoming increasingly ubiquitous in research laboratories (see Note 2). I then present what I consider to be simplest and most valuable method for monitoring Ca^{2+} changes in eggs based upon my experience of using nearly a dozen different fluorescent dyes in hamster, mouse, or human eggs. I will present two alternative dyes for monitoring intracellular Ca^{2+} in mammalian eggs, namely, rhod-dextran and PE3. Rhod dextran is a dextran-linked version of rhod2 and is preferred for several reasons.

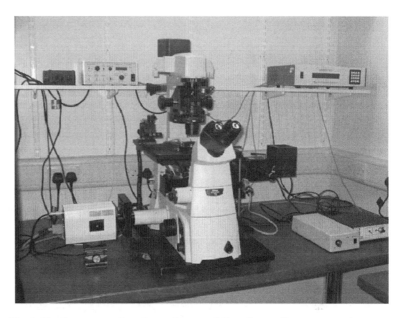

Fig. 1. The fluorescence imaging system consisting of an epifluorescence microscope, with cooled CCD camera, filter wheels, and shutters controlled by a PC (not in picture).

Firstly, it stays indefinitely in the cytosol of eggs since it is dextran coupled. Secondly it has a higher K_d than most other Ca^{2+} dyes (~750 nm) which means it buffers Ca^{2+} less and it is able to detect the amplitude and Ca^{2+} peaks better than the more commonly used fura dyes (14). It is highly recommended if one wants to make a point about differences in the amplitude of Ca^{2+} transients in eggs. Thirdly, rhod dextran has the advantage over some of the other dextran-linked dyes that we have used, such as Oregon green BAPTA dextran, in that it has a larger dynamic range in eggs, and it changes its fluorescence intensity by about fivefold during oscillations. Finally rhod dextran uses green excitation light which is less harmful for mammalian eggs and its fluorescence emission is in the red which is away from the dynamic NADH or FAD driven changes in autofluorescence. Its disadvantage is that it is a single wavelength indicator and hence it prone to small artifacts in signal if eggs move around significantly. PE3 is the best of the many dyes that can be loaded into eggs without using microinjection. It has the advantage over the commonly used fura2 since it does not appear to undergo compartmentalization to any significant degree and is retained in the cytoplasm for prolonged periods (15). Since it uses the same filter set as fura2 it is simple to make the switch to using it. It is also a ratio dye (ratio-metric) and hence small movements and changes in autofluorescence have less impact upon recordings than with single wavelength dyes. However, it has a rather low K_d for Ca^{2+} (~200 nM), requires UV excitation, and needs an imaging system with an excitation filter wheel.

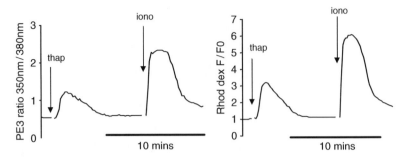

Fig. 2. Recordings of intacellular Ca^{2+} as measured by either PE3 or rhod dextran. This is from a single mouse egg that was both microinjected with rhod dextran and loaded by incubation with PE3-AM. The traces are shown for an egg responding to addition of thapsigargin (20 µM) and the ionomycin (5 µM). The PE3 signals are plotted as the excitation ratio with 350 nm and 380 nm excitation. A XF2043 mirror was used in the dichroic filter block. The rhod dextran traces are plotted as fluorescence intensities divided by the starting fluorescence levels (F/F_0).

Figure 2 shows a mouse egg in which the Ca^{2+} levels were recorded for PE3 and rhod dextran at the same time in the same egg responding to thapsigargin and ionomycin which both release Ca^{2+} from intracellular stores. Both dyes give similar qualitative data.

2. Materials

2.1. Handling Oocytes, Eggs, and Sperm

1. Eggs are obtained from 4- to 6-week-old mice of the MF1 strain.

2. Pregnant mare's serum gonadotrophin (PMSG) and human chorionic gonadotrophin (hCG) to induce super-ovulation (16).

3. Sperm are collected from 10- to 20-week-old F1 hybrid (C57xCBA) male mice.

4. M2 medium (M7167 from Sigma).

5. Hepes buffered KSOM medium: 95 mM NaCl, 2.5 mM KCl, 0.35 mM KH_2PO_4, 0.2 mM Na pyruvate, 1 mM L-glutamine, 0.01 mM EDTA, 0.2 mM $MgSO_4$, 10 mM Na lactate, 4 mM $NaHCO_3$, 1.71 mM $CaCl_2$, 0.2 mM glucose, 20 mM HEPES, 10 mg/l phenol red, pH 7.4 (17), either with or without bovine serum albumin (BSA; 4 mg/ml, Sigma A3311), in cell culture grade water.

6. T6 medium: 99 mM NaCl, 1.42 mM KCl, 0.47 mM $MgCl_2$, 0.36 mM Na_2HPO_4, 1.78 mM $CaCl_2$, 25 mM $NaHCO_3$, Na lactate 24.9 mM, Na pyruvate 0.47 mM, glucose 5.56 mM, phenol red 10 mg/l in cell culture grade water. On the day before use, add 16 mg/ml BSA and place in an incubator to allow equilibration with CO_2.

7. 0.3 mg/ml hyaluronidase in M2 medium. Keep frozen (–20°C) in aliquots of 1 ml until use.

8. Acid Tyrode's solution. Aliquots of ~1 ml kept frozen until use.

9. Mouth pipettes for oocyte handling. Glass pasteur pipettes are pulled in a Bunsen flame to have a very narrow diameter of approximately 100 μm. These are controlled by mouth suction via a piece of silicone tubing connected to the back end of the Pasteur pipette. A small plug of cotton wool is also placed in the tubing. The mouth piece is a 1 ml syringe with the plunger removed and hence is sterile and can be replaced regularly.

10. 35 × 10 mm Plastic petri dishes (Nunc).

11. 14 ml plastic tubes (Falcon) for holding media either in incubator or on the heating block.

12. Embryo tested mineral oil (M8410 from Sigma).

13. Heated block (Grant BT3).

14. CO_2 incubator (Binder).

15. Dissection microscope equipped with diascopic stand (e.g., SMZ1000, Nikon).

2.2. Fluorescent Dyes and Drugs

1. 5 mM Fura PE3-AM in dimethyl sulfoxide (DMSO) containing 5% Pluronic F127. Divide into 1 μl or 0.5 μl aliquots (in 1 ml tubes) and store at –80°C until use. Thaw and dilute for use and do not re-freeze.

2. 1 mM Rhod dextran (Molecular Probes, http://www.probes.com) in KCl Hepes buffer: 120 mM KCl, 20 mM HEPES, pH 7.2 (see Note 3). Divide into 2 μl aliquots and store at –80°C until use. Aliquots can be thawed and re-frozen several times so each aliquot can be used for more than one experiment.

3. 20 mM Thapsigargin in DMSO. Divide into 1 μl aliquots for storage.

4. 5 mM Ionomycin in DMSO. Divide into 1 μl aliquots for storage.

2.3. Microinjecting Dyes

1. Glass for making microinjection pipettes (GC150F and G100 TF [1.5 mm outer diameter and 0.86 mm inner diameter], Harvard Apparatus Ltd.).

2. Micropipette pullers (e.g., Sutter model P-30 and Narashige PN-30).

3. Microforge (Narashige MF-900).

4. Micro-loader pipette tips (20 μl capacity) (Eppendorf).

5. Picopump Pressure injector with foot switch (PV820; World Precision Instruments, www.wpiinc.com).

6. Electrical amplifier (Electrometer IE-251A, Warner Instruments Corporation, www.warneronline.com).

7. Holder for the injection pipette (e.g., MEH2SFW, World Precision Instruments, www.wpiinc.com). This holder has a side-port for pressure pulse application and a wire for making electrical contact with the solution in the micropipette.

8. Syringe system for holding eggs (e.g., Eppendorf Cell Tram Air 5176).

9. Benchtop centrifuge (e.g., Spectrafuge 24D, Jencons).

10. Hydraulic manipulators (3D fine micromanipulator, Narashige).

11. Inverted microscope equipped with 4×, 10×, and 20× long working distance objectives and Hoffman contrast (e.g., TE2000-S, Nikon UK Ltd).

12. Anti-vibration table (Speirs Robertson Corporation).

13. CCD camera for viewing eggs and for demonstration purposes (e.g., Nikon DS-5M, Nikon Instruments).

2.4. Measuring Fluorescence from Eggs

The imaging system we use, as described above, is shown in Fig. 1. It consists of:

1. An epifluorescence microscope equipped with at least a 4× and 20× objectives. The 20× objective used for imaging must be of high numerical aperture (NA) (see Note 4). We use a Nikon Ti-U Eclipse with a plan Apo DIC or S Fluor objective with 0.75 NA.

2. Filter wheels, shutters, and shutter controller (Sutter Instruments, www.sutter.com; the controller is a Sutter Lambda 10–3).

3. Bandpass interference filters (Omega [www.omegafilters.com] or Thorlabs [www.thorlabs.com]). These will be 350 nm and 380 nm filters for excitation of PE3 and a 550 nm bandpass filter for excitation of Rhod dextran. The bandwidths for excitation filters are all 10 nm. For PE3 emission we use a 510–530 nm bandpass filter and for Rhod dextran a 600 nm bandpass filter with 40 nm bandwidth.

4. Dichroic filter blocks can be 505 nm (for PE3) or 580 nm (for rhod dextran) (see Note 5). The filter block only contains the dichroic mirror, and not any bandpass filters, since the wavelength filters are in wheels. Alternatively, and more commonly for rhod dextran we use a dichroic filter block with a triple bandpass filter. For this order an empty filter block and then purchase suitable filters from Horiba Scientific (www.horiba.com) (see Note 6).

5. Excitation light source. Either a halogen lamp with stabilized power supply (Nikon) or else an LED can be used. We use an Optolite LED Lite (Cairn Research Ltd; www.cairn-research.co.uk), with either a 365 nm LED light head for PE3 or a "white" LED light head for rhod dextran. We also have a dual

lamp housing attached to the side of our microscope so that one can switch between lamps. It also has a 50% mirror so that one can combine UV excitation from the LED with the visible range from the halogen lamp (as used in Fig. 2).

6. CCD camera (e.g., Photometrics CoolSNAP HQ²; www.photometrics.com) which is internally cooled to –30°C. The light is projected onto the camera chip with a 1× relay lens.

7. Software for controlling filters and shutters and for data acquisition and analysis (e.g., QED InVivo, Media Cybernetics, www.mediacy.com). Data analysis can be performed using freely available software (e.g., ImageJ, http://rsweb.nih.gov/ij/index.html).

8. Heated stage (e.g., a series 40 Quick Change imaging chamber controlled by a CL-100, Warner Instruments).

9. Anti-vibration table or block. We use a 1 in. steel block covered in plastic that sits on silicone anti-vibration mounts (tennis balls have been used in the past). The camera is supported by a Swiss Boy lab jack (Sigma), with a squash ball as a cushion.

3. Methods

3.1. Collection of Eggs

1. Hormonally prime mice by intraperitoneal injection of PMSG and then hCG (5 or 10 I.U. each).

2. Sacrifice mice 13–15 h after hCG injection and dissect out the oviducts.

3. Place oviducts into a Petri dish of pre-warmed M2 medium containing hyaluronidase and pierce swellings in oviducts to release oocyte cumulus complexes.

4. Remove surrounding cumulus cells by leaving oocyte cumulus complexes in hyaluronidase-treated medium for about 5 min on the heated block until oocytes can be seen to separate from the cumulus cells.

5. Collect oocytes using the finely drawn mouth pipette. It is useful to pipette eggs up and down in the pipette to ensure they are free from cumulus cells.

6. Wash eggs through a series (~6) of drops (~100 µl) of M2 medium to remove hyaluronidase and place in a small (~100 µl) drop of M2 medium under oil in a petri dish in the incubator.

3.2. Loading PE3 into Eggs

If using PE3 to measure Ca²⁺, then eggs can be loaded with the dye without the need to microinject by using PE3-AM.

1. Dilute the stock solution of PE3-AM 1,000-fold into M2 medium to give a final concentration of 5 µM.

2. Make up a Petri dish containing a 100–400 µl drop of 5 µM PE3-AM under oil.

3. Place eggs in this drop for a period of 30 min whilst maintaining the dish either on the warm block away from direct light or else in the incubator (see Note 7). The AM portion of the dye allows it to cross the plasma membrane and once in the cytosol it is cleaved to generate PE3 which is charged and retained in the cytoplasm. PE3 has extra positive charge that makes it different from fura2. As a result it is less subject to compartmentalization and extrusion from cells (15). Otherwise it has almost identical fluorescence properties to fura2.

4. Wash the eggs out of the PE3-AM containing-drop in preparation for imaging. We have held the eggs for several hours after loading with PE3 and not noticed any major loss of fluorescence.

3.3. Microinjecting Rhod Dextran into Eggs

If one is using rhod dextran, it needs to be microinjected since the dye does not cross the plasma membrane. The microinjection is done by pressure injection pulse, but the insertion of the micropipette (needle) makes special use of an electrical method.

1. Pull injection micropipettes just before use with the GC-150F capillary glass and the Sutter P-30 puller. The heat and pull are adjusted to give a short shank but with a tip diameter just less than 1 µm. The estimates of tip size can be done by blowing bubbles in ethanol as described before (18).

2. Pull holding pipettes using the Narashige PN-30 puller and G-100 TF glass. After pulling, break the end of the pipette around the point that it tapers to 100 µm and heat the cut end until it becomes smooth and rounded but still retains an opening.

3. Thaw rhod dextran dye drops and spin in a benchtop centrifuge at 13,000 rpm ($\sim$13,000$\times g$) for 1–2 min. Pick up as small amount of dye as one can see using the micro-loader attached to the end of a 1 ml syringe.

4. Fill glass micropipettes by pipetting a small amount of rhod dextran solution into the back end of the micropipette using the Eppendorf micro-loader. Push the tip of the micro-loader towards the end of the tip before expelling the dye solution into the micropipette. Any solution in the micropipette will run down to the tip because the glass has a filament. The rhod dextran is a distinct purple color and all one needs to see is a small speck of purple in the tip of the pipette. With practice one can use well under 1 µl to load the pipette.

5. Place the dye-loaded micropipette in the MEH2SW holder so that the wire slides into the micropipette through the back end. The wire does not have to make contact with solution but

it is best that it goes most of the way down the inside of the micropipette. Tighten the holder up as much as possible with one's fingers. The MEH2SFW holder should have its "male luer" side-port connected to the output of the Picopump via some stiff silicone tubing.

6. Fit the holder onto the pre-amplifier of the electrometer which is mounted on the Narashige manipulator.

7. Place the eggs to be microinjected in a flat drop of M2 medium covered by a thin layer of mineral oil in an upturned petri dish lid.

8. Manipulate the injection and holding micropipettes so that their tips dip into the medium containing the eggs. Fill the front end of the holding pipette with a small amount of medium. The displacement of the medium is then controlled via the Cell Tram Air device.

9. Dip the silver wire that is electrically connected with the ground of the electrometer into the medium containing the eggs.

10. Switch the electrometer on and a green light indicates when an electrical circuit is successfully made.

11. Stabilize the egg to be microinjected by applying suction through the holding pipette.

12. Insert the injection micropipette through the zona pellucida by advancing it into the egg a distance at least half of the egg diameter using the micromanipulator. The plasma membrane should deform but it will not break.

13. Penetrate the plasma membrane with the micropipette by briefly pressing the yellow "buzz" or "zap" button on the electrometer. This causes an electrical oscillation that will insert the micropipette through the plasma membrane into the egg.

14. Inject solution by applying pressure to the back of the pipette by pressing the foot switch. We use a pressure of ~20 psi and a pulse length of about 0.1–0.5 s. This will cause a brief balloon of solution to become visible in the egg. The diameter of this solution should be around ¼ of the egg diameter (see Note 8).

15. Rapidly withdraw the micropipette from the egg following application of the pressure pulse and use the holding pipette to separate injected eggs away from uninjected ones. Once started this method can be used to microinject several eggs per minute.

3.4. Preparing Eggs for Imaging

After loading eggs with fluorescent dyes, either via AM incubation or via microinjection, transfer them to a recording chamber on the stage of the epifluorescence microscope. The chambers we have used always have a glass coverslip (No. 1) as its base. Whatever heater is adopted, it is recommended that a glass coverslip based chamber is used. A key requirement for effective imaging is to have the eggs stay in one place, or at least move minimally, during the recording period (see Note 9). When making recordings after microinjection of PLCzeta (19), or when using Sr^{2+} medium, since both will induce Ca^{2+} oscillations without the need for any further solutions, the eggs will stay where they are put so that eggs with an intact zona pellucida can be used. In this case, place eggs with intact zona pellucida in a small drop of medium (~10 µl) in the middle of the recording chamber and cover with mineral oil to prevent evaporation. However, if one is carrying out an experiment to measure Ca^{2+} during in vitro fertilization or is adding drugs or chemicals to the medium containing the eggs the best practice is to attach the eggs to the coverslip as described in the following steps.

1. Remove the zona pellucida by pipetting eggs into a small drop of acid Tyrode's solution (usually 100 ml on an upturned petri dish lid). The carryover of M2 medium should be minimal to avoid changing the pH of the acid Tyrode's solution.

2. Examine the eggs continuously with a dissection microscope during the procedure, ensuring that the microscope focus is adjusted so that one can readily see the zona pellucidas of the eggs.

3. Pipette the eggs up and down whilst they are in acid Tyrode's.

4. Remove eggs from acid Tyrode's and transfer to M2 medium as soon as the zona pellucidas starts to be break away from the eggs. This has to be done promptly or else many eggs will lyse. The whole procedure should be over in about 1 min.

5. Pipette the zona-free eggs into a drop of medium (~0.5 ml) under oil in the heated chamber on the microscope stage.

3.5. Inducing Calcium Release in Eggs

To test eggs for Ca^{2+} release, directly pipette either thapsigargin or ionomycin (at 2× or 5× working concentrations in KSOM) into the drop of medium containing the eggs on the microscope stage whilst briefly interrupting fluorescence recordings (see Fig. 2 and Note 10). Although there are a number of other stimuli that can cause Ca^{2+} oscillations in mammalian eggs, such as thimerosal and Sr^{2+} ions, fertilization is the benchmark response. The following steps describe the preparation and use of mouse sperm for IVF experiments.

1. Pierce the cauda epididymis of one culled male and release sperm into warm equilibrated T6 medium.

2. Allow sperm to swim out and then decant into ~5 ml of T6 medium in a 14 ml Falcon tube.

3. Place Falcon tube with sperm in the incubator with the cap loose.

4. Wait 3 h after starting the sperm preparation before adding them to eggs. This is important as the sperm are to be used to fertilize zona-free eggs so they will have to undergo a spontaneous acrosome reaction to be able to fertilize.

5. Add sperm to eggs by pipetting 10 µl of T6 with sperm into 0.5 ml of medium containing the eggs (see Note 11). Eggs should start to generate their first Ca^{2+} transients in about 5–10 min (see Note 12).

3.6. Epifluorescence Imaging

1. Confirm fluorescence of eggs using the 20× objective of the epifluorescence microscope (see Note 13). For PE3 this is best done by just looking at the fluorescence with 380 nm excitation, with rhod dextran there's only one channel to look at. In initial runs it is useful to put some unloaded eggs in the dish. The fluorescence of the loaded eggs should be several times the signal from the unloaded eggs.

2. Adjust the intensity of the excitation light so that it is just strong enough to see the eggs clearly with the CCD camera and aim for 10–15 eggs to be visible within the field of view of the 20× objective (see Note 14).

3. Cover recording chamber with a black plastic lip and keep surroundings darkened during recordings with no overhead lights on.

4. Draw a region of interest around each egg using imaging software. This is important to enable the fluorescence intensities or ratios to be displayed against time during the course of an experiment and will also verify whether the eggs are responding to the sperm or other stimuli.

5. Set up recordings so that one image or one set of ratio images is captured every 10 s with exposure times of 100–300 ms (see Note 15).

6. Store data as a series of images in .tif format files, so that there is a .tif stack for each wavelength for each experiment.

3.7. Analysis of Fluorescence Intensities

At the end of the experiment analyze data offline. Although this can be done with different types of acquisition software, here we describe the technique using ImageJ software.

1. Download the Multi Measure plugin from Bob Dougherty's section of the ImageJ plugins page that enables data from many eggs to be analyzed.

2. Use the plugin to draw a region of interest around individual eggs.

3. Draw a circle the same size as an egg within an egg-free space in the field of view. This will be taken as a measure of nonspecific background fluorescence. Ideally the background should be

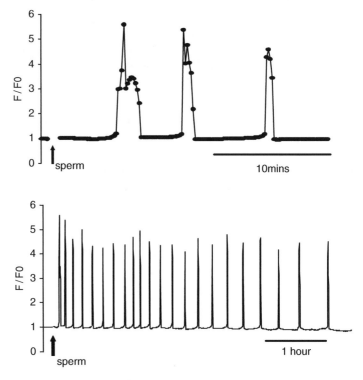

Fig. 3. Ca²⁺ oscillations at fertilization in a mouse egg measured with rhod dextran. A recording of Ca²⁺ levels is plotted as F/F_0 as in Fig. 2, with the top trace showing the initial phase of Ca²⁺ oscillations, with individual data points shown as *filled circles*. The bottom trace shows the full scale of the response that lasts several hours and is characteristic of mammalian eggs in general.

from an unloaded egg in the same field of view but in practice the fluorescence intensities are strong enough that there is little if any difference in signal between an unloaded egg and an empty space with no egg.

4. Use the software to measure the fluorescence intensities within the regions of interests and copy these fluorescence measurements into a suitable spreadsheet program.

5. Obtain background-corrected fluorescence values by subtracting the background intensity from the intensity values of each egg.

6. For rhod dextran one simply plots background-corrected egg fluorescence intensity versus time (see Fig. 2).

7. For PE3, divide the 350 nm fluorescence intensities by those at 380 nm and plot these ratios against time (see Fig. 2). These sort of plots give the raw data for Ca²⁺ levels in eggs and at fertilization should produce data as in Figs. 2 and 3. It is up to the experimenter to decide how far to go in calibrating Ca²⁺ levels in absolute terms (see Note 16). The essential features of

the Ca^{2+} oscillations in mammalian eggs to capture are how many spikes occur, when the spikes occur, for how long they occur, and their relative amplitudes. For the most part these parameters do not require absolute calibration.

4. Notes

1. If culture medium is made in which the ~1.7 mM Ca^{2+} is replaced by 5–10 mM Sr^{2+} then Ca^{2+} oscillations can be triggered for prolonged periods in mouse and rat eggs. This is a simple way of testing that Ca^{2+} oscillations can be detected with the system being used. However, it may only work in rodents. As yet there are no indications that Sr^{2+}-containing medium causes Ca^{2+} oscillations in unfertilized eggs of other species such as pigs, cows, sheep, or humans.

2. Intracellular Ca^{2+} in mammalian eggs can be measured using confocal imaging microscopes. This is very useful if one wants to monitor the rapid waves of Ca^{2+} that cross the egg and constitute the rising phase of each oscillation. However, this is best done to provide a "snapshot" of a few Ca^{2+} transients. For recording the entire Ca^{2+} signal at fertilization one needs to measure for periods of 4–10 h. This is impractical on most confocal-based imaging systems because it requires extensive amounts of hard-drive memory capacity and can be expensive if one is using a core facility with hourly access charges. During long-term recordings, conventional confocal systems also expose mammalian eggs to more light than is ideal (see Note 15).

3. Rhod dextran is the high affinity form of the Ca^{2+} dye Rhod2 linked to 10,000 MW dextran. It should not be confused with rhodamine dextran which is not a Ca^{2+}-sensitive dye.

4. We use an air objective lens because it makes it easier to control the temperature of the incubation chamber. If one uses an oil immersion objective then any slight drift in temperature can also cause the focus to change with time which can lead to a loss or drift of signal. The standard range of a 40× oil immersion objective also tends to offer a small field of view so that only a few (4, 5) eggs can be imaged at the same time. With a 20× objective we can view 10–15 eggs at the same time.

5. For using rhod dextran one can simply use a standard rhodamine filter block that is found on most epifluorescence microscopes. If you just want to measure Ca^{2+} then you do not need the filter wheels and controllers. All that is required is a rhodamine filter block, with its standard excitation and emission filters, a stable excitation source such as halogen lamp, and a cooled CCD camera.

6. The triple bandpass filters have the advantage that one can use a range of fluorescent dyes at the same time. For example fluorescence from CFP and YFP probes or dyes can be monitored at the same time as Ca^{2+} using Rhod dextran. For this type of experiment we use an XF2054 made by Chroma which is purchased through GlenSpectra in the UK (www.horiba.com).

7. The loading conditions for AM dyes in mammalian eggs can vary according to the species. Mouse and human eggs load rather well with different AM dyes. However, hamster eggs load poorly with AM dyes and much longer incubation times are required. Many non-mammalian eggs, for example, sea urchin eggs, do not load significantly at all with the AM method and Ca^{2+} dyes have to be microinjected.

8. Microinjection with this electrical method is the most effective way for MII eggs which are otherwise fragile. It is important to make sure there is an electrical circuit formed between the inside of the pipette and media. The electrometer we use shows a red light if there is no circuit. It can be useful to have an oscilloscope connected to the output of the electrometer to verify that there is a circuit and then that there is good oscillation of the signal when the "buzz" button is pushed. If there is no circuit then the most common problems are that the silver wire is not dipped in the media with the eggs, or else the micropipette tip is blocked. The tip can often be unblocked, or else broken slightly, by touching it on the end of the holding pipette. This is often necessary when microinjecting RNA which can be viscous, but is less often needed for injecting fluorescent dyes by themselves. This may then require one to lower the injection pressure. As a rule, if the electrical oscillation "buzz" method does not allow a micropipette to penetrate the egg membrane then the tip is too blunt and should not be used. The first egg is often used as a practice egg in order to adjust the volume and duration of pressure pulse so as to get an appropriately sized balloon of solution. This egg is then discarded and not used in experiments.

9. The trick to get eggs to stick is to use a drop of medium in the imaging chamber that does not contain BSA. We typically make up HKSOM media without any BSA. The eggs are sticky for several minutes after the acid Tyrode's step and when placed onto the glass coverslip in the absence of albumin they stick down within a few seconds. Once stuck they cannot be removed without a degree of damage so it is worth practicing this step to make sure several eggs can be placed in a tight bunch that can be imaged by the field of view of the CCD camera. Once the eggs have stuck medium containing BSA can be added if desired, but it's worth noting that this is not needed when carrying out an IVF experiment, since one will be adding some medium containing sperm in 15 mg/ml BSA.

10. Before starting a full IVF experiment it is helpful to test the response of some of the loaded eggs to drugs that release Ca^{2+} from intracellular stores. Figure 2 shows the Ca^{2+} response to thapsigargin, which inhibits the Ca^{2+} ATP pump in the endoplasmic reticulum, and then ionomycin which causes Ca^{2+} release from all intracellular organelles. The response to ionomycin should always be larger than that to thapsigargin, and responses like those in Fig. 2 are a good sign that the system and the eggs are suitably responsive.

11. If one uses HKSOM medium to carry out IVF we have found that the supplementation of the medium with glucose (to increase it to 5 mM) helps with IVF success rates since sperm–egg fusion in the mouse requires glucose (20), and HKSOM media otherwise has rather low glucose (0.2 mM).

12. Eggs generally need to be maintained at a temperature above 30°C for fertilization to occur. We typically run experiments at 35–37°C. Whatever heated stage is used it is useful to have a thermocouple-based temperature probe with a fine tip so that one can verify before experiments start that the center of the dish, where the eggs are placed, is genuinely in this temperature range.

13. We use a standard epifluorescence microscope that has the option of the light being directed either to the eyepiece or to a side-port with the CCD camera. It is best to have the option with 100% of the light being directed to the camera to maximize light collection.

14. If the eggs look clear and bright when viewed down the eyepiece with the naked eye, then you are probably using too much excitation light. We use our CCD camera with considerable pixel binning (4×4, or 8×8). This gives a rather pixelated view of the eggs but by binning pixels you reduce the readout noise on the CCD chip and effectively increase the sensitivity of the camera. This in turn means that the excitation light can be turned down as much as possible. In general it is the whole egg Ca^{2+} that is of interest, and so spatial resolution is sacrificed for the sake of measuring signals without harm for many hours. The light level on halogen lamps or LEDs is readily turned up or down. For other light sources you will have to use neutral density filters. The raw light from Xenon or metal halide lamps is far too bright for mammalian eggs and you will need a set of neutral density filters to cut it down.

15. This is sufficient to resolve all the Ca^{2+} spikes whilst keeping overall light exposure down. If recording fertilization in the mouse, then records should be about 4 h as standard. For other eggs they may need to be longer and we have often recorded Ca^{2+} in eggs for up to 20 h. If you want to monitor the Ca^{2+} waves in mammalian eggs at fertilization then you are best

advised to use full frame pixels (no binning) and to capture around 5 images per second. This can be done for periods of ~10 min but is not generally a practical option for several hours. Measuring Ca^{2+} oscillations with fluorescent dyes is generally not conducive to later embryo development. Mammalian eggs are sensitive to light and their development is impaired by conventional fluorescence imaging light (21). However, it is still possible to observe the main events of egg activation such as polar body emission and the formation of pronuclei, and the timing of the first cell division after several hours of imaging (8).

16. Calibrating the fluorescence in terms of absolute Ca^{2+} concentration is nontrivial in mammalian eggs. In principle this is done by obtaining the minimum and maximum fluorescence intensity at the end of the experiment and then converting this to Ca^{2+} using: $Ca^{2+} = K_d ((f - F_{min})/(F_{max} - f))$, where "$f$" is the measured fluorescence and F_{min} and F_{max} are the fluorescence of the dye obtained at zero Ca^{2+} and saturating Ca^{2+} levels (11, 12, 14). The K_d is the dissociation constant for the dye used. For ratio-metric dyes there is an analogous equation with ratios instead of fluorescence intensities. However, the standard method in cells is to add ionomycin and use extracellular buffers to control intracellular Ca^{2+}. This works well in flat cultured cells, but with large spherical eggs it is often difficult to obtain a convincing fluorescence minimum or ratio-metric minimum. Eggs often die and lose their fluorescence before obtaining a convincing fluorescence or ratio-metric maximum. Also it is often desirable to follow the events subsequent to Ca^{2+} oscillations, such as pronuclear formation, and so the destructive use of ionophores is not an option. For fura2 ratiometric dyes one can attempt to calibrate signals by measuring the fluorescence ratio of the dye in drops of solution with high or low Ca^{2+} to obtain F_{max} or F_{min} for your imaging system (11). However, it is never clear whether your test solution fully mimics the cytosol of cells.

Consequently, any traces that show absolute Ca^{2+} levels in eggs have to be interpreted with some caution. For example, calibrated fluorescent traces using fura2, PE3, or fura dextran tend to give rather low values for the amplitude of the Ca^{2+} spikes (e.g., around 500 nM). I suspect this is in error since the luminescence recordings with aequorin, which detects Ca^{2+} more accurately at higher levels, suggest that most of the Ca^{2+} spikes peak at around 2 μM (3). The use of a low affinity Ca^{2+} dye also suggests that the peaks of Ca^{2+} transients at fertilization are above 1 μM (22). Most research studies do not calibrate the Ca^{2+} levels explicitly in eggs and just plot the data as a fluorescence ratio or as fluorescence intensities versus time. One way to at least normalize Ca^{2+} recordings is to plot the fluorescence intensity (F) with respect to the starting level before fertilization (F_0). Fluorescence is converted to "F/F_0" by division of all fluorescence intensities by the mean value at the start of the experiment. The absolute calibration of Ca²⁺ levels in mammalian eggs deserves further investigation.

Acknowledgments

My methods for studying Ca²⁺ in eggs are based upon working in the laboratories of Michael Whitaker and Shun-ichi Miyazaki and upon discussions with Mark Larman who worked in my laboratory. I thank Yuansong Yu for comments in the manuscript. My laboratory has been supported by funds from the WellcomeTrust, the BBSRC, and Cardiff University School of Medicine.

References

1. Swann K, Ozil JP (1994) Dynamics of the calcium signal that triggers mammalian egg activation. Int Rev Cytol 152:183–222

2. Ducibella T, Fissore R (2008) The roles of Ca²⁺, downstream protein kinases and oscillatory signalling in regulating fertilization and the activation of development. Dev Biol 315: 257–279

3. Cuthbertson KS, Cobbold PH (1985) Phorbol ester and sperm activate mouse oocytes by inducing sustained oscillations in cell Ca²⁺. Nature 316:541–542

4. Igusa Y, Miyazaki S (1986) Periodic increases in cytoplasmic free calcium in fertilized hamster eggs measured with calcium-sensitive electrodes. J Physiol 377:193–205

5. Cheek T, McGuinness O, Vincent C, Moreton RB, Berridge MJ, Johnson MH (1993) Fertilisation and thimerosal stimulate similar calcium spiking patterns in mouse oocytes but by separate mechanisms. Development 119: 179–189

6. Dumollard R, Marangos P, Fitzharris G, Swann K, Duchen M, Carroll J (2004) Sperm-triggered [Ca²⁺] oscillations and Ca²⁺ homeostasis in the mouse egg have an absolute requirement for mitochondrial ATP production. Development 131:3057–3067

7. Kline D, Kline JT (1992) Repetitive calcium transients and the role of calcium in exocytosis and cell cycle activation in mouse eggs. Dev Biol 149:80–89

8. Lawrence Y, Ozil JP, Swann K (1998) The effects of a Ca²⁺ chelator and heavy metal ion chelator upon Ca²⁺ oscillations and activation at fertilization in mouse eggs suggest a role for repeptitve Ca²⁺ increases. Biochem J 335: 335–342

9. Nakano Y, Shirakawa H, Mituhashi N, Kuwabara Y, Miyazaki S (1997) Spatiotemporal dynamics of intracellular calcium in the mouse egg injected with a spermatozoa. Mol Human Reprod 3:1087–1093

10. Campbell K, Swann K (2006) Ca²⁺ oscillations stimulate an ATP increase during fertilization of mouse eggs. Dev Biol 298:225–233

11. Mehlmann L, Kline D (1994) Regulation of intracellular calcium in the mouse egg: calcium release in response to sperm or inositol trisphosphate is enhanced after meiotic maturation. Biol Reprod 51:1088–1098

12. Igarashi H, Takahashi T, Takahashi E, Tezuka N, Nakahara K, Takahashi K, Kurachi H (2005) Aged oocytes fail to readjust intracellular adenosine triphosphates at fertilization. Biol Reprod 72:1256–1261

13. Carroll J, Swann K, Whittingham D, Whitaker MJ (1994) Spatiotemporal dynamics of intracellular [Ca²⁺]$_i$ oscillations during growth and meiotic maturation of mouse oocytes. Development 120:3507–3517

14. http://products.invitrogen.com/ivgn/product/R34676

15. Vorndran C, Minta A, Poenie M (1995) New fluorescent calcium indicators designed for cytosolic retention or measuring Ca²⁺ near membranes. Biophys J 69:2112–2124

16. Hogan B, Costantini F, Lacy E (1986) Manipulating the mouse embryo: a laboratory manual. Cold Spring Harbor Lab. Press, Plainview, NY

17. Summers MC, Bhatnagar PR, Lawitts JA, Biggers JD (1995) Fertilization in vitro of mouse ova from inbred and outbred strains: complete preimplantation embryo development in glucose-supplemented KSOM. Biol Reprod 53:431–437

18. Schnorf M, Potrykus I, Neuhaus G (1994) Microinjection technique: routine system for characterizion of microcapillaries by bubble pressure measurement. Exp Cell Res 210: 260–267

19. Saunders CM, Larman MG, Parrington J, Cox LJ, Royse J, Blayney LM, Swann K, Lai FA (2002) PLCζ: a sperm-specific trigger of Ca^{2+} oscillations in eggs and embryo development. Development 129:3533–3544

20. Urner F, Sakkas D (1996) Glucose participates in sperm-oocyte fusion in the mouse. Biol Reprod 55:922

21. Squirrel JM, Wokosin DL, White JG, Bavister BD (1999) Long term two photon fluorescence imaging of mammalian embryos without compromising viability. Nat Biotechnol 12: 763–767

22. Halet G, Tunwell R, Parkinson SJ, Carroll J (2004) Conventional PKCs regulate the temporal pattern of Ca^{2+} oscillations in mouse eggs. J Cell Biol 164:1033–1044

Chapter 17

Counting Chromosomes in Intact Eggs

Teresa Chiang and Michael A. Lampson

Abstract

Chromosomal spreads are an established method to assess ploidy in different cell types. However, many traditional chromosome-spreading techniques require dissolution of the cell and can only be used to assess hyperploidy because of potential chromosome loss inherent in the procedure. Here we describe a method to evaluate chromosome numbers in intact eggs so that both hyperploidy and hypoploidy can be accurately detected.

Key words: Chromosome spread, Chromosome count, Ploidy, Monastrol, Kinetochores

1. Introduction

Traditional chromosome-spreading techniques used to determine chromosome numbers in cells can only detect hyperploidy due to an intrinsic risk of chromosome loss in the procedure. We developed a method that can accurately detect both hyperploidy and hypoploidy in eggs (1). Using monastrol, a kinesin-5 inhibitor, the bipolar spindle collapses into a monopolar spindle and leaves chromosomes dispersed (Fig. 1) (2). After staining DNA and kinetochores, chromosome numbers can be assessed by counting the number of kinetochores in each egg. Several investigators have successfully used this method (3–5).

2. Materials

All materials are prepared with ultrapure water (e.g., using Milli-Q systems from Millipore, Billerica, MA) unless otherwise noted.

Hayden A. Homer (ed.), *Mammalian Oocyte Regulation: Methods and Protocols*, Methods in Molecular Biology, vol. 957,
DOI 10.1007/978-1-62703-191-2_17, © Springer Science+Business Media, LLC 2013

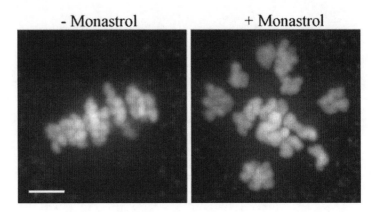

Fig. 1. Monastrol collapses the bipolar spindle and disperses chromosomes. Examples of chromosome configuration in Metaphase II eggs treated either with or without monastrol for 1 h. Images are maximal intensity projections, and scale bar represents 5 μm.

2.1. Oocyte Culture and Maturation

1. CZB maturation medium, modified from ref. 6: 81.62 mM NaCl, 4.83 mM KCl, 1.18 mM KH_2PO_4, 1.18 mM $MgSO_4$, 1.7 mM $CaCl_2$, 25.12 mM $NaHCO_3$, 31.1 mM Na-lactate, 0.27 mM Na-pyruvate, 0.11 mM EDTA, 10 μg/mL Gentamycin, 10 μg/mL Phenol Red, 7 mM Taurine, 3 mg/mL Bovine serum albumin (BSA), and 1 mM L-glutamine (added prior to use). All materials are embryo-culture grade from Sigma-Aldrich (St. Louis, MO).

To make 1 L CZB, measure out 4.77 g NaCl, 360 mg KCl, 160 mg KH_2PO_4, 290 mg $MgSO_4 \cdot 7H_2O$, 250 mg $CaCl_2 \cdot 2H_2O$, 2.11 g $NaHCO_3$, 443 μL Na-lactate (60% syrup), 30 mg Na-pyruvate, 40 mg EDTA-Na_2, 1 mL Gentamycin (10 mg/mL), 1 mL Phenol Red (10 mg/mL), 875 mg Taurine, and 3 g BSA. Add embryo-culture water (Sigma) up to 1 L. Filter-sterilize through 0.22 μm PVDF filters (e.g., Millipore Stericups, Millipore). Store at 4°C.

To prepare 10 mL of 100 mM L-glutamine stock solution, dissolve 146 mg L-glutamine in 10 mL embryo-culture water and filter-sterilize. Store at −20°C.

2.2. Drugs and Fixative

1. Monastrol (Sigma-Aldrich): Prepare stock solution of 100 mM monastrol by dissolving 1 mg monastrol in 34 μL DMSO. Store as instructed.

2. 2% Paraformaldehyde (PFA): Weigh out 0.2 g PFA, add 8 mL water and 100 μL 1 M NaOH (see Note 1). Boil until PFA is dissolved. When solution reaches room temperature, add 500 μL 20× PBS, and adjust pH to 7.4 with 1 M HCl. Bring total volume to 10 mL with water.

2.3. Immunocyto-chemistry

1. Permeabilization solution: 0.3% BSA, 0.1% Triton X-100, 0.02% NaN_3 in PBS. To make a 500 mL solution, weigh out 1.5 g BSA and 0.1 g NaN_3. Add 25 mL 20× PBS and 500 μL

Triton X-100. Bring total volume to 500 mL with water and filter-sterilize. Store at 4°C.

2. Blocking solution: 0.3% BSA, 0.01% Tween-20, 0.02% NaN$_3$ in PBS. To make 1 L solution, weigh out 3 g BSA and 0.2 g NaN$_3$. Add 50 mL 20× PBS and 100 µL Tween-20. Bring total volume to 1 L with water and filter-sterilize. Store at 4°C.

3. Primary antibody: Human antibody against centromeres/CREST (Immunovision, Springdale, AR). Reconstitute with water and store as instructed.

4. Secondary antibody: Alexa fluor 594 goat anti-human (Invitrogen, Carlsbad, CA). Make 2 mg/mL stocks and store as instructed.

5. SYTOX green nucleic acid stain (5 mM in DMSO; Invitrogen). Store as instructed (see Note 2).

3. Methods

1. Add 10 µL stock L-glutamine (100 mM) and 1 µL monastrol (100 mM) to 1 mL CZB, so that the final concentrations are 1 mM and 100 µM, respectively (see Note 3).

2. Pipet 500 µL of CZB + glutamine + monastrol to the inside of an organ culture dish with water in the outer ring (see Note 4). Transfer Metaphase II eggs to center of the dish, with CZB + glutamine + monastrol. Place dish in incubator for 1 h.

3. Prepare 2% PFA and transfer treated eggs to the fixative for 30 min at room temperature.

4. Transfer fixed eggs to Blocking solution (see Note 5).

5. Immunocytochemistry is described below with all solutions in 100 µL volumes in a 96-well dish, but may also be performed in smaller volumes to conserve reagents. Transfer eggs to the well containing 100 µL Permeabilization solution for 15 min, then into 100 µL Blocking solution twice, for 15 min each.

6. To detect kinetochores, transfer eggs into 100 µL Blocking solution containing CREST autoimmune serum (1:40 dilution) and let incubate for 1 h. Wash by transferring eggs through Blocking solution three times, for 15 min each.

7. To detect CREST, transfer into 100 µL Blocking solution containing Alexa fluor 594 goat anti-human secondary antibody (1:100 dilution) and let incubate for 1 h. Wash by transferring eggs through Blocking solution three times, for 15 min each.

DNA   Kinetochores   Merge

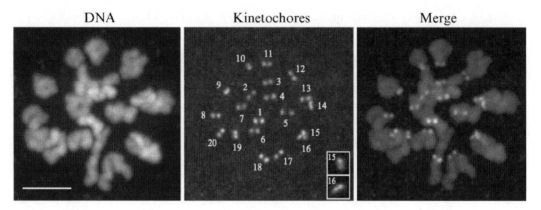

Fig. 2. Assessing ploidy by counting kinetochores. A monastrol-treated Metaphase II egg stained for DNA and kinetochores is shown with numbered kinetochore pairs. Images are maximal intensity projections, with reduced DNA intensity to highlight kinetochores in the merged panel. *Insets* show that kinetochore pairs 15 and 16 are distinct on different z-planes. Note that there are 20 kinetochore pairs in mouse eggs and chromosomes are telocentric. Scale bar represents 5 μm.

8. To stain DNA, transfer eggs to Blocking solution containing SYTOX green nucleic acid stain (1:5,000 dilution) for 10 min.

9. Transfer eggs to mounting medium (e.g., VECTASHIELD, Vector Laboratories, Burlingame, CA) on microscope slides. Image to cover all chromosomes and count the number of kinetochores (Fig. 2) (see Note 6).

4. Notes

1. We find that it is best to prepare this fixative fresh each time.

2. Other stains can be used to detect DNA, such as 4′-6-diamidino-2-phenylindole (DAPI).

3. Prior to use, warm and equilibrate CZB medium in a 37°C incubator with 5-5-90 gas (5% O_2-5% CO_2-90% N_2). Since CZB is not buffered to maintain pH, avoid removing the medium from incubator for extended periods of time; CZB will turn pink in color at high pH. L-glutamine should be added to CZB fresh each time.

4. Other dishes can also be used, but avoid contact with oil if possible since DMSO is oil-soluble.

5. The eggs can be stored overnight at 4°C at this stage.

6. We use a confocal microscope with high magnification (e.g., 100×, NA 1.4). Make sure all chromosomes and especially kinetochores are imaged with fine z-sections (e.g., 0.5 μm intervals).

Acknowledgements

This work was supported by a grant from the National Institutes of Health (HD 058730) and a Searle Scholar award to M.A.L.

References

1. Duncan FE, Chiang T, Schultz RM, Lampson MA (2009) Evidence that a defective spindle assembly checkpoint is not the primary cause of maternal age-associated aneuploidy in mouse eggs. Biol Reprod 81:768–776

2. Mayer TU, Kapoor TM, Haggarty SJ, King RW, Schreiber SL, Mitchison TJ (1999) Small molecule inhibitor of mitotic spindle bipolarity identified in a phenotype-based screen. Science 286:971–974

3. Chiang T, Duncan FE, Schindler K, Schultz RM, Lampson MA (2010) Evidence that weakened centromere cohesion is a leading cause of age-related aneuploidy in oocytes. Curr Biol 20:1522–1528

4. Lane SI, Chang HY, Jennings PC, Jones KT (2010) The Aurora kinase inhibitor ZM447439 accelerates first meiosis in mouse oocytes by overriding the spindle assembly checkpoint. Reproduction 140:521–530

5. Illingworth C, Pirmadjid N, Serhal P, Howe K, Fitzharris G (2010) MCAK regulates chromosome alignment but is not necessary for preventing aneuploidy in mouse oocyte meiosis I. Development 137:2133–2138

6. Chatot CL, Ziomek CA, Bavister BD, Lewis JL, Torres I (1989) An improved culture medium supports development of random-bred 1-cell mouse embryos in vitro. J Reprod Fertil 86:679–688

Chapter 18

Free-Hand Bisection of Mouse Oocytes and Embryos

Zbigniew Polanski and Jacek Z. Kubiak

Abstract

Mouse oocytes and zygotes are semitransparent and large cells approximately 80 μm in diameter. Bisection is one of the easiest ways for performing micromanipulations on such cells. It allows living sister halves or smaller fragments to be obtained, which can be cultured and observed for long periods of time. Bisection can be used for different kinds of experiments such as analysis of nucleo-cytoplasmic interactions, the relationship between different cellular structures or between different parts of embryos, eventually for analyzing the developmental potential of embryonic fragments. Oocyte or embryo halves can be examined by immunostaining, by measuring different cellular functions and by Western blot and genetic analysis (e.g., RT-PCR). Here we describe a detailed protocol for the free-hand bisection of mouse *zona pellucida*-free oocytes and embryos on an agar layer using a glass needle.

Key words: Bisection, Cell cycle, Early embryo development, Micromanipulations, Mouse, Nucleo-cytoplasmic interactions, Oocyte maturation

1. Introduction

Mouse oocyte and embryo bisection was routinely used in the series of analyses of nucleo-cytoplasmic interactions during oocyte maturation and early embryo development performed by Andrzej K. Tarkowski in the 1970s and 1980s. The idea for cutting mouse eggs by hand evolved, as it often happens, from the need to do such experiments and the lack of specialized equipment. In communist Poland, limited access to precise and complex equipment for research forced researchers to invent "homemade" methods. Free-hand bisection of cells was one such method. With time, it became clear that the shortcoming turned out to be an advantage.

In earliest days, bisected female germ cells were used by Tarkowski to determine the developmental potential of the separate half-oocytes or embryos during in vitro culture (1, 2) and to observe the behavior of sperm heads that were introduced into the oocyte halves by in vitro fertilization (3–6). Later on, cell fusion

Hayden A. Homer (ed.), *Mammalian Oocyte Regulation: Methods and Protocols*, Methods in Molecular Biology, vol. 957, DOI 10.1007/978-1-62703-191-2_18, © Springer Science+Business Media, LLC 2013

was used to test the activities present in nuclear and anuclear oocyte and embryo halves (7–17). Cell fusion techniques adapted to mouse embryos (18) may also be used to combine nuclear and anuclear fragments of eggs that have been obtained from different strains of mice thereby creating reconstituted cells having modified proportions of different mouse strains (15, 16). Blastomeres obtained for instance from embryos at the two-cell stage can also be bisected by the free-hand method described here (e.g., (19)). It should also be possible to adapt this method to 4-cell and 8-cell blastomeres, if required, by using a cutting fiber of smaller diameter.

The development of modern methods including two-dimensional protein electrophoresis (20, 21), immunocytochemistry (19, 22), Western blotting and immunoprecipitation (12, 19, 23), kinase activity assays (12, 19), live-imaging of tubulin or cyclin B (22, 24) as well as RT-PCR (12) allowed the proteins and nucleic acids in bisected oocyte or embryo fragments to be analyzed. At present therefore, there is the possibility to combine free-hand bisection with other sophisticated techniques. For example, studies that were performed on bisected egg fragments with the aim of looking for a putative role of localized information in mammalian development (25, 26) might be nicely complemented by protein mass spectrometry and multiplex Real-Time PCR.

The use of micromanipulators allowed for the enucleation of mammalian oocytes and embryos (27). This early technique, however, had its limitation since, once removed, the nucleus invariably underwent destruction. By 1983, McGrath and Solter (28) developed the micromanipulator-based procedure of mouse one-cell embryo enucleation, adapted later for mammalian cloning (29), in which one obtained a viable nucleus within a minimal volume of cytoplasm (karyoplast) and a large cell devoid of a nucleus (cytoplast) having a volume almost equal to that of a whole oocyte or 1-cell embryo. Thus, until 1983, Tarkowski's method of oocyte/embryo bisection remained the only technique that allowed the production of viable karyoplasts and cytoplasts from a single egg. Experiments involving free-hand bisection were instrumental for gaining insight into the regulation of the activities controlling the meiotic cell cycle in oocytes, fertilization and egg activation as well as early embryo development. This knowledge was indispensable for the success of mammalian cloning (29). Notably, the bisection procedure was later adapted to enable nuclear transfer without the use of micromanipulators (30).

2. Materials

Prepare all solutions using ultrapure water and analytical grade reagents and store at room temperature (unless indicated otherwise).

2.1. Preparation of Oocytes or One Cell Embryos

1. 10–12 week-old female mice of a chosen strain (e.g., Swiss, CBA, C57Bl) for isolation of ovaries to obtain oocytes arrested in meiotic prophase I (Germinal vesicle (GV) stage) or for injection with five to ten units of PMSG and 48 h later with five to ten units of hCG to obtain ovulated oocytes arrested in MII. If embryos are needed, PMSG/hCG-injected females should be coupled with males (e.g., from the strains listed above).

2. Scissors, tweezers, and needles for ovary or oviduct dissection and manipulation.

3. Pre-warmed (30–37°C) M2 medium (containing 0.4% BSA) for oocyte and embryo manipulations in open air (see Note 1).

4. Laboratory stereo-microscope.

5. Watch slides or small Petri dishes for isolating oocytes and embryos (see Note 2).

6. Glass Pasteur pipettes for preparing handling pipettes used for oocyte collection/transfer.

7. 100–300 IU/ml hyaluronidase solution in PBS or M2 medium (see Note 3).

8. Tyrode's acid solution, pH 2.5–3 for *zona pellucida* dissolution; acidified M2 (pH 2.5–3) works equally well (see Note 4).

9. 5 μg/ml Cytochalsin B (or cytochalasin D).

10. Heating plate set at 30°C.

2.2. Oocyte/One Cell Embryo Bisection

1. Pasteur pipettes or glass capillaries for preparation of a cutting fiber. Capillaries of 1 μm outer diameter such as those supplied by Drummond, Sutter Instrument or Clark Electronics are suitable.

2. Bunsen burner with adjustable pilot flame.

3. 1% Agar in PBS.

4. 60 mm plastic Petri dishes.

5. Petri dishes coated with a layer of 1% agar in PBS. Dishes are prepared by pouring liquefied 1% agar into a plastic Petri dish (60 mm size) so that it forms a layer ~2–3 μm thick. Leave to stand for a few minutes until the agar solidifies (see Note 5).

6. Laboratory stereo-microscope (see Note 6).

7. Micropipette puller (optional, for preparing the cutting fiber).

3. Methods

All procedures are carried out at room temperature unless specified otherwise. There is no strictly defined method for bisection. Below we describe the basic procedures, which can be modified to best fit the preferences of individual investigators.

3.1. Isolation of
Oocytes and Embryos

1. Sacrifice female mouse by cervical dislocation. Use non-stimulated mice if GV-stage oocytes required; PMSG/hCG-treated mice for ovulated oocytes, and; PMSG/hCG-treated mice that have been coupled with males for zygotes and early embryos.

2. Open abdominal cavity. For experiments involving GV-stage oocytes, dissect out the ovaries using tweezers and scissors and proceed to step 3. For experiments involving MII-arrested oocytes or embryos, dissect out the oviducts and proceed to step 5.

3. For obtaining oocytes at the GV-stage, place dissected ovaries into either a Petri dish or a watch slide containing M2 medium (see Note 7).

4. Using a needle, liberate oocytes by puncturing the ovarian follicles, which are visible as clear bulges at the ovarian surface. Pipette them vigorously to remove follicular cells and transfer cumulus-free oocytes to clean M2 medium using a mouth pipette and proceed to step 9.

5. For obtaining ovulated MII oocytes and one-cell embryos, place the dissected oviduct into pre-warmed PBS containing hyaluronidase (see Note 8).

6. Puncture the ampullary region of the oviduct using a needle to liberate the ovulated oocytes or embryos into hyaluronidase-containing PBS.

7. Maintain cumulus-surrounded MII oocytes/embryos for 3–6 min at 30°C on a heating plate. During this time the surrounding follicular cells become detached and the oocytes or embryos become easily visible with the stereomicroscope.

8. Transfer cumulus-free oocytes or embryos to hyaluronidase-free M2 medium using a mouth pipette.

9. Remove the *zonae pellucidae* by incubation of oocytes in a drop of acid Tyrode's medium. Because the treatment with acid medium dissolves the *zona pellucida* within a relatively short time, the process should be continuously monitored under the stereomicroscope. The *zona pellucida* becomes gradually thinner before disappearing completely (see Note 9).

10. Wash zona-free oocytes/embryos through a series of large droplets (~100 µL) of standard M2 medium. After the removal of the *zona pellucida* the oocytes become more sticky and fragile, thus more care should be taken to avoid injuring the cells during further handling.

Optionally pronase solution in M2 may be used in which the oocytes should be incubated for 3–5 min at 37°C (see Note 10).

3.2. Preparation of a
Cutting Glass Fiber

There are two options for preparing the cutting fiber involving either the flame of a Bunsen burner or a micropipette puller.

1. Heat the glass capillary or Pasteur pipette over the flame of a Bunsen burner.

2. When the glass begins to soften in the flame, pull it into a fine diameter. This is the same method as used for preparing pipettes for oocyte collection/handling. However, obtaining good quality (thin enough) glass fibers usually requires many attempts (see Note 11).

1. Adjust the parameters of the puller to obtain fibers with the desired length and diameter.

2. Load glass capillary into puller and activate.

3. Readjust puller settings until fibers are obtained that are most effective at bisecting oocytes/embryos.

Option 1 does not require highly specialized equipment but does not consistently produce good fibers. Option 2 requires a relatively expensive puller and some trial and error before the best settings are established. However, once established, the same parameters may then be used repeatedly to obtain fibers of the desired characteristics, which suit best the individual needs of an investigator.

3.3. Bisection of Oocytes/Embryos

1. Place oocytes for at least 15 min in cytochalasin B (or D)-containing M2 medium. This treatment disrupts the cell actin cytoskeleton and prevents oocytes from lysing either during or immediately after the bisection. Cytochalasin treatment eliminates the need for oocyte cooling during the bisection, which was described in the original papers by Tarkowski (1, 2). However, cooling (see Note 12) may still be applied and should further increase the efficiency of bisection (see Note 13).

2. Place a group of oocytes in a row on the agar-coated Petri dish filled with cytochalasin B (or D)-containing M2 medium. The distance between oocytes should be at least five times the oocyte diameter. The bigger distance between the oocytes reduces also the risk of mixing up the halves obtained from different oocytes.

3. Place the cutting fiber over the first oocyte in the row and press gently (the procedure is schematically shown in Fig. 1). The oocyte should flatten under the pressure and will tend to separate into two hemispheres emerging at both sides of the cutting fiber. If it is intended that the oocyte be cut into two equal halves, adjust the position the cutting device (still having the oocyte pressed gently by the fiber) such that equal volumes of cytoplasm emerge on either side of the fiber (see Note 14).

4. Press the cutting fiber downwards to divide the oocyte into two parts by "squishing" out the two separated parts on both sides of the fiber. Usually the two oocyte-halves remain connected by a cytoplasmic bridge (sometimes very thin, thus difficult to

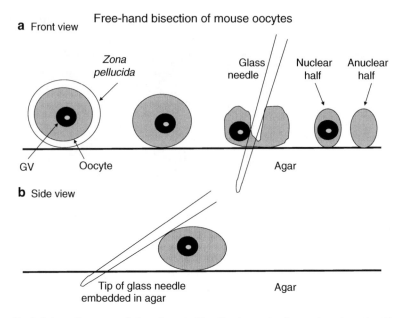

Fig. 1. Schematic representation of oocyte bisection (example of a prophase I oocyte with GV). The same procedure is applied for oocytes during maturation, MII-arrested oocytes, one-cell embryos or blastomeres. (**a**) Front view showing the sequence of proceedings. (**b**) Side view showing how the tip of the glass needle should be immobilized in the agar to facilitate cutting.

see and easily overlooked). Leaving the fragments connected by cytoplasmic bridges will result in fast (even within 1 min) reunification of the two fragments into a single round cell.

5. Soon after cutting, separate both fragments completely by tearing any cytoplasmic bridges against the agar surface perpendicularly to the fiber axis. In this way the both parts are pushed apart from each other and the cytoplasmic bridge becomes narrower, and finally breaks after which the cell membranes seal at the breaking point.

6. Repeat the procedure with subsequent oocytes by gently pushing the Petri dish under the stereomicroscope to bring the next oocyte into the field of vision. Avoid rapid movements, which may displace the already obtained oocyte halves.

7. Collect the oocyte halves/fragments for further use (see Note 15). The outcome of the procedure is shown in Fig. 2.

4. Notes

1. Mouse oocytes at the germinal vesicle (GV)-stage are isolated from ovaries, whereas ovulated oocytes (arrested at metaphase of the second meiotic division or MII) and early-stage embryos

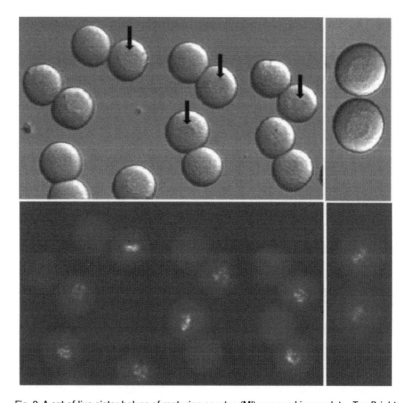

Fig. 2. A set of live sister halves of maturing oocytes (MI) arranged in couplets. *Top*: Bright field. Each couplet is composed of the nuclear and anuclear half obtained from a single oocyte. The nuclear halves are distinguished by the presence of the meiotic spindle (*arrows*). In the remaining couplets (*without arrows*) the nuclear halves can be identified by modifying their orientation to a position in which the spindle becomes visible. On the right: two intact oocytes are shown for the comparison of sizes (*arrows* mark the spindles as in the couplets on the left). *Bottom*: Dark field. The cells were live-labeled with Hoechst (which stains the DNA) to demonstrate the co-localization of the spindles and chromosomes. The images were acquired using a Zeiss Axioplan 300 M inverted microscope: bright field images were obtained with Nomarski contrast, whilst dark field images were obtained using UV fluorescence.

(zygotes) are isolated from the oviduct. The pH of M2 medium is stabilized by HEPES, so permitting manipulations to be made in air. In contrast, non-HEPES containing media must be CO_2-equilibrated and cannot be used for such manipulations since their pH changes when exposed to air.

2. Glass slides are better since they are not scratched by forceps or needles and allow easier collection of oocytes.

3. Ovulated oocytes and early zygotes are surrounded by the *cumulus oophorus* composed of a mass of very sticky follicular cells that pose difficulties during handling using pipettes (for more details see Note 8).

4. Alternatively, 0.25% pronase (grade B, Calbiochem) in M2 medium may be used for *zona pellucida* dissolution.

5. The agar-coated Petri dishes can be prepared in advance and stored for 1–2 weeks at 4°C. They should not be stored longer because water evaporates from the agar making the coating overly hard and uneven. A stock of 1% agar in PBS may be stored for a longer time in a glass flask at 4°C and when needed, can be liquefied by heating in a microwave oven and used to prepare agar coated dishes. It is important to be aware, however, that after repeated cycles of liquefaction, the agar gel becomes harder (through evaporation at each heating), which in turn may affect the efficiency of bisection. Thus, we recommend heating the stock of agar no more than three times.

6. The optics quality of the steromicroscope is crucial when oocytes or embryos at the M-phase are bisected. In some microscopes the halves containing the spindle/chromosomes are easily distinguished because of the presence of a darker area that appears as a slight concavity indicating the position of the spindle (Fig. 2). In some stereomicroscopes, however, this area is barely visible making the identification of nuclear and anuclear halves problematic. In cases in which oocytes or embryos with interphase nuclei are used (e.g., GV-containing prophase I-arrested oocytes, activated eggs at the pronuclear stage or interphase 2-cell embryos) the identification of nuclear and anucelar fragments is relatively easy since the visibility of the nuclei is much better.

7. If an electric bulb is used with the stereo microscope, the heating that is produced on the stage maintains a suitable temperature for oocyte and embryo manipulations. If a cold light is used, all solutions should be placed on a heating plate next to the microscope to keep them at approximately 30°C.

8. Cumulus-surrounded oocytes/embryos should be liberated from oviductal *ampullae* in the presence of hyaluronidase. In the absence of hyaluronidase, the cumulus-surrounded oocytes/embryos easily stick to the pipette, forceps, or needles and can be easily lost.

9. The time required for *zona pellucida* removal may be increased if oocytes are kept for long periods in culture prior to zona removal. The prolongation is caused by *zona pellucida* hardening, which occurs when oocytes are cultured in vitro in media without serum (31).

10. In contrast to the treatment with the acidified medium, the pronase solution digests *zona pellucida* enzymatically. It becomes detached from the oocytes progressively. It is not necessary to wait until the complete disappearance of the *zona pellucida*, because it can be easily removed by gentle pippeting during the transfer of oocytes to M2 medium. The choice which method is applied for *zona pellucida* removal may depend on the further scheme of the experiment. For example,

fertilization in vitro of zona-free oocytes is significantly reduced after pronase treatment (32). On the other hand, if the oocyte fragments are planned to be used for cell fusion the pronase treatment may improve the fusion efficiency (9).

11. The width of glass fiber used for bisection should be around one-third to one-fifth of the diameter of the oocyte/embryo. Based on our experience, thinner diameter fibers allow more precise control of the cutting line and thus the volume of the fragments obtained by bisection. Our preference is for the thinnest section of the cutting fiber to be 1–2 cm in length. In cases in which smaller glass capillaries are used for making the cutting fiber, attachment to a thicker rod (made of glass or plastic, etc.) makes it more convenient for handling.

12. Cooling may be achieved either by using an agar-coated dish filled with medium which was pre-cooled in a refrigerator or by using a cooling plate placed on the stereomicroscope stage.

13. In GV-stage oocytes, the cytochalasin treatment results initially in the formation of numerous protrusions (blebs) on the oocyte surface. This may simplify bisection as such oocytes have reduced tendency to escape from under the cutting fiber during bisection. On the other hand, it is more difficult to control the volume of the bisected fragments. After longer culture in the presence of cytochalasin (usually 30 min) the blebs disappear.

14. Fixing the tip of the cutting fiber in the agar at some distance from the oocyte may help in bisection because it allows better control of the grip and reduces hand-shaking. The GV oocyte bisection is usually trickier than maturing oocytes or embryos. Most probably the GV makes the oocyte very rigid. Since the GV is rarely centrally placed, it is better, right before cutting, to turn the oocyte gently to a position in which the GV is slightly eccentric. Alternatively, one can gently aspirate the oocyte into a pipette with a slightly smaller diameter than the oocyte. This deforms the oocyte into a sausage-like shape, making it much easier to cut. However, because this requires each oocyte to be manipulated just before cutting the time spent on cutting is increased and may become tiresome for the experimenter. This method is also very useful for unequal bisection especially when much smaller nuclear fragments are required (e.g., (15, 16)).

15. If after cutting each oocyte, sister halves (the two halves originating from a single oocyte) are required for further analysis, they can be transferred either singly or in pairs into separate drops of medium under oil in another Petri dish under a second stereomicroscope. Alternatively, fragments can be transferred after cutting the whole series of oocytes. If a higher number of oocyte/embryo halves is required without distinction of sister halves, the selection of nuclear and anuclear halves should be done after cutting of the whole series of oocytes.

Acknowledgments

We thank Malgorzata Kloc for valuable discussions and critical reading of the manuscript. This work was supported by grants from ARC and LCC to JZK.

References

1. Tarkowski AK, Rossant J (1976) Haploid mouse blastocysts developed from bisected zygotes. Nature 259:663–665
2. Tarkowski AK (1977) In vitro development of haploid mouse embryos produced by bisection of one-cell fertilized eggs. J Embryol Exp Morphol 38:187–202
3. Bałakier H, Tarkowski AK (1980) The role of germinal vesicle karyoplasm in the development of male pronucleus in the mouse. Exp Cell Res 128:79–85
4. Tarkowski AK (1980) Fertilization of nucleate and anucleate egg fragments in the mouse. Exp Cell Res 128:73–77
5. Borsuk E, Mańka R (1988) Behavior of sperm nuclei in intact and bisected metaphase II mouse oocytes fertilized in the presence of colcemid. Gamete Res 20:365–376
6. Borsuk E, Tarkowski AK (1989) Transformation of sperm nuclei into male pronuclei in nucleate and anucleate fragments of parthenogenetic mouse eggs. Gamete Res 24:471–481
7. Balakier H, Czolowska R (1977) Cytoplasmic control of nuclear maturation in mouse oocytes. Exp Cell Res 110:466–469
8. Bałakier H, Masui Y (1986) Chromosome condensation activity in the cytoplasm of anucleate and nucleate fragments of mouse oocytes. Dev Biol 113:155–159
9. Clarke HJ, Masui Y (1987) Dose-dependent relationship between oocyte cytoplasmic volume and transformation of sperm nuclei to metaphase chromosomes. J Cell Biol 104: 831–840
10. Borsuk E (1991) Anucleate fragments of parthenogenetic eggs and of maturing oocytes contain complementary factors required for development of a male pronucleus. Mol Reprod Dev 29:150–156
11. Fulka J Jr, Ouhibi N, Fulka J, Kanka J, Moor RM (1995) Chromosome condensation activity (CCA) in bisected C57BL/6JxCBA mouse oocytes. Reprod Fertil Dev 7:1123–1127
12. Hoffmann S, Tsurumi C, Kubiak JZ, Polanski Z (2006) Germinal vesicle material drives meiotic cell cycle of mouse oocyte through the 3'UTR-dependent control of cyclin B1 synthesis. Dev Biol 292:46–54
13. Kárníková L, Urban F, Moor R, Fulka J Jr (1998) Mouse oocyte maturation: the effect of modified nucleocytoplasmic ratio. Reprod Nutr Dev 38:665–670
14. Kubiak JZ, Weber M, de Pennart H, Winston NJ, Maro B (1993) The metaphase II arrest in mouse oocytes is controlled through microtubule-dependent destruction of cyclin B in the presence of CSF. EMBO J 12:3773–3778
15. Polanski Z (1997) Strain difference in the timing of meiosis resumption in mouse oocytes: involvement of a cytoplasmic factor(s) acting presumably upstream of the dephosphorylation of p34cdc2 kinase. Zygote 5:105–109
16. Polanski Z, Ledan E, Brunet S, Louvet S, Verlhac MH, Kubiak JZ, Maro B (1998) Cyclin synthesis controls the progression of meiotic maturation in mouse oocytes. Development 125:4989–4997
17. Zernicka-Goetz M, Ciemerych MA, Kubiak JZ, Tarkowski AK, Maro B (1995) Cytostatic factor inactivation is induced by a calcium-dependent mechanism present until the second cell cycle in fertilized but not in parthenogenetically activated mouse eggs. J Cell Sci 108:469–474
18. Kubiak JZ, Tarkowski AK (1985) Electrofusion of mouse blastomeres. Exp Cell Res 157:561–566
19. Ciemerych MA, Tarkowski AK, Kubiak JZ (1998) Autonomous activation of histone H1 kinase, cortical activity and microtubule organization in one- and two-cell mouse embryos. Biol Cell 90:557–564
20. Petzoldt U (1990) Survival of maternal mRNA in anucleate and unfertilized mouse eggs. Eur J Cell Biol 52:123–128
21. Petzoldt U, Muggleton-Harris A (1987) The effect of the nucleocytoplasmic ratio on protein synthesis and expression of a stage-specific antigen in early cleaving mouse embryos. Development 99:481–491

22. Brunet S, Polanski Z, Verlhac MH, Kubiak JZ, Maro B (1998) Bipolar meiotic spindle formation without chromatin. Curr Biol 8:1231–1234

23. Pahlavan G, Polanski Z, Kalab P, Golsteyn R, Nigg EA, Maro B (2000) Characterization of polo-like kinase 1 during meiotic maturation of the mouse oocyte. Dev Biol 220:392–400

24. Hoffmann S, Maro B, Kubiak JZ, Polanski Z (2011) A single bivalent efficiently inhibits cyclin B1 degradation and polar body extrusion in mouse oocytes indicating robust SAC during female meiosis I. PLoS One 6:e27143

25. Ciemerych MA, Mesnard D, Zernicka-Goetz M (2000) Animal and vegetal poles of the mouse egg predict the polarity of the embryonic axis, yet are nonessential for development. Development 127:3467–3474

26. Zernicka-Goetz M (1998) Fertile offspring derived from mammalian eggs lacking either animal or vegetal poles. Development 125: 4803–4808

27. Modlinski JA (1978) Transfer of embryonic nuclei to fertilised mouse eggs and development of tetraploid blastocysts. Nature 273:466–467

28. McGrath J, Solter D (1983) Nuclear transplantation in the mouse embryo by microsurgery and cell fusion. Science 220:1300–1302

29. Wilmut I, Schnieke AE, McWhir J, Kind AJ, Campbell KH (1997) Viable offspring derived from fetal and adult mammalian cells. Nature 385:810–813

30. Vajta G, Lewis IM, Hyttel P, Thouas GA, Trounson AO (2001) Somatic cell cloning without micromanipulators. Cloning 3:89–95

31. De Felici M, Siracusa G (1982) Spontaneous hardening of the zona pellucida of mouse oocytes during in vitro culture. Gamete Res 6:107–113

32. Komorowski S, Szczepanska K, Maleszewski M (2003) Distinct mechanisms underlie sperm-induced and protease-induced oolemma block to sperm penetration. Int J Dev Biol 47:65–69

Chapter 19

Microarray-CGH for the Assessment of Aneuploidy in Human Polar Bodies and Oocytes

Souraya Jaroudi and Dagan Wells

Abstract

The cytogenetic analysis of single cells, such as oocytes and polar bodies, is extremely challenging. The main problem is low probability of obtaining a metaphase preparation in which all of the chromosomes are sufficiently well spread to permit accurate analysis (no overlapping chromosomes, no chromosomes lost). As a result, a high proportion of the oocytes subjected to cytogenetic analysis are not suitable for traditional chromosome banding studies or for molecular cytogenetic methods such as spectral karyotyping (SKY) or multiplex fluorescence in situ hybridization (M-FISH). Fortunately, recent innovations in whole genome amplification and microarray technologies have provided a means to analyze the copy number of every chromosome in single cells with high accuracy. Here we describe the use of such methods for the investigation of chromosome and chromatid abnormalities in human oocytes and polar bodies.

Key words: Whole genome amplification, Microarray, Comparative genomic hybridization, Aneuploidy, Polar body, Oocyte

1. Introduction

The cytogenetic assessment of human oocytes has attracted a great deal of attention from scientists and clinicians alike. Not only is the meiotic process of fundamental biological interest, but the outcome of abnormal chromosome segregation has profound medical consequences. Aneuploidy, mostly oocyte derived, is the leading cause of first trimester miscarriage, congenital abnormalities, and mental retardation (e.g., Down syndrome). It is also believed to be the underlying reason for many unsuccessful infertility treatments.

Loss or gain of chromosomes or chromatids during the first and/or second meiotic divisions (i.e., meiosis I and meiosis II) is extremely common in our species. Indeed, the frequency at which errors of this type occur is an order of magnitude higher in human oocytes compared with most other mammalian species.

Hayden A. Homer (ed.), *Mammalian Oocyte Regulation: Methods and Protocols*, Methods in Molecular Biology, vol. 957, DOI 10.1007/978-1-62703-191-2_19, © Springer Science+Business Media, LLC 2013

Malsegregation of chromosomes and chromatids is associated with female age, increasing slowly during early adulthood, but accelerating from the mid-30s onwards. Typically, about 10% of oocytes from women under 30 are found to be abnormal, while more than two-thirds derived from women in their 40s are aneuploid (1–4). Most abnormalities involve premature separation of sister chromatids (these should normally remain attached to each other until anaphase of meiosis II). Failure to maintain sister chromatid cohesion leads to separation during the first meiotic division, followed by segregation to either spindle pole at random (2, 4, 5). In young women most errors occur during meiosis I, but for women towards the end of their reproductive life the second meiotic division may be a more important contributor to abnormality (4).

A great deal of research into oocyte chromosomes has been undertaken using conventional cytogenetic techniques such as G-banding and R-banding (1). However, such approaches are limited by the difficulty in obtaining a good quality chromosome spread from a single oocyte. Many of the oocytes assessed cannot be analyzed due to overlapping chromosomes, whereas other spreads are found to be missing large numbers of chromosomes. It can be difficult to be sure if a single chromosome loss is a consequence of a true meiotic error or an artifact of the spreading procedure.

An alternative to conventional chromosome banding approaches is to use fluorescent in situ hybridization (FISH). This method has the advantage that each chromosome is identified by a fluorescent probe with a distinct color, so chromosomes that are overlapping or of poor morphology can still be readily identified (2). However, the main drawback of the FISH approach is that only a handful of chromosomes can be assessed in each oocyte, due to the limited range of spectrally distinct fluorochromes (i.e., few colors) available for probe labeling. Although it is possible to use molecular cytogenetic methods related to FISH, such as Multiplex-FISH (M-FISH) or spectral karyotyping (SKY), to identify each of the oocyte chromosomes, in practice this runs into the same problems as chromosome banding approaches, since good quality metaphase spreads are needed (6).

In terms of a simple assessment of chromosome/chromatid losses and gains, microarray comparative genomic hybridization (microarray-CGH) is the best method currently available for oocyte studies. Microarray-CGH is a DNA-based technique, so there is no need to obtain chromosome spreads from the sample; rather the cell is placed inside a microcentrifuge tube. The oocyte DNA is amplified using a whole genome amplification technique; the DNA is then labeled and applied to a microarray comprising numerous probes for specific chromosomal regions affixed to a solid support (usually a glass slide); a "reference" DNA sample derived from a chromosomally normal individual, labeled in a different color is applied to the microarray at the same time. Detection of loss or gain of chromosomal material involves computer assisted analysis of the color of each probe on the microarray. An equal quantity of fluorescence corresponding to the oocyte DNA and the reference DNA is indicative of normality for the chromosomal region corresponding to the probe; a chromatid/chromosome gain is revealed by an excess of fluorescence attributable to the oocyte; a deficiency of fluorescence from the oocyte relative to the reference DNA demonstrates a loss of chromosomal material in the oocyte. This approach has the potential to provide a highly accurate analysis of every chromosome. However, most commercially available microarray platforms are not suitable for single cell analysis, so it is advised to only use those that have been validated at the single cell level.

This chapter describes the use of the 24Sure microarray manufactured by BlueGnome (Cambridge, UK). This microarray has been highly optimized and validated for use with single cells, including oocytes and polar bodies (PBs) (4, 7–9). We routinely use this microarray for analyzing polar bodies biopsied from human oocytes as part of our preimplantation genetic screening (PGS) program. Given that the oocyte and PB are "daughter cells" of meiosis I, any abnormality detected in a PB is expected to be reciprocated in the corresponding oocyte (e.g., a loss of a chromosome in the first polar body is associated with a gain of the same chromosome in the oocyte after completion of meiosis I). The aim of PGS is to identify chromosomally normal oocytes, produced for the purpose of in vitro fertilization (IVF) treatment. Embryos derived from the chromosomally normal oocytes are then prioritized for transfer to the uterus with the hope of increasing pregnancy rates and decreasing the risks of miscarriage and Down syndrome. Embryos derived from abnormal oocytes are not transferred.

The 24Sure microarray is produced by "spotting" of more than 3,200 bacterial artificial chromosomes (BACs) onto a glass slide. The BACs have been carefully selected on the basis of having little variation in over 5,000 hybridizations, an absence of known copy-number variations (CNVs) and the highest levels of reproducibility and sensitivity in conjunction with whole genome amplification. Purpose-designed software (BlueFuse Multi) smooth results, allowing robust detection of chromosome losses and gains. If analysis of smaller chromosomal regions is desired, such as chromosome fragments associated with reciprocal translocations, it is recommended to use the higher density 24Sure+ microarray (7, 9). The 24Sure+ microarray has an average resolution of 2–5 Mb compared to 10 Mb for standard 24Sure microarrays. In particular, 24Sure+ has greater coverage of sub-telomeric and peri-centromeric regions, enabling accurate characterization of arm level aneuploidy and other large-scale structural abnormalities.

2. Materials

The SurePlex protocol should be carried out in a sterile (DNA-free, UV-irradiated) laminar flow cabinet in a separate room to where amplified products are handled, in order to prevent sample contamination. All amplification and labeling reactions should be set up on ice (unless indicated otherwise). Prepare washing and hybridization solutions using ultrapure water (18 MΩ cm at 25°C) and molecular biology grade reagents. Prepare and store all solutions at room temperature unless indicated otherwise by the manufacturer. Follow all waste disposal regulations carefully when disposing of materials.

2.1. Cell Preparation, Lysis, and Whole Genome Amplification

1. Collection of oocytes or PBs: Dulbecco's phosphate buffered saline (DPBS) with 0.1% polyvinyl alcohol (PVA) (Sigma-Aldrich).

2. SurePlex DNA Amplification System (Rubicon Genomics Inc., Ann Arbor, Michigan, USA), including: Cell Extraction Buffer, Extraction Enzyme Dilution Buffer, Cell Extraction Enzyme, Pre-amp Buffer, Pre-amp Enzyme, Amplification Buffer, Amplification Enzyme, and Nuclease-free water.

3. 1× Tris–Borate–EDTA (TBE) Buffer: 89 mM Tris base, 89 mM Boric acid, 2 mM EDTA, pH 8.3±0.1.

4. One percent agarose gel: Add 1 g agarose to 100 ml 1× TBE buffer and heat in microwave until dissolved. Allow to cool to room temperature then add 2 μl of Ethidium Bromide (10 mg/ml).

5. 10× Blue Juice gel loading buffer (Invitrogen, CA, USA).

6. 100 base pairs (bp) Lower Scale DNA Ladder (Fisher Scientific Ltd, UK).

7. Benchtop microcentrifuge (appropriate size for 0.2 ml tubes).

8. Thermal cycler.

9. UV transilluminator/gel imaging system.

2.2. Labeling and Hybridization

1. Fluorescent Labeling System (BlueGnome Ltd, Cambridge, UK), including: dCTP-labeling reagents, SureRef Reference Male DNA, and COT Human DNA.

2. 24Sure microarrays or 24Sure+ microarrays (BlueGnome Ltd, Cambridge, UK).

3. Formamide.

4. Centrifugal evaporator.

5. 20× Saline-Sodium Citrate (SSC) buffer and Tween 20.

6. 2× SSC/50% formamide: Prepare 6 ml by mixing 3 ml of formamide with 600 μl of 20× SSC and 2.4 ml of water.

7. Humidified box to be used as a hybridization chamber: Use a box that can be closed off completely such as a slide box or a sandwich box. Line the inside with tissue saturated with 6 ml of 2× SSC/50% formamide.

8. Parafilm.

9. Lidded non-circulating water bath at 47°C.

10. Square coverslips.

11. Wash 1 (2× SSC/0.05% Tween20): Prepare 1,000 ml by adding 100 ml 20× SSC (pH 7.0) and 0.5 ml Tween 20 to 899.5 ml ultrapure water. Prepare all washing solutions fresh on the day of use.

12. Wash 2 (1× SSC): Prepare 500 ml by adding 25 ml 20× SSC (pH 7.0) to 475 ml ultrapure water.

13. Wash 3 (0.1× SSC): Prepare 1,000 ml by adding 5 ml 20× SSC (pH 7.0) to 995 ml ultrapure water.

14. Coplin jars.

15. Stainless steel slide rack.

16. Square glass staining dish.

17. Magnetic stirrer.

18. 2.5 cm magnetic stir bar.

19. Water bath with good temperature stability. We recommend the Hybex Microarray Incubation System with water bath insert (BlueGnome 4303-1).

2.3. Microarray Scanning and Analysis

1. A two-channel microarray scanner (e.g., InnoScan 700, Innopsys, France) for capturing the Cy3 and Cy5 signals produced by the independently labeled test and control DNA samples.

2. MAPIX software for scanning the microarray slides.

3. BlueFuse Multi data analysis software (BlueGnome Ltd, Cambridge, UK) for analyzing scanned images.

3. Methods

The 24Sure protocol described in this chapter involves whole genome amplification using the SurePlex DNA Amplification System and takes a minimum of 8 h and 40 min to complete; however, certain steps (indicated with *) can be extended to allow the procedure to fit into a standard working day. The approximate timings for each step are shown in Table 1. The protocol used for the high-resolution 24Sure+ microarrays is essentially identical.

Table 1
Summary of steps involved in the microarray-CGH protocol

Protocol step	Approximate time	Subheadings
Sample preparation	30 min	3.1
Amplification	2 h	3.1
Labeling*	1.5–4 h (can be left overnight)	3.2
Combination and volume reduction	1 h	3.3
Hybridization*	3–16 h (can be left overnight)	3.4
Post-hybridization washes	30 min	3.5
Scanning and analysis	10 min (for a single slide, 2 samples)	3.6

3.1. Sample Preparation

1. Collect the oocytes or PBs in DNase-free, thin-walled 0.2 ml PCR tubes in 1–2 μl of PBS (0.1% PVA) (see Notes 1 and 2).

2. Wash oocytes through at least three 5 μl droplets of PBS pipetted onto a clean petri dish before transfer to the PCR tube. Washing is recommended in order to remove any DNA contaminants that may have been present in the medium in which the sample had been cultured.

3. If not used immediately, samples should be stored at –80°C and processed within 6 months. Once thawed, keep the oocytes or PBs at 4°C or on ice at all times. Avoid repeated cycles of freezing and thawing (see Note 3).

4. Prepare the oocytes or PBs for lysis by centrifuging the tube(s) at $200 \times g$ for 3 min or pulse centrifuge briefly.

5. Add 3 μl of Cell Extraction Buffer to each sample including any possible controls (see Note 4). The total volume in the tube should be 4–5 μl.

6. Prepare the extraction mastermix by adding 0.2 μl Cell Extraction Enzyme to 4.8 μl Extraction Enzyme Dilution Buffer per sample. Mix by flicking the mastermix tube and centrifuging briefly to collect the fluid at the bottom of the tube. The mastermix is prepared for an extra 10% of tubes, e.g., if analysis of two samples is to take place, the mastermix is prepared for 2.2 tubes.

7. Add 5 μl of freshly prepared extraction mastermix to each of the 0.2 ml PCR tubes containing the 4–5 μl sample.

8. "Pulse" centrifuge the tubes for approximately 10 s to ensure collection of the extraction mix and sample at the bottom of the tube.

Table 2
Thermal cycler program for the SurePlex lysis step

Number of cycles	Temperature (°C)	Time (min)
1 Cycle	75	10
1 Cycle	95	4
1 Cycle	Room temperature	Hold

Table 3
Thermal cycler program for the SurePlex pre-amplification step

Number of cycles	Temperature (°C)	Time
1 Cycle	95	2 min
12 Cycles	95	15 s
	15	50 s
	25	40 s
	35	30 s
	65	40 s
	75	40 s
1 Cycle	4	Hold

9. Incubate the tubes in a thermal cycler using the program shown in Table 2 (see Notes 5 and 6).

10. Place the samples on ice while setting up and adding the pre-amplification mastermix.

3.2. Pre-amplification

1. Prepare the Pre-Amp mastermix by adding 0.2 µl SurePlex pre-amp enzyme to 4.8 µl SurePlex pre-amp buffer per sample and mixing well. As with the lysis step, the mastermix is prepared for an extra 10% of tubes to allow for loss of fluid during pipetting.

2. Add 5 µl of SurePlex Pre-Amp mix to 10 µl of sample. Briefly centrifuge the tubes to ensure that all contents have collected at the bottom of the tube.

3. Incubate the tubes in a thermal cycler using the program shown in Table 3 (see Note 2).

4. Place the pre-amplified products on ice.

Table 4
Thermal cycler program for the SurePlex amplification step

Number of cycles	Temperature (°C)	Time
1 Cycle	95	2 min
14 Cycles	95	15 s
	65	1 min
	75	1 min
1 Cycle	4	Hold

3.3. Amplification

1. Prepare the amplification mastermix by adding 0.8 μl SurePlex amplification enzyme to 25 μl SurePlex amplification buffer and 34.2 μl nuclease-free water per sample and mixing well by vortexing for 5 s. The mastermix is prepared for an extra 10% of tubes, e.g., if analysis of two samples is to take place, then the mastermix is prepared for 2.2 tubes.

2. Add 60 μl of SurePlex Amplification mix to the 15 μl of synthesis reaction product and mix well by flicking and briefly pulse centrifuging for 10 s.

3. Incubate the tubes in a thermal cycler using the program shown in Table 4 (see Note 7).

3.4. Checking Amplification Success Using Agarose Electrophoresis

If whole genome amplification has not been successful, microarray analysis will not yield usable data (see Note 8 and Fig. 1). As microarrays are relatively expensive, it is desirable to avoid wasting them on samples that have not amplified.

1. Mix 5 μl of each amplified sample with 0.5 μl gel loading buffer.

2. Load the samples on a 1% agarose gel alongside 5 μl of 100 bp Lower Scale DNA Ladder (or equivalent).

3. Electrophorese at 175–190 V for 10 min.

4. Visualize the products using a UV transilluminator/gel imaging system. Successful amplification should result in a smear that ranges between 100 and 2,000 base pairs (bp) with strongest intensity around 500 bp (see Fig. 1).

3.5. Labeling and Preparation of Amplified Test and Reference DNA Samples

The amplified test and reference DNAs are labeled with Cy3 (green) and Cy5 (red) fluorophores, respectively, using random primers. The SureRef, normal male (46,XY) reference DNA, is well matched to amplified single cells or PBs and is used as a hybridization control (see Note 9).

1. Thaw, vortex, and centrifuge briefly the components of the Fluorescent Labeling System and keep on ice. Prepare

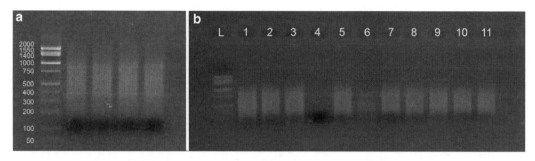

Fig. 1. SurePlex amplified DNA detected on a 1% agarose 1× TBE gel. The 100 bp Lower Scale DNA Ladder (L) was loaded in the first lane followed by the amplified polar bodies. (**a**) The smears observed for the four polar bodies between 200 and 2,000 bp indicate successful whole genome amplification. (**b**) The smears observed for samples 1, 2, 3, 5, 7, 8, 9, 10, and 11 demonstrate successful amplification. The weaker smear observed for polar body 6 indicates poor amplification. This sample is likely to give noisy microarray results that may be difficult to interpret. No amplification was detected for the PB in lane 4.

the labeling mixes by adding the following components in the order they are listed.

2. Prepare mastermix for the test samples (Cy3 labeling mix) by adding the following components in the order they are listed: 5 μl Reaction Buffer, 5 μl Primer Solution, 5 μl dCTP-Labeling Mix, and 1 μl Cy3 dCTP. The mastermix is prepared for an extra 10% of tubes.

3. Prepare mastermix for the reference samples (Cy5 labeling mix) by adding the following components in the order they are listed: 5 μl Reaction Buffer, 5 μl Primer Solution, 5 μl dCTP-Labeling Mix, 1 μl Cy3 dCTP, and 8 μl SureRef DNA.

4. Aliquot 16 μl of the Cy3 labeling mix into clearly labeled PCR tubes. One tube is required per test sample.

5. Add 8 μl of the amplified DNA sample into the appropriate tube and mix well by pipetting up and down.

6. Aliquot 24 μl of Cy5 labeling mix into an equal number of "reference" PCR tubes.

7. Place all of the tubes in a thermal cycler (with a preheated lid) at 94°C for 5 min to denature the DNA.

8. Immediately transfer tubes onto ice or a pre-cooled thermal cycler at 4°C for another 5 min.

9. Add 1 μl of Klenow enzyme to each test and reference sample.

10. Mix well by flicking and then pulse centrifuge for 5 s to collect all of the fluid in the bottom of the tubes. Place the tubes in a thermal cycler (with a preheated lid) for 2–4 h at 37°C (see Note 10). Incubation is generally performed for 3 h, but labeling can be completed in as little as 1.5 h, or left overnight if more convenient.

11. Once the incubation is complete, place all tubes on ice.

3.6. Combination of the Fluorescently Labeled Products and Volume Reduction

In these steps, the labeled test and reference DNAs are combined and the volume reduced using a centrifugal evaporator. Alternatively, ethanol precipitation can be used to reduce the volume (see Note 11). The labeled DNAs will be hybridized onto the probes affixed to the microarray slides.

1. Preheat the centrifugal evaporator to 75°C for 30 min.

2. Combine the labeled test and reference DNA samples by transferring the 25 μl of Cy5-labeled reference DNA to the 0.2 ml PCR tube containing the Cy3-labeled test sample.

3. Add 25 μl COT Human DNA to each tube containing the combined Cy3/Cy5 products.

4. Centrifuge the tubes (the centrifugal evaporator only has a single speed) at 75°C, with the lids open, for approximately 40 min or until most of the liquid volume is evaporated and only roughly 3 μl remain in each tube (see Note 12).

3.7. Probe Preparation, Denaturation, and Hybridization

1. Resuspend each pellet in 21 μl of preheated DS hybridization buffer.

2. Mix by flicking the tube and/or vortexing, then pulse centrifuge for approximately 5 s to collect the fluid in the bottom of the tube.

3. Denature the probes in a thermal cycler at 75°C for 10 min, pulse centrifuge again for approximately 5 s, and allow to cool to room temperature.

4. Assign a sample to each hybridization area. The *24Sure technology* supports the hybridization of two samples on each slide.

5. Load 18 μl of each probe onto a square coverslip for each hybridization area (a paper hybridization template is provided with the microarrays to help with the positioning of the coverslips).

6. Lower the slide facing downwards (barcode at the bottom) until it comes into contact with the coverslip.

7. After the probe mixture spreads out between the slide and the coverslip it is possible to lift the slide and turn it face-up.

8. Place the 24Sure slides in the prepared hybridization chamber, close the lid firmly, seal it with parafilm, and float the chamber in a non-circulating lidded water bath for 3–16 h at 47°C. This incubation is generally carried out overnight.

3.8. Post-hybridization Washes

24Sure slides are washed to remove any un-hybridized DNA. A high temperature, formamide-free wash is used to deliver the correct levels of stringency.

1. Briefly wash the 24Sure slides, agitating them manually in a coplin jar containing Wash 1 solution at room temperature.

Ideally, the coverslips will float off during this wash, but in some cases they will need to be gently removed by hand.

2. After removal of coverslips, immediately transfer each slide to a stainless steel slide rack placed inside a square glass staining dish containing 400 ml of Wash 1 and a 2.5 cm magnetic stir bar (room temperature).

3. Once the rack is fully loaded, cover the dish and wash the slides for 10 min using a magnetic stirrer.

4. Transfer the slides in the rack to another square glass staining dish containing 400 ml of Wash 2 and wash for another 10 min on the magnetic stirrer (room temperature).

5. Transfer the slides to the Hybex Microarray Incubation System containing 400 ml of Wash 3 preheated for 30 min to 59°C and incubate for 5 min.

6. Finally, transfer the slides to a square glass staining dish containing 400 ml of Wash 3 and wash for 1 min using the magnetic stirrer.

7. Dry the slides by centrifuging at $170 \times g$ for 3 min and then store in a slide box at room temperature (see Note 13).

3.9. Microarray Scanning and Analysis

1. Connect the MAPIX software to the scanner.

2. Place the slide in the scanner with the barcode (and hybridization area) facing upwards.

3. Select "Preview" under the Tools menu to pre-scan and obtain a low resolution image of the entire slide allowing the identification of the hybridization areas and detection of the barcode.

4. Once the pre-scan is complete, select a hybridization area and scan it at high resolution (see Note 14).

5. Save the resulting TIFF image and name the file including the barcode and sample number/sample name.

6. Using the BlueFuse Multi data analysis software, open the "Sample Editor" and add the details of each sample (e.g., date and method of extraction) to analyze. It is recommended to enter sample details into the BlueFuse Multi database in order to facilitate information retrieval and future analysis.

7. Click the "Run Batch Processing" button to complete the analysis of all unprocessed samples. The obtained results can be viewed and interpreted (see Note 15).

3.10. Interpretation Criteria

The following is a list of the important criteria used for determining chromosomal constitution:

1. The data obtained from the fluorescence ratios of the reference DNA and the test DNA are presented as a deviation from the

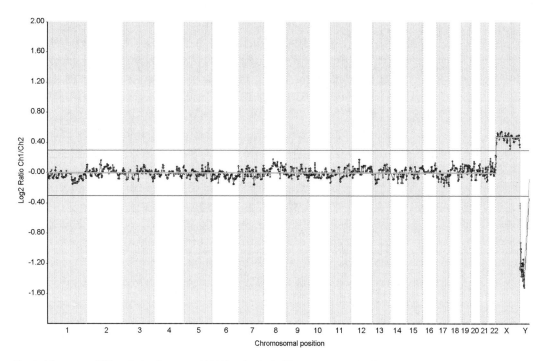

Fig. 2. Microarray-CGH analysis of a normal (23,X) polar body. The apparent gain of X-chromosome material and loss of Y-chromosome material is seen because a male reference DNA was used (SureRef, BlueGnome). If a female reference DNA had been used no losses or gains would have been observed.

Log2 ratio between Cy3 and Cy5 signals for each target in the BAC microarray.

2. The BlueFuse Multi software establishes the normal ranges at Log2 ratio of +0.38 and −0.38.

3. Values for normal chromosome copy number should fall between +0.38 and −0.38 for the DNA sequences of the whole chromosome (Fig. 2).

4. A Log2 ratio greater than +0.38 denotes gain of DNA and a Log2 ratio below −0.38 denotes loss of DNA.

5. In most cases, a profile shift resulting in Log2 ratios above +0.4 for the whole chromosome indicates a gain of that chromosome (see Note 16); whereas a shift resulting in Log2 ratios between +0.25 and +0.4 indicates the gain of a single chromatid (Figs. 3 and 4).

6. A profile shift resulting in Log2 ratios of −0.9 or below indicates loss of the entire chromosome (see Note 17); whereas a shift between −0.40 and −0.80 indicates the loss of a single chromatid (Fig. 3).

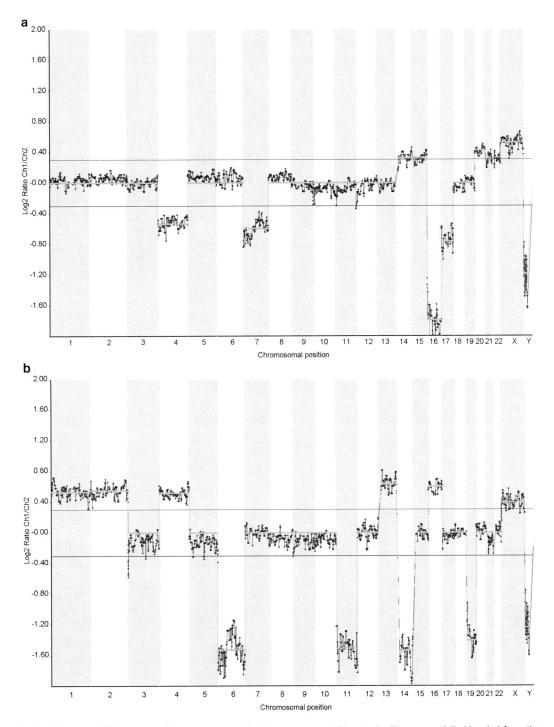

Fig. 3. Microarray-CGH analysis of highly abnormal first (**a**) and second (**b**) polar bodies sequentially biopsied from the same oocyte. (**a**) The first polar body has lost one chromatid belonging to chromosomes 4, 7, and 17 and has one extra chromatid for chromosomes 14, 15, 20, 21, and 22. The polar body has also lost the entire chromosome 16 (both chromatids). (**b**) The second polar body has gains of chromatids 1, 2, 4, 13, and 16 and losses of chromatids 6, 11, 14, and 19. The corresponding oocyte was therefore aneuploid (22,X,−1,−2,+6,+7,+11,−13,−15,+16,+17,+19,−20,−21,−22).

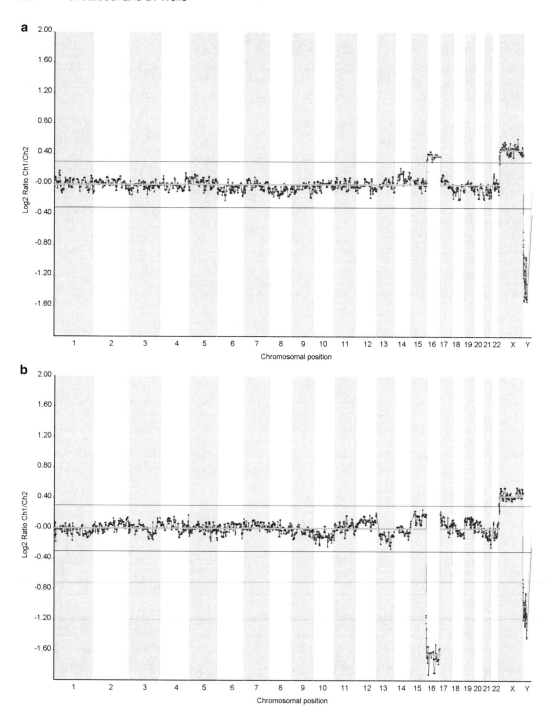

Fig. 4. Microarray-CGH analysis of aneuploid first (**a**) and second (**b**) polar bodies sequentially biopsied from the same oocyte. (**a**) The first polar body has one extra chromatid for chromosome16. (**b**) The second polar body has lost chromatid 16. The corresponding oocyte was therefore euploid (23, X).

4. Notes

1. If analyzing oocytes the zona pellucida should be removed prior to whole genome amplification. This is because somatic cells (cumulus cells) and surplus spermatozoa are often stuck to the outer surface of the zona pellucida after IVF and their DNA is likely to be amplified along with the oocyte DNA. This contamination will prevent accurate analysis of the oocyte. The zona pellucida can be removed by brief immersion in pronase or acidified Tyrode solution.

2. The volume of PBS should not exceed 2 µl or the lysis and amplification reagents will be diluted to an unacceptable level. A volume of 1–1.5 µl is ideal.

3. For storage of samples for more than 6 months it is recommended that the cell is lysed and the DNA amplified. The resulting SurePlex products can be frozen and stored for extended periods (>12 months).

4. It is critical to use good laboratory practice and minimize pipetting errors throughout the protocol. This is particularly important during the amplification step, as any contaminants may also be amplified and could affect the results.

5. Make sure that the pipette tip does not touch the bottom of the tube when adding the lysis buffer or the extraction mastermix.

6. Most "whole genome amplification" techniques do not truly amplify the entire genome. Some sequences may be absent or under-represented, while others may be relatively over-amplified. Consequently it is important to use a microarray that has been optimized for use with the chosen WGA method and consists of probes situated in regions of the genome that are represented after amplification. The 24Sure microarray has been designed to be compatible with SurePlex DNA amplification.

7. Always use a heated lid on the thermal cycler. Do not carry out reactions under oil.

8. Typically, 5% of biopsied cells fail to amplify. The most likely causes of amplification failure are the presence of degraded DNA in the biopsied PB and failure to transfer.

9. It is essential that the reference DNA is prepared in the same way as the sample DNA. It is especially important for both DNAs to be amplified using the same method (i.e., SurePlex). The SureRef DNA, supplied by the manufacturer, is derived from a chromosomally normal male and has been subjected to SurePlex amplification.

10. Half an hour before the labeling incubation is complete, switch on the centrifugal evaporator and set the temperature to 75°C.

11. Standard ethanol precipitation: Spin down the labeled sample and reference tubes. Combine the labeled test and reference DNA samples by transferring the 25 µl of Cy5-labeled reference DNA to the 0.2 ml PCR tube containing the Cy3-labeled test sample. Add 25 µl COT Human DNA and 7.5 µl 3 M sodium acetate to each tube and vortex. Add 187.5 µl of absolute ethanol to each tube and invert two or three times to mix. Place tubes in a −80°C freezer and incubate for 10 min. Centrifuge the probes at 16,000 × g for 10 min and discard the supernatant. Add 500 µl of 70% ethanol, mix by inversion, and centrifuge at full speed for 5 min. Decant the supernatant. Keeping tube inverted, gently tap out any remaining droplets onto a folded tissue. Spin tubes in centrifuge briefly and remove the remaining ethanol with a micropipette. Allow the pellet to air-dry for 2 min at room temperature.

12. While the probes are being centrifuged to reduce the volume, preheat an aliquot of DS Hybridization Buffer at 75°C for 10 min. Make sure a sufficient volume of DS Hybridization Buffer is prepared, taking into account the viscosity of the liquid.

13. The microarrays are reasonably light resistant, but if problems with faint signals are encountered post-hybridization washes can be undertaken in the dark. Microarrays should be stored in a light-proof container. High levels of ozone can also lead to loss of fluorescence and can be a problem in regions with warm climates. Apparatus to remove ozone from the air are available.

14. It is important to include all the clones while minimizing the area around the microarray in order to reduce background pixels.

15. BlueFuse Multi provides fully automated analysis of all 24Sure experiments. This includes grid finding, signal estimation, normalization, exclusion of poor quality results, and combination of replicates, region detection, region classification, and region reporting. The microarray type is used to select algorithms optimized to the format of the microarray.

16. Gains of entire chromosomes should result in profile shifts (0.40–0.50) that are comparable to the profile shift observed for chromosome X. It is important to compare the shifts to chromosome X in order to determine whether it is a gain of a chromatid or an entire chromosome.

17. Losses of entire chromosomes should result in profile shifts to −0.90 and lower that are comparable to the profile shift observed for chromosome Y. Generally, loss of chromosomal material results in more dramatic shifts than gain. It is important to compare those shifts to chromosome Y in order to determine whether it is a loss of a chromatid or an entire chromosome.

Acknowledgement

D.W. is funded by the NIHR Biomedical Research Centre, Oxford.

References

1. Pellestor F, Andreo B, Arnal F, Humaeu C, Demaille J (2003) Maternal ageing and chromosomal abnormalities: new data drawn from in vitro unfertilized human oocytes. Hum Genet 112:195–203

2. Kuliev A, Cieslak J, Ilkevitch Y, Verlinsky Y (2003) Chromosomal abnormalities in a series of 6,733 human oocytes in preimplantation diagnosis for age-related aneuploidies. Reprod Biomed Online 6:54–59

3. Fragouli E, Escalona A, Gutiérrez-Mateo C, Tormasi S, Alfarawati S, Sepulveda S, Noriega L, Garcia J, Wells D, Munné S (2009) Comparative genomic hybridization of oocytes and first polar bodies from young donors. Reprod Biomed Online 19:228–237

4. Fragouli E, Alfarawati S, Goodall NN, Sánchez-García JF, Colls P, Wells D (2011) The cytogenetics of polar bodies: insights into female meiosis and the diagnosis of aneuploidy. Mol Hum Reprod. Apr 14. [Epub ahead of print] PMID: 21493685

5. Angell RR (1991) Predivision in human oocytes at meiosis I: a mechanism for trisomy formation in man. Hum Genet 86:383–387

6. Sandalinas M, Marquez C, Munné S (2002) Spectral karyotyping of fresh, non-inseminated oocytes. Mol Hum Reprod 8:580–585

7. Alfarawati S, Fragouli E, Colls P, Wells D (2011) First births after preimplantation genetic diagnosis of structural chromosome abnormalities using comparative genomic hybridization and microarray analysis. Hum Reprod 26:1560–1574

8. Gutiérrez-Mateo C, Colls P, Sánchez-García J, Escudero T, Prates R, Wells D, Munné S (2010) Validation of microarray comparative genomic hybridization for comprehensive chromosome analysis of embryos. Fertil Steril 95:953–958

9. Fiorentino F, Spizzichino L, Bono S, Biricik A, Kokkali G, Rienzi L, Ubaldi FM, Iammarrone E, Gordon A, Pantos K (2011) PGD for reciprocal and Robertsonian translocations using array comparative genomic hybridization. Hum Reprod 26:1925–1935

Chapter 20

Nuclear Transfer in the Mouse Oocyte

Eiji Mizutani, Atsuo Ogura, and Teruhiko Wakayama

Abstract

The nuclear transfer (NT) technique in the mouse has enabled us to generate cloned mice and to establish NT embryonic stem (ntES) cells. Direct nuclear injection into mouse oocytes with a piezo impact drive unit can aid in the bypass of several steps of the original cell fusion procedure. It is important to note that only the NT approach can reveal dynamic and global modifications in the epigenome without using genetic modification as well as generating live animals from single cells. Thus, these techniques could also be applied to the preservation of genetic material from any mouse strain instead of preserving embryos or gametes. Moreover, with this technique, we can use not only living cells but also the nuclei of dead cells from frozen mouse carcasses for NT. This chapter describes our most recent protocols of NT into the mouse oocyte for cloning mice and for the establishment of ntES cells from cloned embryos.

Key words: Nuclear transfer, Reprogramming, ntES cell, Cloning, Embryo, Stem cell, Mouse oocyte

1. Introduction

There are two ways to achieve NT in the mouse oocyte: by cell fusion or by nuclear injection. Although the cell fusion method is beneficial for NT with large donor nuclei such as haploid pronuclei or diploid nuclei from blastomeres (1, 2), it requires live intact donor cells and many steps. In the usual electrofusion procedure, a donor cell is inserted into the perivitelline space of an enucleated oocyte after cutting the zona pellucida, moving the sets of donor cells and recipient oocytes to an electrofusion machine, applying electronic pulses, and later confirming cell fusion. On the other hand, nuclear injection using a piezo impact drive unit ("piezo unit") (3) can bypass several steps of the original cell fusion procedure. Importantly, it does not require intact donor cells because their nuclei will be intermingled with the recipient ooplasm. Surprisingly, we could produce normal cloned mice from dead cells or even from frozen mouse bodies stored in a standard laboratory

Hayden A. Homer (ed.), *Mammalian Oocyte Regulation: Methods and Protocols*, Methods in Molecular Biology, vol. 957, DOI 10.1007/978-1-62703-191-2_20, © Springer Science+Business Media, LLC 2013

freezer for long periods. The piezo unit will greatly help not only in NT but also in other forms of micromanipulation, such as intra-cytoplasmic sperm injection (ICSI) into oocytes (4, 5) or in the production of chimeric mice carrying embryonic stem (ES) cells by subzonal or blastocyst injection (6–8).

Since the first cloned mouse was generated from cumulus cell nuclei (3), many types of cells have been used to produce such clones (9, 10). These include tail-tip fibroblasts (11), Sertoli cells (12), fetal cells (2, 13), ES cells (14), natural killer T cells (15), primordial germ cells (16), hematopoietic stem cells (17), keratinocyte stem cells (18), fetal neuronal cells (19), and newborn neuronal stem cells (7, 20). The success rate of mouse cloning has varied with the genetic background and types of donor cell (13, 21). Thus, carefully choosing the donor cell type is important in producing cloned mice.

There are several approaches to improve the efficiency of producing cloned mice. For example, Inoue et al. succeeded in producing cloned mice at a rate of 19% based on transferred embryos by correcting X inactivation in cloned embryos using *Xist* knock-out mice as the donor source for NT. This is the highest success rate of NT in mice to date (22). We also succeeded in increasing the efficiency of mouse cloning by addition of histone deacetylase inhibitors (HDACi), such as trichostatin A (TSA), scriptaid (SCR), and suberoylanilide hydroxamic acid (SAHA) to the oocyte activation medium (23–27). Although it is almost impossible to generate cloned mice from inbred mouse strain donor cells without any special treatment, this new protocol has allowed us to generate cloned mice from such "unclonable" strains (26, 28).

The NT technique applied to mouse oocytes has been used not only to generate cloned mice but also to create ES cell lines from adult somatic cells (ntES cell lines) (29, 30). These ntES cell lines have the same potential as ES cells from fertilized blastocysts (31). Of course, we can also use these cells as donor cells for NT. The ntES cells have the same genome as the donor somatic cell, so they can be used as a backup for the donor cell genome and help increase the overall production in mouse cloning (30). In addition, this technique potentially may also be applied to the preservation of genetic material from any mouse strain instead of preserving embryos or gametes (30, 32). By combining the ntES technique with NT or chimera formation, we can preserve and propagate valuable genetic resources from mutant mice that are infertile or old, or even recovered from freeze-dried cells or carcasses, without the use of germ cells (33–35). Moreover, we succeeded recently in generating normal live mice from frozen carcasses by using these techniques (36). These results suggest that mouse cloning can be performed using donor cells in any condition as long as their genome remains intact.

2. Materials

2.1. Mouse Strains for NT Experiments

1. 2- to 3-month-old F1 hybrid mouse strains such as B6D2F1 (C57BL/6 × DBA/2) or B6C3F1 (C57BL/6 × C3H/He) are used for recipient metaphase II oocyte collection (see Note 1). Usually hybrid strains, such as B6D2F1, B6C3F1, or B6 × 129F1 can be used as donors, because these provide much better donor cells for the production of cloned mice. Inbred stains, such as C57BL/6 or C3H/He, are possible to use, but with a lower success rate for producing cloned mice than with hybrid strains. The ICR (CD-1) strain of mice is used for pseudopregnant surrogate mothers or foster mothers and for providing vasectomized males.

2. Hormones for superovulation: 50 IU/ml equine chorionic gonadotropin (eCG), 50 IU/ml human chorionic gonadotropin (hCG). eCG and hCG are dissolved in saline and stored at –40°C until use.

2.2. Chemicals and Sources

1. Cytochalasin B (CB) 100× stock solution: 500 µg/ml CB. CB is dissolved in dimethyl sulfoxide (DMSO) (37). Divide into small tubes (10–20 µL) and store at –40°C.

2. $SrCl_2$ 20× stock solution: 100 mM $SrCl_2 \cdot 6H_2O$. $SrCl_2 \cdot 6H_2O$ is dissolved in ultra pure water and stored in aliquots at room temperature (10).

3. Ethylene glycol tetraacetic acid (EGTA) 100× stock solution: 200 mM EGTA. EGTA is dissolved in ultra pure water and stored in aliquots at 4°C.

4. HDACi 200× stock solutions: 1 µM TSA, 25 µM SCR, 100 µM SAHA, 100 µM oxamflatin. A 200× stock solution of each HDACi is prepared by dissolving it in DMSO at a suitable concentration and stored at –80°C until use.

5. Hyaluronidase 100× stock solution: 10% hyaluronidase. Hyaluronidase is dissolved in M2 medium and stored at –40°C.

6. PBS (–): Ca/Mg free phosphate buffered saline.

7. Trypsin–EDTA: 0.25% trypsin, 1 mM ethylenediaminetetraacetic acid (EDTA). Trypsin and EDTA are diluted in PBS (–) and stored at –40°C until used.

8. Modified EGTA solution: 50 mM EGTA, 100 mM Tris–HCl, pH 8.0.

9. Mitomycin C 100× stock solution: 1 mg/ml mitomycin C. Mitomycin C is dissolved in PBS (–) and stored at –40°C until used.

10. Mineral oil.

11. Mercury.

12. Tyrode's solution acidic.

13. Homogenizer pestle.

14. Cell strainer.

2.3. Media

1. KSOM: 0.5 mg/ml PVA, 95 mM NaCl, 2.5 mM KCl, 0.3 mM KH$_2$PO$_4$, 0.2 mM MgSO$_4$·7H$_2$O, 0.2 mM Na pyruvate, 25 mM NaHCO$_3$, 10 mM Na lactate, 0.01 mM EDTA, 2.8 mM Glucose, 1.0 mM Glutamine, 1.7 mM CaCl$_2$·2H$_2$O, 0.5× Non-essential amino acids, 0.5× Essential amino acids, 1 mg/ml bovine serum albumin. All components are dissolved in ultra pure water.

2. M2 medium: 95 mM NaCl, 4.8 mM KCl, 1.2 mM KH$_2$PO$_4$, 1.2 mM MgSO$_4$·7H$_2$O, 23 mM Na lactate, 0.3 mM Na pyruvate, 5.6 mM Glucose, 4.15 mM NaHCO$_3$, 1.7 mM CaCl$_2$·2H$_2$O, 20.9 mM Hepes. All reagents are dissolved in ultra pure water. pH is adjusted to 7.3–7.4.

3. M2 medium containing 0.1% hyaluronidase: 2 μl of hyaluronidase 100× stock solution, 198 μl of M2 medium.

4. Nuclear isolation medium (NIM): 123.0 mM KCl, 2.6 mM NaCl, 7.8 mM NaH$_2$PO$_4$, 1.4 mM KH$_2$PO$_4$, 3 mM EDTA disodium salt, and 0.5 mM phenylmethylsulfonyl fluoride (PMSF). Reagents are dissolved in ultra pure water and the pH is adjusted to 7.2 by addition of a small quantity of 1 M KOH.

5. Polyvinylpyrrolidone (PVP) medium. PVP (360 kD) is dissolved in M2 medium at 10–12%. Divide into small tubes (50–100 μL) and store at –40°C.

6. PVP-NIM medium: PVP (360 kD), NIM. PVP is dissolved in NIM at 4%. Divide into small tubes and store at –40°C.

7. EF medium: Dulbecco's modified Eagle's medium (DMEM), fetal calf serum (FCS). Add FCS at 10% to DMEM.

8. ES cell maintenance medium: Glasgow minimal essential medium (GMEM), 10% FCS, 100 μM 2-mercaptoethanol, 1×non-essential amino acids, 1 mM of sodium pyruvate, 1,000 U/ml LIF (see Note 2).

9. M2+CB medium: 2 μL of CB 100× stock solution, 198 μL of M2 medium.

10. Activation medium 1: 10 μL of SrCl$_2$ 20× stock solution, 2 μL of EGTA 100× stock solution, 2 μL of CB 100× stock solution, 1 μL of HDACi 200× stock solution, 185 μL of KSOM (see Note 3).

11. Activation medium 2: 2 μL of CB 100× stock solution, 1 μL of HDACi 200× stock solution, 197 μL of KSOM (see Note 3).

12. HDACi medium: 1 μL of HDACi 200× stock solution, 199 μL of KSOM.

13. Mitomycin C medium: mitomycin C 100× stock solution, EF medium. Mitomycin C 100× stock solution is diluted 100 times with EF medium.

14. ntES cell establishing medium: ES cell medium without FCS, 20% Knockout serum replacement (KSR), 0.1 mg/ml ACTH (fragments 1–24) (see Note 2).

2.4. Tools for NT

1. Inverted microscope with Hoffman modulation contrast optics from Olympus (Tokyo, Japan, #IX71) (see Note 4).

2. Micromanipulator set (Narishige, Tokyo, Japan; #MMO-202, http://www.narishige.co.jp/).

3. Microforge (Narishige #MF-900).

4. Pipette puller (Sutter Instrument, Novato, CA, USA; #P-97, http://www.sutter.com/index.html) and glass pipette (Sutter Instrument #B100-75-10) (see Note 5).

5. Piezo impact drive system (Prime Tech Ltd, Ibaraki, Japan; #PMM-150FU, http://www.primetech-jp.com/index.html).

6. Warming plate (Tokai Hit Co., Shizuoka, Japan; Thermo Plate, http://www.ivf.net/ivf/tokai_hit_thermo_plate-o738.html).

7. 37°C, 5% CO_2, humidified incubator.

8. Tissue culture hood.

9. Centrifuge.

3. Methods

3.1. Outline of the NT Procedure (Fig. 1)

1. Start the donor cell culture if cultured cells are used as donor nuclei (e.g., tail-tip fibroblasts 2 weeks before injection, ES cells 2–3 days before injection).

2. Day 3: hormone administration to oocyte donors (eCG).

3. Day 1: hormone administration to oocyte donors (hCG) and mating foster mothers.

4. Day 0: day of NT experiment.

5. Oocyte collection.

6. Enucleation.

7. Preparing donor cells.

8. Injection of donor nuclei.

9. Activation of cloned oocytes, HDACi treatment, and culture.

10. Mating recipient females with vasectomized male mice.

11. Subsequent days: embryo transfer; establishment of ntES cell lines; Caesarian section.

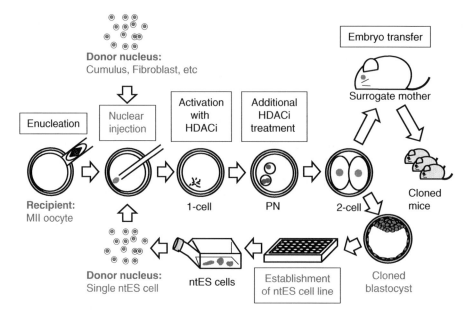

Fig. 1. Outline of the basic procedure of NT experiments in the mouse. Ovulated mature oocytes have the MII spindle removed and the donor nuclei are inserted. They are then activated with activation medium containing $SrCl_2$, CB, and HDACi for 6 h (*see* Subheading 3.7). Additional HDACi treatment for 2–4 h can improve the success rate for generating cloned pups or for establishing ntES cell lines. If the cloned embryos are transferred to surrogate mothers, a few can develop to term. ntES cell lines can also be established from cloned embryos reaching the blastocyst stage. In some cases, newly established ntES cells are used as donor cells for a second round of NT (*bottom left*).

3.2. Donor Cell Preparation

3.2.1. Cultured Cells

1. Prepare cells in culture for use, remove culture medium from the dish, and wash them in PBS (−).

2. Remove the PBS (−) and add trypsin–EDTA

3. Incubate for 5–20 min at 37°C in the incubator.

4. Add culture medium (including serum) and triturate the cells to produce a single cell suspension.

5. Spin down the cells in a centrifuge at $200 \times g$ for 10 min.

6. Wash cells with PBS (−) by centrifugation at least three times.

7. After the final washing, make a very concentrated cell suspension in EF medium. The final volume should be less than 10 μL (see Note 6).

3.2.2. Primary Fresh Cells

1. Wash in PBS (−) after collecting the cells using effective methods for each cell type.

2. Centrifuge at $200 \times g$ for 10 min, at least three times, to remove any enzymes that might be present. Cumulus cells are exceptional and can be used without washing.

3. Prepare concentrated cell suspension as described above (see Notes 6 and 7).

3.2.3. Frozen Dead Cells

1. Collect tissue samples from the frozen carcass and break them into small pieces (1–2 mm³) on dry ice.

2. Homogenize under 500 µL of NIM in 1.5 ml tubes using a homogenizer pestle, then filter using a cell strainer and collect the supernatant. This suspension contains many naked nuclei.

3. Introduce a few microliters of suspension into NIM on the manipulation chamber instead of into PVP medium.

3.2.4. Freeze-Dried Cells

1. Collect live single cells as described above.

2. Wash cells with modified EGTA solution.

3. Place about 50 µL of cell suspension into vials and place into a freezer at –30°C for 3 h.

4. Remove vials from the freezer and connect to a lyophilizer. After drying in a vacuum for more than 5 h, vials are stored at 4°C. For NT, cells are resuspended in PVP-NIM medium.

3.3. Superovulation and Collection of Oocytes

1. Superovulate female mice by injecting 5 IU eCG into the abdominal cavity 3 days prior to the planned day of the experiment.

2. Inject 5 IU hCG intra-peritoneally 48 h later (1 day before an experiment). Usually we inject mice at 5–6 pm.

3. Sacrifice mice 14–15 h after hCG injection and dissect out both oviducts.

4. On the day of the experiment, make up separate culture dishes containing ~15 µl micro-drops of M2 medium, M2 medium containing 0.1% hyaluronidase, or KSOM under mineral oil and pre-warm in the incubator for ~1–2 h.

5. Retrieve oocyte–cumulus cell complexes from the oviductal ampullae (usually we collect oocytes at 8–9 am) and transfer them into a droplet of M2 medium containing 0.1% hyaluronidase for removing the surrounding cumulus cells.

6. After 5 min, select the good quality cumulus-free oocytes and wash through M2 medium three times.

7. Transfer oocytes to drops of KSOM and incubate until used.

3.4. Setting up the Micromanipulator

1. Prepare holding, microinjection, and enucleation micropipettes. All micropipettes for NT can be made in the lab by using a pipette puller and microforge. For the holding pipette, the outside diameter (OD) should be equal or a little smaller than that of oocyte (e.g., OD 70 µm, inner diameter (ID) 10 µm). The ID of the enucleation pipette is 7–9 µm. The ID of the injection pipette depends on donor cell type (e.g., cumulus cells 5–6 µm, ES cells, and fibroblasts 6–7 µm). The distal section of the pipettes (extending backwards from the tip for about 300 µm) is given an angle of 15–20°.

2. Prepare dishes for the micromanipulation chamber: Place droplets (~10 μL) of the three media (M2 for injection, M2 + CB for enucleation, PVP for donor cell suspension, and washing needles) on the lid of a 10 cm dish and cover with mineral oil. To distinguish each medium, draw lines on the back of the dish (Fig. 2c).

3. Backload a small amount of mercury into the enucleation and injection pipettes before use (see Note 8).

4. Attach enucleation pipette to the pipette holder that is connected with the piezo unit (Fig. 2a, b). The top of the pipette holder must be screwed in tightly. Position the piezo unit on the micromanipulator.

5. Expel any air and oil and a few drops of mercury from the enucleation pipette into the droplet of PVP medium in the dish. Wash both the inside and outside of the pipette using PVP medium. While expelling the air and mercury from the pipette in the PVP droplet, activate the piezo unit with high power and high speed continuously for at least 1 min.

6. Attach a holding pipette to the pipette holder that is not attached to the piezo unit. Expel any air and oil into the M2 medium. Ensure that PVP does not become adherent to the holding pipette.

3.5. Enucleation of Oocytes

1. Place one group of oocytes (about 10–25/group) in the droplet of M2 + CB medium in the dish on the micromanipulation chamber and wait at least 5 min.

2. Rotate the oocyte using the enucleation pipette so that the metaphase II (MII) spindle (a clear zone in the oocyte identifiable using Nomarski or Hoffman optics) is located between 2 and 4 o'clock or between 8 and 10 o'clock and immobilize it with the holding pipette (Fig. 2d).

3. Penetrate the zona pellucida with the enucleation pipette by applying a few piezo pulses (see Note 9).

4. Without breaking the oolemma, adjust the enucleation pipette so that it comes to lie adjacent to the MII spindle in the oocyte (Fig. 2d').

5. Aspirate the MII spindle with a minimal volume of cytoplasm. Then, slowly pinch off the oocyte membrane and spindle by carefully withdrawing the pipette. The MII spindle is more rigid than the cytoplasm (Fig. 2d'', d''').

6. Wash the enucleated oocytes more than three times in KSOM to remove CB completely and return them to the incubator in droplets of KSOM for at least 30 min until donor cell injection (see Note 10).

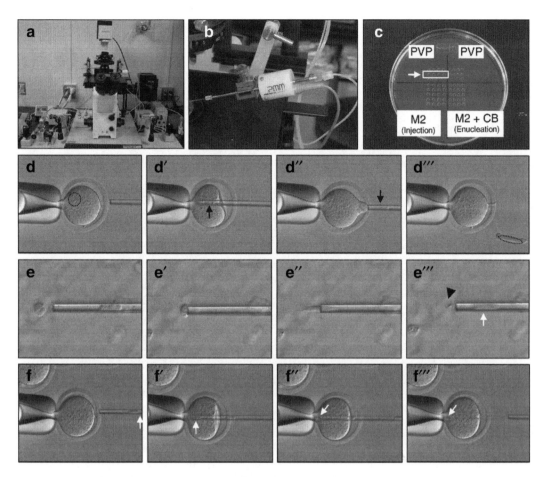

Fig. 2. Illustrations of each step in micromanipulation for carrying out NT. (**a**) Inverted microscope with a piezo impact drive unit. (**b**) The piezo unit. (**c**) Preparation of the droplets used for micromanipulation. The drops on the right half of the dish (M2 + CB medium) are used for enucleation. The left half (M2 medium) is used for microinjection. Donor cells are suspended in the enclosed PVP medium drops. Other PVP medium drops are used for washing needles (*see* Subheading 3.4). (**d–d'''**) The enucleation procedure (*see* Subheading 3.5). (**d**) The *black circle* indicates an MII spindle. It can be distinguished as a clear region in the ooplasm. (**d', d''**) After cutting the zona pellucida, the MII spindle is sucked into the enucleation pipette and drawn out without applying a piezo pulse. *Arrows* indicate the MII spindle. (**d'''**) The removed MII spindle is circled. It is harder than the oocyte cytoplasm. (**e–e'''**) Collection of a donor nucleus using an injection pipette is shown. The donor cells and the injection pipette are in a drop of PVP-containing medium. The inner diameter of the injection pipette is smaller than that of the donor cells. (**e'', e'''**) Aspirate the donor cells into and out of the injection pipette to break the donor cell membrane. The *arrowhead* indicates the removed donor cell membrane and cytoplasm. The *white arrow* indicates a denuded donor cell nucleus. (**f–f'''**) The injection procedure (*see* Subheading 3.6); *white arrows* indicate the donor nucleus. (**f**) Cutting the zona pellucida by applying piezo pulses. (**f'**) The injection pipette inserted almost as far as the holding pipette. Do not apply any piezo pulses in this step. (**f''**) Apply a single weak piezo pulse to break the oocyte membrane, then expel a donor nucleus. (**f'''**) Remove the injection pipette from the oocyte gently.

3.6. Injection of Donor Nuclei

1. Set the injection pipette on the micromanipulator and wash it in a droplet of PVP medium as described in steps 4 and 5 in Subheading 3.4.

2. Place one group of enucleated oocytes into M2 medium.

3. Add a small volume of donor cell suspension into the PVP medium and mix completely (Fig. 2e). If donor nuclei from

frozen carcasses or freeze-dried cells are used, use NIM medium or PVP-NIM medium instead of PVP medium, respectively.

4. Break the membrane of each donor cell by gently aspirating it in and out of the injection pipette until any visible cytoplasmic material is isolated (Fig. 2e′, e″, e‴). You can pick up a few donor nuclei in the same way in one injection procedure.

5. Stabilize the enucleated oocytes using the holding pipette.

6. Penetrate the zona pellucida by applying a few piezo pulses (Fig. 2f).

7. Push one donor nucleus to near the tip of the injection pipette and advance the injection pipette until it almost reaches the opposite side of the oocyte cortex (near the holding pipette, Fig. 2f; see Note 11).

8. Apply one weak piezo pulse (speed 1, intensity 1) to penetrate the oolemma, and then immediately expel the donor nucleus into the oocyte cytoplasm (Fig. 2f′).

9. After discharging the donor nucleus, gently withdraw the injection pipette from the oocyte (Fig. 2f″; see Note 12).

10. Wash the injection pipette thoroughly with PVP medium.

11. Keep the injected oocytes in the injection droplet for at least 10 min before returning them to KSOM in the incubator (see Note 13).

3.7. Oocyte Activation and Embryo Culture

1. Prepare dishes for oocyte activation: Place ~15 μL droplets of activation medium 1 and 2, and 15 μL droplets of HDACi medium on a 6 cm dish and cover with mineral oil; draw lines on the back of the dish to distinguish these media.

2. Transfer the reconstituted oocytes to droplets of activation medium 1, wash twice, and culture for 1 h in the incubator at 37°C under 5% CO_2.

3. Transfer oocytes to activation medium 2 and culture for 5 h in the incubator (see Notes 14 and 15).

4. After 5 h culture in activation medium 2, transfer the resulting embryos to droplets of HDACi medium and wash twice. If NT and oocyte activation are performed properly, a few pseudo-pronuclei can be formed in reconstituted embryos at this time (see Note 16).

5. Culture embryos in HDACi medium for up to 3–4 h (TSA, SAHA, Oxamflatin) or 18 h (SCR) to enhance genomic reprogramming. If cloned embryos are ES or ntES cell origin, omit this step (see Note 3).

6. Wash the cloned embryos in KSOM and transfer to new droplets of the same medium prepared in another dish, then continue culturing (see Note 17).

3.8. Embryo Transfer

1. Transfer two-cell stage cloned embryos (24 h after activation) or 4- to 8-cell embryos (48 h after activation) to oviducts of pseudo-pregnant female mice at 0.5 days postcopulation (dpc). Transfer morulae/blastocysts (72 h after activation) into the uterus of 2.5 dpc pseudopregnant female mice. Thus, ICR females have to be mated with vasectomized male mice at the appropriate times depending on the purpose of your experiment.

2. At 18.5 or 19.5 dpc, euthanize surrogate mothers, remove the uterus from the abdomen, and dissect out the cloned pups with their placentas. Wipe away the amniotic fluid from the skin, mouth, and nostrils and stimulate the pups to breathe by rubbing their backs or pinching them gently with blunt forceps and warm to 37°C (see Note 18).

3. After allowing time for recovery, transfer the cloned pups to the cage of a naturally delivered foster mother. Take some soiled bedding from the cage and nestle the cloned pups in the bedding material so that they take on the odor of the bedding. Remove some pups from the foster mother's litter and then mix the cloned pups with the foster female's pups.

3.9. Establishment of ntES Cell Lines from Cloned Embryos

1. To establish embryonic feeder cells, collect fetuses at 12.5–13.5 dpc from the pregnant mother and remove the head and internal organs into PBS (−).

2. Place the embryos into a new 10 cm dish and mince them into very small pieces with sterile scissors.

3. Add 25 ml of EF medium and plate into 175 cm² culture flasks.

4. One or two days later, split the cells 1:5 by trypsinization and allow them to grow to confluence.

5. When the cells become confluent, replace EF medium to mito-mycin C medium and culture for 2 h in the incubator, then wash with PBS (−) at least twice, and trypsinize to detach the cells from the flask. These cells can be preserved in freezing medium (final concentration about 1×10^6 cells/ml) at −80°C until used.

6. Prepare 96-well plates with mitotically inactivated feeder cells at least 1 day before ntES cell establishment.

7. Change the medium of feeder cells from EF medium to ntES cell establishing medium. Alternatively, you can use the new ES cell establishment "3i medium," which inhibits GSK3, MEK, and FGF receptor tyrosine kinases and enhances the ntES cell establishment rate significantly (see Note 19).

8. Remove the zona pellucida from cloned morulae/blastocysts by treating them with acid Tyrode's solution and wash three times in M2 medium.

9. Place the zona-free cloned embryos on the feeder cells of the 96-well plate one by one and culture 10–14 days in an incubator at 37°C under 5% CO_2 in air without changing the medium.

10. After clumps of inner cell mass derived cells appear and grow sufficiently, treat them with trypsin and disaggregate the cells by pipetting.

11. Place the cell suspension into another well with feeder cells (see Note 20). After 2 or 3 days of culture, ES-like colonies appear.

12. Trypsinize the cells and disaggregate them by pipetting.

13. Resuspend cells in ES cell maintenance medium.

14. Place the cell suspension into 0.1% gelatin-coated 48-well dishes without feeder cells.

15. Passage the cells to gelatin-coated 24-well, 12-well, and 12.5-cm^2 flask gradually to expand. After counting the numbers of ntES cells, they can be cryopreserved (38).

3.10. Production of Offspring from Established ntES Cells

The ntES cells can be used as donor cells for NT or as donor cells for producing chimeric mice in the same way as fertilization-derived ES cells. This technique can be applied to production of cloned mice by a second round of NT. It can also be used to generate pups from chimeric mice produced from donor cells in situations where it is difficult to generate cloned mice by direct NT, such as from infertile very aged mice or cells from frozen mouse carcasses (Fig. 3).

4. Notes

1. The mouse strain used for recipient oocytes is also important for successful NT experiments. In B6D2F1 oocytes, it is easy to find the MII spindle and they are robust enough to withstand in vitro manipulation and culture.

2. Media for ES cell maintenance and establishment can be obtained from several companies. For instance, ES cell medium (Millipore R-ES-101) is used for maintenance of ES cells and CultiCell medium for ES cells (Stem Cell Sciences KK, Saitama, Japan; http://www.scskk.com/index.php) is used for establishment of ntES cell lines.

3. There is no need to treat cloned oocytes from ES or ntES cell nuclei with HDACis. In this case, activation media 1 and 2 are prepared without HDACis.

4. An inverted microscope can also be used with Nomarski differential interference optics. In this case, a glass-bottomed dish must be used to obtain the best resolution.

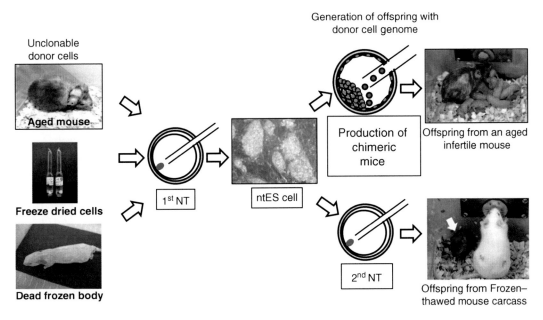

Fig. 3. Generating offspring from "unclonable" donor cells. Usually, it is very difficult or almost impossible to produce cloned mice from the somatic cells of very old mice, freeze-dried cells, or cells from frozen mouse carcasses. To generate offspring from such "unclonable" cells, a combination of the NT technique with ntES cell and chimera generation is very useful. For this, the first round of NT is carried out with "unclonable" donor cells, then ntES cell lines are established. These ntES cells possess the genome of the donor cell. The success rate of mouse cloning from such cells is better than from differentiated somatic cells. Thus to generate clone mice, the second round of NT is performed with established ntES cells used as donor nuclei. Pups with donor cell genomes can also be produced via chimeric mice, because ntES cells have pluripotency and their genomes can be transmitted to the germ cells of chimeric mice.

5. Micropipettes for micromanipulation can be ordered from several companies (e.g., Prime Tech Ltd., http://www.primetech-jp.com/en/01products/psk.html).

6. Trypsin is very toxic for oocytes, so the donor cells must be washed by centrifugation at least three times. If the final concentration of the cell suspension is too low, it is difficult to find proper donor cells at the time of the microinjection procedure.

7. If Sertoli cells are used, adult Sertoli cells are inappropriate as NT donors because of their large size, but those collected from neonatal mouse testes (immature Sertoli cells) are small enough for injection and usually give better results than cumulus cells (12, 21). Thus, the age of donors will affect the success rate. We recommend the use of newborn males younger than 6 days of age.

8. Mercury is volatile and is toxic if absorbed by breathing or through the skin. Wear appropriate gloves and always handle mercury in a working fume hood. Use appropriate safety handling conditions, as recommended by your Institutional Safety Office.

9. If you cannot cut the zona pellucida, check the connection between the pipette and the pipette holder. The top of the pipette holder must be screwed on tightly. Expel all oil inside the pipette, as this can reduce the piezo power output. There should be a slight negative pressure inside the pipette to enhance the piezo effect.

10. Contamination with CB affects the survival rate of oocytes after injection. If the CB is not washed out completely, many oocytes will lyse during microinjection.

11. Do not apply the piezo pulse until the tip of the injection pipette reaches the opposite wall of the oocyte. This is important to enable the oocytes to survive after injection.

12. If it is difficult to release the donor nuclei from the injection pipette, the tip of the injection pipette might be dirty. To clean it, wash in PVP medium with several piezo pulses.

13. This step is also important for the survival of injected oocytes. The oocyte membrane must be allowed to recover before returning it to the incubator.

14. If there is no time to change from activation medium 1 to activation medium 2, this step can be omitted and culturing can continue in activation medium 1 for an additional 5 h.

15. Activation with medium containing CB is important for using G0/G1 phase somatic cells as donor nuclei. If G2/M phase ES cells are used as donor cells, CB must be omitted from the activation medium to allow the oocyte to eject the polar body and the embryos can remain diploid.

16. The most probable reason for failure to form pseudopronuclei is a failure to break the donor cell membranes when they are injected into oocytes. The inner diameter of the injection pipette must be smaller than the donor cell. Apply a few piezo pulses to break the donor cell membrane when you pick it up.

17. To avoid carry-over of the chemicals contained in activation media into the medium for culture, all embryos should be washed at least three times and moved to different culture dishes for long-term culture.

18. If you have no success in generating cloned mice, change the donor cell type or use other hybrid mouse strains. In addition, check the media used in the experiments. However, the most important thing is your technical skill. The way to overcome the failure of mouse cloning is to keep practicing. Do not give up!

19. ntES cell establishing medium (CultiCell medium) does not contain FCS, which contains potential differentiation factors (39). Therefore, it is important to use this medium for establishing new ntES cell lines.

20. ntES cells in this step are very easy to differentiate. It is sufficient to separate clumps to small aggregates with 5–10 cells with gentle pipetting. Do not dissociate the ntES cell clumps completely into single cells.

References

1. Ono Y, Kono T (2006) Irreversible barrier to the reprogramming of donor cells in cloning with mouse embryos and embryonic stem cells. Biol Reprod 75:210–216
2. Ono Y, Shimozawa N, Ito M, Kono T (2001) Cloned mice from fetal fibroblast cells arrested at metaphase by a serial nuclear transfer. Biol Reprod 64:44–50
3. Wakayama T, Perry AC, Zuccotti M, Johnson KR, Yanagimachi R (1998) Full-term development of mice from enucleated oocytes injected with cumulus cell nuclei. Nature 394:369–374
4. Kimura Y, Yanagimachi R (1995) Intracytoplasmic sperm injection in the mouse. Biol Reprod 52:709–720
5. Wakayama T, Yanagimachi R (1998) Development of normal mice from oocytes injected with freeze-dried spermatozoa. Nat Biotechnol 16:639–641
6. Kawase Y, Iwata T, Watanabe M, Kamada N, Ueda O, Suzuki H (2001) Application of the piezo-micromanipulator for injection of embryonic stem cells into mouse blastocysts. Contemp Top Lab Anim Sci 40:31–34
7. Mizutani E, Ohta H, Kishigami S, Van Thuan N, Hikichi T, Wakayama S et al (2006) Developmental ability of cloned embryos from neural stem cells. Reproduction 132:849–857
8. Mizutani E, Ohta H, Kishigami S, Van Thuan N, Hikichi T, Wakayama S et al (2005) Generation of progeny from embryonic stem cells by microinsemination of male germ cells from chimeric mice. Genesis 43:34–42
9. Wakayama T (2007) Production of cloned mice and ES cells from adult somatic cells by nuclear transfer: how to improve cloning efficiency? J Reprod Dev 53:13–26
10. Thuan NV, Kishigami S, Wakayama T (2010) How to improve the success rate of mouse cloning technology. J Reprod Dev 56:20–30
11. Wakayama T, Yanagimachi R (1999) Cloning of male mice from adult tail-tip cells. Nat Genet 22:127–128
12. Ogura A, Inoue K, Ogonuki N, Noguchi A, Takano K, Nagano R et al (2000) Production of male cloned mice from fresh, cultured, and cryopreserved immature Sertoli cells. Biol Reprod 62:1579–1584
13. Wakayama T, Yanagimachi R (2001) Mouse cloning with nucleus donor cells of different age and type. Mol Reprod Dev 58:376–383
14. Wakayama T, Rodriguez I, Perry AC, Yanagimachi R, Mombaerts P (1999) Mice cloned from embryonic stem cells. Proc Natl Acad Sci USA 96:14984–14989
15. Inoue K, Wakao H, Ogonuki N, Miki H, Seino K, Nambu-Wakao R et al (2005) Generation of cloned mice by direct nuclear transfer from natural killer T cells. Curr Biol 15:1114–1118
16. Miki H, Inoue K, Kohda T, Honda A, Ogonuki N, Yuzuriha M et al (2005) Birth of mice produced by germ cell nuclear transfer. Genesis 41:81–86
17. Inoue K, Ogonuki N, Miki H, Hirose M, Noda S, Kim JM et al (2006) Inefficient reprogramming of the hematopoietic stem cell genome following nuclear transfer. J Cell Sci 119:1985–1991
18. Li J, Greco V, Guasch G, Fuchs E, Mombaerts P (2007) Mice cloned from skin cells. Proc Natl Acad Sci USA 104:2738–2743
19. Yamazaki Y, Makino H, Hamaguchi-Hamada K, Hamada S, Sugino H, Kawase E et al (2001) Assessment of the developmental totipotency of neural cells in the cerebral cortex of mouse embryo by nuclear transfer. Proc Natl Acad Sci USA 98:14022–14026
20. Inoue K, Noda S, Ogonuki N, Miki H, Inoue S, Katayama K et al (2007) Differential developmental ability of embryos cloned from tissue-specific stem cells. Stem Cells 25:1279–1285
21. Inoue K, Ogonuki N, Mochida K, Yamamoto Y, Takano K, Kohda T et al (2003) Effects of donor cell type and genotype on the efficiency of mouse somatic cell cloning. Biol Reprod 69:1394–1400
22. Inoue K, Kohda T, Sugimoto M, Sado T, Ogonuki N, Matoba S et al (2010) Impeding Xist expression from the active X chromosome improves mouse somatic cell nuclear transfer. Science 330:496–499
23. Kishigami S, Mizutani E, Ohta H, Hikichi T, Thuan NV, Wakayama S et al (2006) Significant improvement of mouse cloning technique by treatment with trichostatin A after somatic nuclear transfer. Biochem Biophys Res Commun 340:183–189

24. Kishigami S, Van Thuan N, Hikichi T, Ohta H, Wakayama S, Mizutani E et al (2006) Epigenetic abnormalities of the mouse paternal zygotic genome associated with microinsemination of round spermatids. Dev Biol 289:195–205

25. Rybouchkin A, Kato Y, Tsunoda Y (2006) Role of histone acetylation in reprogramming of somatic nuclei following nuclear transfer. Biol Reprod 74:1083–1089

26. Van Thuan N, Bui HT, Kim JH, Hikichi T, Wakayama S, Kishigami S et al (2009) The histone deacetylase inhibitor scriptaid enhances nascent mRNA production and rescues full-term development in cloned inbred mice. Reproduction 138:309–317

27. Ono T, Li C, Mizutani E, Terashita Y, Yamagata K, Wakayama T (2010) Inhibition of class IIb histone deacetylase significantly improves cloning efficiency in mice. Biol Reprod 83:929–937

28. Kishigami S, Bui HT, Wakayama S, Tokunaga K, Van Thuan N, Hikichi T et al (2007) Successful mouse cloning of an outbred strain by trichostatin A treatment after somatic nuclear transfer. J Reprod Dev 53:165–170

29. Wakayama T, Tabar V, Rodriguez I, Perry AC, Studer L, Mombaerts P (2001) Differentiation of embryonic stem cell lines generated from adult somatic cells by nuclear transfer. Science 292:740–743

30. Wakayama S, Ohta H, Kishigami S, Thuan NV, Hikichi T, Mizutani E et al (2005) Establishment of male and female nuclear transfer embryonic stem cell lines from different mouse strains and tissues. Biol Reprod 72:932–936

31. Wakayama S, Jakt ML, Suzuki M, Araki R, Hikichi T, Kishigami S et al (2006) Equivalency of nuclear transfer-derived embryonic stem cells to those derived from fertilized mouse blastocysts. Stem Cells 24:2023–2033

32. Wakayama S, Mizutani E, Kishigami S, Thuan NV, Ohta H, Hikichi T et al (2005) Mice cloned by nuclear transfer from somatic and ntES cells derived from the same individuals. J Reprod Dev 51:765–772

33. Ono T, Mizutani E, Li C, Wakayama T (2008) Nuclear transfer preserves the nuclear genome of freeze-dried mouse cells. J Reprod Dev 54:486–491

34. Mizutani E, Ono T, Li C, Maki-Suetsugu R, Wakayama T (2008) Propagation of senescent mice using nuclear transfer embryonic stem cell lines. Genesis 46:478–483

35. Wakayama S, Kishigami S, Van Thuan N, Ohta H, Hikichi T, Mizutani E et al (2005) Propagation of an infertile hermaphrodite mouse lacking germ cells by using nuclear transfer and embryonic stem cell technology. Proc Natl Acad Sci USA 102:29–33

36. Wakayama S, Ohta H, Hikichi T, Mizutani E, Iwaki T, Kanagawa O et al (2008) Production of healthy cloned mice from bodies frozen at –20 degrees C for 16 years. Proc Natl Acad Sci USA 105:17318–17322

37. Wakayama T, Yanagimachi R (2001) Effect of cytokinesis inhibitors, DMSO and the timing of oocyte activation on mouse cloning using cumulus cell nuclei. Reproduction 122:49–60

38. Schatten G, Smith J, Navara C, Park JH, Pedersen R (2005) Culture of human embryonic stem cells. Nat Methods 2:455–463

39. Ogawa K, Matsui H, Ohtsuka S, Niwa H (2004) A novel mechanism for regulating clonal propagation of mouse ES cells. Genes Cells 9:471–477

INDEX

Hayden A. Homer (ed.), *Mammalian Oocyte Regulation: Methods and Protocols*, Methods in Molecular Biology, vol. 957,
DOI 10.1007/978-1-62703-191-2, © Springer Science+Business Media, LLC 2013